Food Engineering Aspects
of Baking Sweet Goods

Contemporary Food Engineering

Series Editor

Professor Da-Wen Sun, Director

Food Refrigeration & Computerized Food Technology
National University of Ireland, Dublin
(University College Dublin)
Dublin, Ireland
http://www.ucd.ie/sun/

Food Engineering Aspects of Baking Sweet Goods, *edited by Servet Gülüm Şumnu and Serpil Sahin* (2008)

Computational Fluid Dynamics in Food Processing, *edited by Da-Wen Sun* (2007)

Contemporary Food
Engineering Series
Da-Wen Sun, Series Editor

Food Engineering Aspects
of Baking Sweet Goods

Edited by

Servet Gülüm Şumnu
Serpil Sahin

CRC Press
Taylor & Francis Group
Boca Raton London New York

CRC Press is an imprint of the
Taylor & Francis Group, an **informa** business

CRC Press
Taylor & Francis Group
6000 Broken Sound Parkway NW, Suite 300
Boca Raton, FL 33487-2742

First issued in paperback 2019

ISBN-13: 978-1-4200-5274-9 (hbk)
ISBN-13: 978-0-367-38761-7 (pbk)

Library of Congress Cataloging-in-Publication Data

Food engineering aspects of baking sweet goods / editors, Servet Gulum Sumnu, Serpil Sahin.
    p. cm. -- (Contemporary food engineering)
    Includes bibliographical references and index.
    ISBN 978-1-4200-5274-9 (hardback : alk. paper)
    1. Baked products--Analysis. 2. Food--Analysis. I. Sumnu, Servit Gulum. II. Sahin, Serpil. III. Title. IV. Series.

TP431.F66 2007
664'.752--dc22   2007049013

Visit the Taylor & Francis Web site at
http://www.taylorandfrancis.com

and the CRC Press Web site at
http://www.crcpress.com

To

Erin Bora, Melisa Defne, and Devin Kerem Dindoruk

and

Tuğçe and Gökçe Özkan and Kaan Demirezen

Contents

Series Editor's Preface

CONTEMPORARY FOOD ENGINEERING

Food engineering is the multidisciplinary field of applied physical sciences combined with the knowledge of product properties. Food engineers provide the technological knowledge transfer essential to the cost-effective production and commercialization of food products and services. In particular, food engineers develop and design processes and equipment in order to convert raw agricultural materials and ingredients into safe, convenient, and nutritious consumer food products. However, food engineering topics are continuously undergoing changes to meet diverse consumer demands, and the subject is being rapidly developed to reflect market needs.

In the development of food engineering, one of the many challenges is to employ modern tools and knowledge, such as computational materials science and nanotechnology, to develop new products and processes. Simultaneously, improving food quality, safety, and security remain critical issues in food engineering study. New packaging materials and techniques are being developed to provide more protection to foods, and novel preservation technologies are emerging to enhance food security and defense. Additionally, process control and automation regularly appear among the top priorities identified in food engineering. Advanced monitoring and control systems are developed to facilitate automation and flexible food manufacturing. Furthermore, energy saving and minimization of environmental problems continue to be an important food engineering issue and significant progress is being made in waste management, efficient utilization of energy, and reduction of effluents and emissions in food production.

Consisting of edited books, the *Contemporary Food Engineering* book series attempts to address some of the recent developments in food engineering. Advances in classical unit operations in engineering applied to food manufacturing are covered as well as such topics as progress in the transport and storage of liquid and solid foods; heating, chilling, and freezing of foods; mass transfer in foods; chemical and biochemical aspects of food engineering and the use of kinetic analysis; dehydration, thermal processing, nonthermal processing, extrusion, liquid food concentration, membrane processes and applications of membranes in food processing; shelf-life, electronic indicators in inventory management, and sustainable technologies in food processing; and packaging, cleaning, and sanitation. The books aim at professional food scientists, academics researching food engineering problems, and graduate-level students.

The editors of the books are leading engineers and scientists from many parts of the world. All the editors were asked to present their books in a manner that will address the market need and pinpoint the cutting-edge technologies in food engineering. Furthermore, all contributions are written by internationally renowned experts who have both academic and professional credentials. All authors have attempted to provide critical, comprehensive, and readily accessible information on the art and science of a relevant topic in each chapter, with reference lists to be used by readers for further information. Therefore, each book can serve as an essential reference source to students and researchers in universities and research institutions.

Da-Wen Sun, Series Editor

Preface

Baking is a complex food process that involves simultaneous heat and mass transfer. Understanding the baking process is necessary for process and product development. The books dealing with this topic mainly concentrate on different product formulations and functions of ingredients in cookies, cakes, and so forth. To date, baking has been a field of trial and error. In this book, we chose to look at the topic of baking from a different perspective. We aimed to combine engineering and science aspects of baking, as this is lacking in baking books.

Rheological and emulsion properties of dough and batter, physical properties of sweet goods, and heat and mass transfer during baking are important in understanding the baking process. For this reason, in this book we include chapters on the rheology of cake batter and cookie dough, physical and thermal properties of sweet goods, cake emulsions, and heat and mass transfer during baking. In addition, information is presented on the food science aspects of soft wheat products, including quality of soft wheat, functions of ingredients in the baking of sweet goods, and chemical reactions during processing. Moreover, information on cake and cookie technology is provided. The principles of recent technologies for baking soft wheat products, such as jet impingement, microwave and hybrid ovens, and recent studies in this area are also summarized. Presented in the last chapter of this book is a summary of the nutritional issues regarding the consumption of fats and sugars and general strategies of substituting fats and sugars in baked products, because the recent trend among consumers is to consume low-calorie products.

Various experts in different fields from different countries contributed to this book. The editors believe that this book will be helpful for undergraduate or graduate students who are working in the field of baking, food science, and food engineering, and also people from the food industry.

Servet Gülüm Sumnu
Serpil Sahin

About the Editors

Servet Gülüm Sumnu is an associate professor in the Department of Food Engineering, Middle East Technical University, Ankara, Turkey. She has authored or coauthored 55 journal articles and book chapters. She is one of the authors of *Physical Properties of Foods* (2006, Springer). She received BS (1991), MS (1994), and PhD (1997) degrees from the Department of Food Engineering, Middle East Technical University. Sumnu was a visiting scholar in the Department of Food Science and Technology at the Ohio State University for one year (1996). She is working on microwave food processes, especially microwave baking and frying. Her research also focuses on physicochemical properties of hydrocolloids and the determination of physical properties of foods.

Serpil Sahin is an associate professor in the Department of Food Engineering, Middle East Technical University, Ankara, Turkey. She has authored or coauthored about 40 journal articles and book chapters. She is one of the authors of *Physical Properties of Foods* (2006, Springer). She received BS (1989), MS (1992), and PhD (1997) degrees from the Department of Food Engineering, Middle East Technical University. Sahin was a visiting scholar in the Department of Food Science and Technology at the Ohio State University for one year (1996). She is working on food processes, especially frying, baking, separation processes, and applications of the microwave.

Contributors

Vural Gökmen
Department of Food Engineering
Hacettepe University
Ankara, Turkey

Manuel Gómez
E.T.S. Ingenierías Agrarias. Avda
Madrid, Spain

Dasappa Indrani
Flour Milling, Baking, and
  Confectionery Technology
Central Food Technological Research
  Institute
Mysore, India

Mukund V. Karwe
Department of Food Science
Rutgers, The State University of New
  Jersey
New Brunswick, New Jersey

Dilek Kocer
National Food Starch Innovation
Bridgewater, New Jersey

Hamit Köksel
Department of Food Engineering
Hacettepe University
Ankara, Turkey

Perry K.W. Ng
Department of Food Science and
  Human Nutrition
Michigan State University
East Lansing, Michigan

Gandham Venkateswara Rao
Flour Milling, Baking, and
  Confectionery Technology
Central Food Technological Research
  Institute
Mysore, India

Shyam S. Sablani
Department of Biological Systems
  Engineering
Washington State University
Pullman, Washington

Sarabjit S. Sahi
Cereals Processing and Bakery Science
Campden and Chorleywood Food
  Research Association
Gloucestershire, United Kingdom

Serpil Sahin
Department of Food Engineering
Middle East Technical University
Ankara, Turkey

Servet Gulum Sumnu
Department of Food Engineering
Middle East Technical University
Ankara, Turkey

Edmund J. Tanhehco
Mennel Milling Company
Fostoria, Ohio

Nantawan Therdthai
Department of Product Development,
  Faculty of Agro-Industry
Kasetsart University
Bangkok, Thailand

Suzan Tireki
ETI Group of Companies
Eskisehir, Turkey

Meryem Esra Yener
Department of Food Engineering
Middle East Technical University
Ankara, Turkey

Weibiao Zhou
Food Science and Technology Program
National University of Singapore
Singapore

1 Soft Wheat Quality

Edmund J. Tanhehco, Perry K.W. Ng

CONTENTS

1.1   INTRODUCTION

The category of sweet goods made from wheat flour encompasses a wide variety of products with different appearances, textures, flavors, nutritional values, and shelf lives. These include different types of cakes, cookies, doughnuts, pastries, and many more items. The quality of these goods begins with that of the soft wheat flour used to produce them. Flour quality is, in turn, affected by the wheat genotype, growing environment, and processing. The genotype and growing environment determine the amount and characteristics of the wheat components, including proteins, carbohydrates, and lipids. To produce high-quality flour, wheat must be properly milled; postmilling processing such as chlorination is also sometimes utilized for its benefits. Quality testing assures that a flour meets any necessary standards and gives valuable information to those seeking to improve it. These tests include the determination of proximate composition along with various chemical, rheological, and baking tests. The following sections of this chapter describe the milling of soft wheat into flour, composition of flour, quality testing, and how flour properties relate to the quality of products such as cookies and cakes.

1.2   WHEAT PRODUCTION, CLASSIFICATION, AND USAGE

Wheat is one of the major crops grown in the world, with over 620 million metric tons (MMT) produced worldwide in 2005 (USDA Foreign Agriculture Service 2007). U.S. and Canadian wheat production accounted for over 57 and 26 MMT, respectively. Common wheat, *Triticum aestivum*, is used for a wide range of products including breads, cakes, cookies, crackers, noodles, breakfast cereals, and much more. When describing wheat varieties, classification can be based on texture, color, and growth habit.

1.2.1   TEXTURE

Wheat is categorized as hard or soft based on kernel texture, one of the major determinants of end use. Compared to wheat with a softer texture, hard wheat requires more energy to be milled into flour and produces a coarser flour, and also one with more starch damage. Conversely, wheat kernels with softer texture produce finer flour with less starch damage, both important attributes of high-quality soft wheat flour. The majority of wheat grown worldwide is hard. In the United States, soft wheat accounts for about 25% of wheat production.

1.2.2   COLOR

Wheat can also be classified as red or white depending on the color of the bran covering the wheat kernel. The major difference between the two, other than appearance, is the greater susceptibility of white wheat to sprouting under favorable (moist and warm) conditions. This makes the use of white wheat undesirable for some food-processing applications such as thickening. However, there are advantages to white wheat, such as the bran being less bitter in flavor. Milling yields (or extraction rates) can also be higher in some cases because the bran of white wheat does not darken flour as much as red wheat bran (Lin and Vocke 2004).

1.2.3   GROWTH

Wheat planted in the spring and harvested in late summer in the same year is referred to as spring wheat. Winter wheat is usually planted in late summer or early fall and harvested the following summer. Soft red winter wheat accounts for the majority of soft wheat planted in the United States. Major soft wheat producing areas lie around the Mississippi River, Ohio, and some areas on the east coast (USDA Economic Research Service 2006). States that grow soft white wheat include those in the Pacific Northwest (Washington, Oregon, and Idaho), along with Michigan and New York. The provinces of Ontario and Alberta, Canada, account for much of the Canadian soft wheat production.

Hard wheats are generally bred to have higher protein content than soft wheats, although protein content and hardness are not necessarily linked. This reflects the different end-use requirements of hard (>11% protein) and soft wheat flours (8 to 10% protein). The main use of hard wheat flours is in bread, where strong and high levels of protein are needed. Soft wheat flours on the other hand are used in products where weaker protein (i.e., weaker dough strength and weaker viscoelastic properties) is desired, including products such as cakes and cookies. However, soft wheat flours are also used for a wide range of goods, some requiring higher levels of proteins, although not necessarily "strong" proteins. Crackers and noodles fall into this category.

1.3   FLOUR MILLING

The major components of the wheat kernel are the outer covering of bran, the embryo or germ, and the endosperm. The goal of flour milling is to separate these three as cleanly as possible, along with reducing the endosperm into flour particles. Higher extraction rates of flour, while economically desirable, may result in flour with excessive bran contamination (and thereby higher ash content) as well as increased starch damage. Therefore, a proper balance needs to be achieved, depending on the desired end use of the flour.

After cleaning the wheat of any debris, the first step in milling is the addition of water, referred to as tempering. The purpose of tempering is to toughen the bran, keeping it in larger flakes, thereby reducing the amount of small bran particles later contaminating the flour. An additional benefit of tempering is that it softens the endosperm, further reducing breakage of the bran when it is crushed against the endosperm during milling. The amount of water added in tempering varies depending on the hardness of the wheat and the flour mill machinery being used. Soft wheat is commonly tempered to around 14 to 15% moisture; hard wheat requires higher levels. The time needed for tempering can range from a few hours up to 24 h, again depending on the wheat. Detailed information on wheat tempering and preparation for experimental milling can be found in the American Association of Cereal Chemists International (AACCI) Method 26-10A (AACC International 2000).

The production of flour is achieved through roller milling which involves sets of two steel rolls spinning in opposite directions, between which the wheat falls to be ground. The flour mill is composed of two main systems: the break and the reduction. The purpose of the break system is to rip open the wheat kernel and separate the endosperm from the bran as cleanly as possible. This is achieved with corrugated rolls turning with a differential in speed. The slower roll serves to hold the kernel while the faster roll breaks it open. Multiple passes through different sets of break rolls with different gaps and corrugations are used to achieve a gradual separation of the bran and endosperm, while keeping the bran as intact as possible. In addition to separating the bran and endosperm, some flour is also produced after each pass through the break rolls and is sifted out. Wheat with a softer kernel texture fractures more easily and produces more flour in the break system than does wheat with a harder texture (Finney 1989). Flour obtained in the break system is called "break flour" and has a smaller particle size than the flour produced later on during milling in the reduction system (i.e., reduction flour). The remaining endosperm freed by the break rolls requires further milling and goes on to the reduction system.

The reduction system is similar to the break system with the main difference being that the reduction rolls are smooth. Multiple passes through different reduction rolls with sieving in between each pass are used to gradually reduce the endosperm to flour of the desired particle size. This gradual reduction is done to control the level of starch damage. Adjusting the pressure between the rolls and changing the level of tempering can also help to control the amount of starch damage in the milled flour.

Further downstream in the milling process, ash content is higher due to increases in fine bran contamination, and starch damage is higher due to the narrower reduction roll gaps. Therefore, the different break and reduction flour streams need to be selectively blended together to produce flour with the desired characteristics. Straight-grade flours are a combination of all of the flour streams. Patent flours consist of higher-grade streams with less bran (lighter in color) and consequently less ash, and clear flours have higher bran contamination (darker in color) and higher ash. Detailed information regarding milling can be found in the literature (Posner and Hibbs 1997).

1.4   MAJOR CONSTITUENTS OF SOFT WHEAT FLOUR

The constituents of wheat flour vary due to the genotype and the growing environment. These, in turn, determine the end-use characteristics, with certain varieties of wheat being better suited to specific types of products. The most important flour constituents in relation to flour functionality include the proteins, starches, pentosans (the largest portion of nonstarch polysaccharides), and lipids.

1.4.1   PROTEINS

Osborne (1907) fractionated wheat proteins into four classes based on their solubility in different solvents. By his classification, albumins were proteins soluble in water, and globulins were soluble in salt solutions. Prolamins were found to be soluble in 70 to 85% ethanol, and glutelins were soluble in dilute acid. Over the decades, further work was done to fractionate the proteins, as there is some overlap between the different classes and because still further fractionation can be done with different solvents (Chen and Bushuk 1970; Kreis et al. 1985).

Wheat proteins have the unique ability to form a viscoelastic network that allows for the production of products such as bread. The proteins mainly responsible for the viscoelastic properties of flour are the gliadins (prolamins) and glutenins (glutelins). Glutenins are large polymeric proteins held together by disulfide bonds. These proteins give dough strength and elasticity. Gliadins are smaller monomeric proteins that are responsible for dough extensibility. Together these proteins form the gluten proteins. Both the quantity (amount) and quality (type) of protein are important to flour characteristics. The strong gluten proteins found in hard wheat flour are able to form a network with good gas-retaining properties vital for yeast-leavened products.

Soft wheat flours are typically low in protein content (8 to 10%) and the proteins are weak in strength, characteristics better suited to making more tender products such as cakes and cookies. Most research has been focused on understanding the more obvious role of proteins in hard wheat products, with less focus on the role of proteins in soft wheat products. However, studies have shown that in addition to quantity, protein composition is important in soft wheat products, making its study necessary (Finney and Bains 1999; Hou et al. 1996a, 1996b; Huebner et al. 1999; Souza et al. 1994).

1.4.2   STARCH

In general, wheat flour contains over 70% starch (Sollars and Rubenthaler 1971) that is composed of approximately 25% amylose and 75% amylopectin. Amylose is a primarily straight-chain polymer of α-1,4-linked D-glucopyranose molecules. Amylopectin is a branched polymer of α-1,4-linked glucose connected by α-1,6-linked branch points. Amylose and amylopectin are organized in starch granules ranging from 1 to 45 µm in diameter. Wheat starch granules come in two forms: oval type A granules about 35 µm in diameter, and round type B granules approximately 3 µm in diameter (Alexander 1995). One of the most important properties of starch is its ability to swell and absorb water when it is heated in excess water. As starch granules swell, they cause an increase in the viscosity of the starch–water slurry, until eventually the granules break down, releasing primarily amylose, followed by amylopectin. Upon cooling, the starch molecules, especially amylose, can reassociate, forming a gel. The processes of granule swelling and breakdown are referred to as gelatinization and pasting, respectively, and can be visualized in wheat flour by measuring the viscosity of a flour–water slurry as it is heated and cooled (Figure 1.1). These properties of starch are important in many aspects relating to flour quality because they influence the interactions of starch and water in a food system.

Starch granules can be physically damaged during flour milling, increasing their water-holding ability and susceptibility to attack from the enzyme α-amylase. Greer and Steward (1959) found that 2 g of water was absorbed by each gram of damaged starch, compared to only 0.44 g of water absorbed by each gram of native starch. Soft wheat flour, in general, is lower in damaged starch content than hard wheat flour, due to the softer kernel texture and higher break flour yield. In bread flour, a controlled amount of damaged starch is needed because the enzymatic breakdown of starch provides some food for the yeast. However, in soft wheat products, the increased water absorption associated with increased levels of damaged starch can be detrimental to product quality.

1.4.3 PENTOSANS

Pentosans are carbohydrates of interest due to their ability to absorb ten times their own weight in water (D'Appolonia and Kim 1976; Kulp 1968). They are found in the

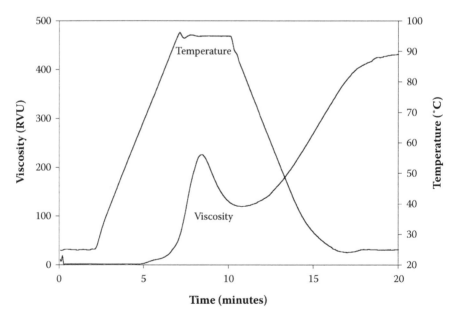

FIGURE 1.1 Rapid Visco Analyzer pasting curve of 3.5 g of soft wheat flour in 25 ml of water. (RVU: viscosity in Rapid Visco units.)

cell walls of wheat endosperm and bran and are composed mainly of arabinoxylan, a polymer with a β-(1-4)-linked D-xylopyranose backbone and branches of L-arabino-furanose residues (Cole 1967; Gruppen et al. 1992; Perlin 1951a, 1951b; Wang et al. 2006). Pentosans exist as both water-insoluble and water-soluble forms, depending on the degree of branching of the arabinose side chains. A higher degree of arabinose substitution is associated with higher water solubility (Hoseney 1984; Medcalf and Gilles 1968; Wang et al. 2006). Wang et al. (2006) measured the total pentosan content in six varieties of hard spring wheat and found it to range from 5.45 to 7.32% of the whole grain and from 1.88 to 2.04% of the straight-grade flours produced from this grain. The ratio of water-soluble to water-insoluble pentosans in this flour was 0.36:0.37. Pentosan content was also found to be higher in the lower-grade streams of flour, an important fact to consider when blending milling streams. Finnie et al. (2006) specifically measured the arabinoxylan content in soft white winter wheat flour and found variation among cultivars to be greatest in the water-soluble fraction, ranging from 3.23 to 5.74 mg xylose equivalents per gram sample. Water-insoluble arabinoxylan ranged from about 7 to 10 and total arabinoxylan from about 11 to 13.5 mg xylose equivalents per gram sample of soft white winter wheat flour.

1.4.4 LIPIDS

Flour lipids are important for quality attributes of soft wheat products such as cookie spread and cake volume. Whole grain wheat contains approximately 2 to 4% and the endosperm about 1 to 2% crude fat (Morrison 1978a). In flour, lipids exist as either nonstarch lipids or starch lipids that are held in amylose-inclusion complexes in starch granules (Acker and Becker 1971). Starch lipids are deemed to be less functionally important than nonstarch lipids due to their protected environment. Supporting evidence of this is that chlorination of flour (see Section 1.8) affects nonstarch lipids but not starch lipids (Morrison 1978b). The nonstarch lipids can be characterized as two types: free lipids extractable with petroleum or diethyl ether, and bound lipids extractable with cold polar solvent mixtures (Morrison 1978a). The free lipids can be further fractionated into nonpolar lipids (triglycerides, diglycerides, monoglycerides, fatty acids, sterols, and hydrocarbons) and polar lipids (glycolipids and phospholipids). The bound polar lipids consist of phospholipids and glycolipids (Pomeranz 1988).

1.5 QUALITY EVALUATION OF WHEAT GRAIN AND FLOUR

Characterization of wheat grain and wheat flour is necessary for both commercial and research purposes. Potential buyers need to know if what they will be getting will meet their needs, and researchers use these methods to better understand how flour affects end-use quality.

Quality tests on wheat grain include determining the test weight, milling yield, and kernel hardness. Flour is typically tested for proximate composition along with various chemical, rheological, and baking tests.

1.5.1 WHEAT GRAIN

1.5.1.1 Test Weight

Test weight is a measure of the weight of grain per unit volume in kilograms per hectoliter (kg/hl) or pounds per bushel (lb/bu) (AACCI Method 55-10). Higher test weights are generally correlated with greater milling flour yield; lower test weights resulting from shriveled and less sound kernels result in lower flour yields (Gaines et al. 1997).

1.5.1.2 Experimental Milling

Flour yield is dependent on the amount of endosperm in the kernel and how well it can be separated from the bran. As mentioned previously, flour yield must be balanced with flour quality characteristics such as starch damage and ash content. The milling characteristics of small quantities of wheat (<1 kg) can be evaluated by laboratory scale experimental milling. AACCI Methods 26-30A, 26-31, and 26-32 describe procedures for milling soft wheat flour with a Bühler MLU-202 experimental mill (Bühler Inc., Uzwil, Switzerland). This mill produces three break and three reduction flour streams. The Brabender Quadrumat Jr. experimental mill (C.W. Brabender Instruments, Inc., South Hackensack, NJ) is suited for smaller samples than the Bühler MLU-202 and produces flour and bran after passing wheat through a fixed set of three breaks (AACCI Method 26-50). The USDA-ARS Soft Wheat Quality Laboratory in Wooster, OH, has also developed and modified experimental milling methods to better evaluate the milling quality of soft wheat (Finney and Andrews 1986; Gaines et al. 2000; Yamazaki and Andrews 1982). In addition to valuable information regarding milling quality, the flour produced by these mills is important for use in flour quality evaluation.

1.5.1.3 Break Flour Yield

The break flour yield, expressed as a percent, is the weight of the flour produced by the break rolls relative to the weight of all products obtained from the combined break and reduction rolls (all streams of flour, bran, and germ). It is an excellent indicator of wheat hardness, because softer wheat produces more break flour. For soft wheat products, higher break flour yields are particularly important because of the desire for flour with finer particle size and lower starch damage. Typical break flour yields from a Bühler experimental mill used in the Michigan State University Wheat Quality Testing Program (milling soft white winter wheat) are around 30% of the total products recovered from milling (Figure 1.2; Ng et al. 2007), with harder wheats giving a lower percentage of break flour, typically less than 25%.

1.5.1.4 Kernel Texture

In addition to comparing break flour yields from milling, standardized methods exist to measure kernel hardness. Particle size index (AACCI Method 55-30) is measured by using a standardized grinder to mill grain into meal followed by weighing what meal passes through a U.S. No. 75 sieve. A softer wheat passes more of the meal

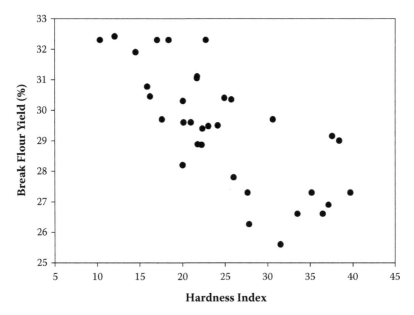

FIGURE 1.2   Scatter plot of hardness index measured by the Single Kernel Characterization System and break flour yield of Michigan soft white winter wheat milled in a Bühler MLU-202 flour mill. Wheat varieties were grown in the years 2001 to 2005.

through the sieve. A near-infrared (NIR) instrument can also be calibrated to measure hardness of a sample of ground wheat (AACCI Method 39-70A).

A more convenient and increasingly utilized way of measuring kernel hardness is with the Single Kernel Characterization System (SKCS; Perten Instruments, Huddinge, Sweden). Hardness is measured by assigning a hardness index value to the sample based on the force needed to crush the individual kernels (AACCI Method 55-31; Martin et al. 1993). There has been some limited information reported on the use of the SKCS for assessing soft wheats. Gaines et al. (1996a) reported a relationship between SKCS hardness values and softness equivalent (which is a measure of break flour yield used by the USDA-ARS Soft Wheat Quality Laboratory in Wooster, OH) for a group of soft wheat cultivars. However, Hazen et al. (1997) did not find a significant relationship between SKCS hardness values and softness equivalent for their group of tested soft wheat cultivars. This could be due to the fact that the SKCS was developed initially for a hard-wheat-growing region and perhaps the sensitivity of the measured values requires some adjustment for very soft wheat cultivars. Nevertheless, it appears that the SKCS can still be used with soft wheats for evaluation of hardness, in relative terms.

1.5.2   WHEAT FLOUR

1.5.2.1 Moisture

The moisture content of flour is most easily determined from the difference in weight of a sample before and after drying in an air oven (AACCI Methods 44-15A and 44-16). Moisture content can also be determined with a properly calibrated NIR spectrophotometer or with moisture-measuring instruments made by various manufacturers. Results of flour analysis are usually adjusted to 14% moisture basis as a way of expressing results on a constant solids basis between samples that may have different moisture contents.

1.5.2.2 Ash

Ash or mineral content of flour is often measured as an indicator of the quality of milling. As it is higher in the bran than the endosperm, ash content indicates the degree of bran contamination in flour. However, it should be noted that the endosperm ash content varies among wheat genotypes; therefore, ash levels may not completely correlate with the degree of bran contamination (Greffeuille et al. 2005). Ash content is also of interest because it is correlated with flour color (Kim and Flores 1999), an attribute that affects marketability of a flour. Flour ash contents are typically below 0.5% and can be determined by incinerating a flour sample in a muffle furnace, leaving only the ash (AACCI Methods 08-01 and 08-02).

1.5.2.3 Protein

Protein content is typically determined indirectly through measuring nitrogen content by methods such as Kjeldahl (AACCI Method 46-11A) and combustion (AACCI Method 46-30). A correction factor accounting for amino acid composition and non-protein nitrogen (×5.7) is then applied to calculate the protein content. Calibration of a NIR spectrophotometer using either of the previously mentioned methods can also be done to provide a rapid way of determining protein content that does not require chemicals or reagents (AACCI Method 39-11).

1.5.2.4 Sprout Damage

Sprout damage, caused by increased amounts of α-amylase activity, is a problem in products where a high hot paste viscosity of the wheat flour is needed, as in soup thickeners. High levels of α-amylase are found in grain that has begun to germinate because of exposure to moisture before harvest. This enzyme, while necessary in a germinating kernel, reduces soft wheat flour quality by hydrolyzing the α-1,4-linked glucose molecules of starch.

The α-amylase activity in grain or flour can be measured colorimetrically by incubating it with dyed and cross-linked amylose tablets (AACCI Method 22-05). Due to their ease, however, methods that measure the effects of α-amylase activity on heated flour-water slurries are more commonly used. The Falling Number System (Perten Instruments, Huddinge, Sweden) provides a rapid method of assessing sprout damage by measuring the time it takes for a stirrer to fall through a heated wheat meal and water or flour and water gel. Higher levels of α-amylase decrease the viscosity of the gel, causing the stirrer to fall faster. Wheat with a Falling Number value below 300 is suspected to have some sprout damage (Kaldy and Rubenthaler 1987).

Instruments that record viscosity while heating and stirring a flour–water slurry include the Amylograph (C.W. Brabender Instruments, Inc., South Hackensack, NJ) and the Rapid Visco Analyzer (RVA; Newport Scientific Pty. Ltd., Warriewood, Australia) (AACCI Methods 22-10 and 76-21, respectively). Higher α-amylase activity results in a curve with lower peak viscosity. The resulting curve can also give information about starch pasting characteristics not related to sprouting. Noodles are an example where texture has been correlated with Amylograph and RVA pasting properties such as the pasting temperature and peak viscosity (Batey et al. 1997; Morris et al. 1997; Oda et al. 1980).

1.5.2.5 Damaged Starch

The level of damaged starch can be measured by incubating a flour sample with α-amylase, followed by measurement of the reducing sugars or glucose that are produced (AACCI Methods 76-30A and 76-31). In soft wheat flour, damaged starch typically is below 3%. Levels as low as possible are preferred due to the increased susceptibility of damaged starch to the action of amylases during food processing.

1.5.2.6 Polyphenol Oxidase

Polyphenol oxidase (PPO), an enzyme that causes the formation of colored compounds (melanins) from phenols (Bettge 2004; Fuerst et al. 2006), is mostly removed with the bran during milling. However, some does make its way in with the flour, especially at higher flour extraction rates. This enzyme activity is especially detrimental to the quality of Asian noodles due to its darkening and discoloring effects (Kruger et al. 1992, 1994). PPO has also been reported to discolor batters, pie crusts, and refrigerated doughs (Gajderowicz 1979). Levels of PPO in wheat differ due to both genotype and growth environment (Baik 1994a; Park et al. 1997). AACCI Method 22-85 was developed as a rapid and small-scale test for PPO activity that can be used by both breeders and industry (Bettge 2004). This method measures PPO activity by incubating wheat or flour with a substrate (L-DOPA) and monitoring the color change spectrophotometrically.

1.5.2.7 Alkaline Water Retention Capacity of Flour

Alkaline water retention capacity (AWRC) is a test developed to simulate the alkaline conditions of the formula for evaluating sugar-snap cookie-making potential of a wheat flour (Finney and Yamazaki 1953). The test is defined as the amount of alkaline water held by the flour against a centrifugal force. Flour that binds alkaline water poorly is considered to be of good quality (AACCI Method 56-10). Yamazaki (1953) found a negative relationship between the amount of alkaline water held by the flour and cookie diameter. However, the relationship is not as clear for more recently developed soft wheat varieties (Finney 1994) and for distinguishing among flours within a softness or hardness class (Kitterman and Rubenthaler 1971). Breeders, though, are still selecting for low AWRC in their soft wheat lines.

1.5.2.8 Solvent Retention Capacity of Flour

More recently, a method for measuring the solvent retention capacity (SRC) of wheat flour was established to predict commercial flour properties (AACCI Method 56-

11). This method uses four solvents independently—water, 50% sucrose, 5% sodium carbonate, and 5% lactic acid—and measures a flour's ability to hold them after centrifugation. In general, water SRC is affected by all flour constituents, sucrose SRC is associated with pentosan characteristics, sodium carbonate SRC is associated with the level of damaged starch, and lactic acid SRC is associated with glutenin characteristics (Bettge et al. 2002; Gaines 2000).

The use of different solvents for SRC allows the separation of effects of different flour components, and the combined pattern of the four SRC profiles provides a practical flour quality assessment for predicting baking performance (Bettge et al. 2002; Guttieri et al. 2004; Slade and Levine 1994). In a collaborative study by Gaines (2000), lactic acid SRC was found to correlate with Mixograph number (protein content multiplied by peak height and peak time), and sodium carbonate SRC with damaged starch, softness equivalent, AWRC, and sugar-snap cookie spread. Sucrose SRC correlated with damaged starch, AWRC, and cookie spread (Table 1.1). SRC tests are currently used in a number of soft wheat breeding programs, including the Michigan State University Wheat Quality Testing Program (Ng et al. 2007). Variations on the SRC methods using smaller quantities of material and wheat meal instead of flour have also been developed, allowing for rapid screening of early generation breeder lines of wheat (Bettge et al. 2002; Guttieri et al. 2004).

1.5.3 DOUGH RHEOLOGY

With regard to wheat flour, rheology is the measure of the flow and deformation of doughs. These dough properties can affect product qualities such as geometry (e.g., cookie spread or cake volume), texture, and handling during processing. Dough rheological instruments were originally designed for use with materials such as bread doughs, where strength and elasticity are valued. Soft wheat flour products, however, generally require doughs that are weaker. Results obtained from these rheological instruments should not be interpreted using the same cri-

TABLE 1.1

Correlation Coefficients between Solvent Retention Capacity and Various Flour Quality Parameters

	Water	50% Sucrose	5% Sodium Carbonate	5% Lactic Acid
Protein content	0.33a	0.39a	0.31a	0.39a
Damaged starch	0.94a	0.77a	0.95a	0.23
Flour yield	0.51a	0.41a	0.54a	−0.06
AWRC	0.97a	0.81a	0.97a	0.33a
SSC diameter	−0.88a	−0.76a	−0.86a	−0.33a
Mixograph number	0.50a	0.49a	0.43a	0.69a

Notes: AWRC, alkaline water retention capacity; SSC, sugar-snap cookie.

[a] Significant at the 1% level.

Source: Adapted from Gaines, C.S., *Cereal Foods World*, 45, 303–306, 2000..

teria as results from hard wheat flours, as the rheological properties of soft and hard wheat flours are not simply opposites (Hoseney et al. 1988). Dough-forming properties of flours are commonly evaluated using the Alveograph, Mixograph, and the Farinograph.

1.5.3.1 Alveograph

The Alveograph (Chopin Technologies, Villeneuve-la-Garenne Cedex, France) measures air pressure inside of a dough bubble as it is inflated until it bursts (AACCI Method 54-30A). This biaxial extension is meant to simulate the deformation of a dough during fermentation and oven spring during baking. It allows for the measurement of the maximum overpressure (P), which relates to the resistance of dough to deformation, and the average length of the curve baseline at rupture (L), which is a measure of dough extensibility. The deformation energy (W) is a measure of the energy needed to inflate the dough and is derived from the area under the curve. W is related to the flour strength (Faridi and Rasper 1987). Bettge et al. (1989) investigated the ability of the Alveograph to evaluate soft wheat varieties for cookies and found that the parameter best able to predict cookie diameter was P in combination with the flour protein content. Nemeth et al. (1994) found that P and P/L were significantly correlated with sugar-snap cookie spread and score. Yamamoto et al. (1996) found that Alveograph P was negatively correlated and L positively correlated with Japanese sponge cake volume (Table 1.2).

TABLE 1.2

Correlation Coefficients between Rheological Properties and Qualities of Japanese Sponge Cakes and Sugar-Snap Cookies Made from Soft Wheat Flour Grown in the United States

Quality Parameter	Japanese Sponge Cake Volume	Sugar-Snap Cookie Diameter
P	−0.639[a]	ns
L	0.492[b]	0.522[b]
MPT	ns	0.577[b]
MPH	−0.692[a]	−0.590[b]
FWA	ns	−0.667[a]
FPT	−0.490[b]	ns

Notes: P, Alveograph maximum overpressure; L, Alveograph length; MPT, Mixograph peak time; MPH, Mixograph peak height; FWA, Farinograph water absorption; FPT, Farinograph peak time; ns, not significant.

[a] Significant at the 1% level.

[b] Significant at the 5% level.

Source: Adapted from Yamamoto, H., Worthington, S.T., Hou, G., and Ng, P.K.W., *Cereal Chemistry*, 73, 215–221, 1996.

1.5.3.2 Mixograph and Farinograph

The Mixograph (National Manufacturing, Lincoln, NE) and Farinograph (C.W. Brabender Instruments, Inc., South Hackensack, NJ) are both mixers that record changes in dough properties over time (AACCI Methods 54-40A and 54-21, respectively). These instruments are able to give information regarding optimum dough water absorption, strength, mixing time, and tolerance to overmixing. The main difference between the two is in the geometry of the mixers. The Mixograph uses vertically oriented pins that move in a planetary motion, and the Farinograph uses sigmoid-shaped mixing paddles. The Mixograph was developed to provide the more intensive mixing that North American wheats require. It is therefore mainly used there as well as in Australia. The Farinograph is widely used around the world (Ingelin 1997). Hazen et al. (1997) reported significant relationships between wire-cut cookie diameters and Mixograph peak time, and peak height. Uriyo et al. (2004) found significant negative correlations between Farinograph water absorption and cookie diameter, and with cake volume, in products made from soft red winter wheat. Cake tenderness was correlated with Farinograph departure time and mixing stability. Yamamoto and coworkers (1996) also reported a negative correlation between cookie diameter and Farinograph water absorption along with a negative correlation of cookie diameter to Mixograph peak height (Table 1.2); there was a positive correlation between cookie diameter and Mixograph peak time. Other workers have also correlated Farinograph and Mixograph measurements with various cake and cookie qualities (Finney and Bains 1999; Nemeth et al. 1994; Uriyo et al. 2004).

1.5.4   Products Requiring Weaker Proteins

Much research in the past has been focused on developing procedures for soft wheat quality evaluation. However, no test has proven more satisfactory than a baking test, which is an all-inclusive test. Most U.S. Eastern soft wheats have been tested for cake and cookie-making qualities. Most of these tests have followed standard AACCI methods.

1.5.4.1 Cookies

The sugar-snap cookie baking test (AACCI Method 10-52) was considered "the standard" cookie test for many years and has been used to evaluate flour for products such as cookies, crackers, cakes, and pies (Gaines 2004). Flours that produce cookies with larger spread and softer texture are favored. As there are fewer sugar-snap-type cookies on the market, the wire-cut cookie baking test was developed which utilizes a cookie formulation that more closely reflects the commercial wire-cut cookie formulation (AACCI Method 10-54; Slade and Levine 1994) (Table 1.3). Gaines et al. (1996b) compared the sugar-snap and wire-cut cookie formulations and found that even though both tests were capable of evaluating spread, the wire-cut cookies better reflected differences in cookie texture based on instrumental hardness evaluated using an Instron universal testing machine.

1.5.4.2 High-Ratio Cakes

The high-ratio (more sugar than flour, wt:wt) cake baking test (AACCI Method 10-90) is commonly used to evaluate soft wheat flours for cake products. The important characteristic of the flour used in these cakes is that they must be able to carry 1.3 to 1.4 times their weight in sugar (see Table 1.4 for formula). To accomplish this, cake flour is chlorinated to modify the flour components. Baked cakes are scored based on their volume, contour (symmetry), cell structure, grain, texture, color, and flavor.

TABLE 1.3
Comparison of Micro Wire-Cut (AACCI Method 10-54) and Micro Sugar-Snap (AACCI Method 10-52) Cookie Formulations

	Formulation	
Ingredient	Wire-Cut (g)	Sugar-Snap (g)
Sucrose	12.8	24
Brownulated sugar	4.0	—
Nonfat dry milk	0.4	1.2
NaCl	0.5	0.18
Sodium bicarbonate	0.4	0.4
Solution A[a]	na	0.32
Solution B[b]	na	0.20
Shortening	16.0	12
High-fructose corn syrup	0.6	—
Ammonium carbonate	0.2	—
Water	Variable	Variable
Flour	40.0 (13% m.b.)	40 (14% m.b.)

[a] Solution A: 7.98% sodium bicarbonate in water.

[b] Solution B: 10.16% ammonium chloride and 8.88% NaCl in water.

TABLE 1.4
High-Ratio White Layer Cake Formulation (AACCI Method 10-90)

Ingredient	Weights (g)	Weight Percent (Flour Basis)
Flour (14% m.b.)	200.0	100.0
Sugar	280.0	140.0
Shortening	100.0	50.0
Nonfat dry milk	24.0	12.0
Dried egg whites	18.0	9.0
NaCl	6.0	3.0
Baking powder and water	Variable	Variable

1.5.5   Products Requiring Stronger Proteins

Crackers and noodles are economically significant categories of products that are made with soft wheat flour, though they are not sweet goods. Crackers referred to in the following section are produced by fermentation with yeast to make saltine and similar crackers. Flour quality for noodle production is especially important for export of wheat to countries in the Far East.

1.5.5.1 Crackers

Crackers require stronger gluten than other soft wheat products and are often made from blends of both hard and soft wheat flours. This stronger gluten is necessary to give structure to crackers as they are fermented and sheeted. There is still no official test method for evaluating flours for cracker-baking potential, although there are published procedures using a two-stage sponge and dough approach to making crackers (Doescher and Hoseney 1985; Pizzinatto and Hoseney 1980). This involves fermentation of a sponge (containing yeast, water, and 60 to 70% of the flour) for 16 to 18 h followed by addition of the remaining ingredients and fermentation of the dough for another 6 h (Creighton and Hoseney 1990; Doescher and Hoseney 1985; Ranhotra and Gelroth 1988). However, duration of these test procedures limits the number of samples that can be evaluated by an operator in a given time. Lee et al. (2002) developed a practical one-stage procedure that enables an operator to evaluate 15 samples, as compared to about 6 samples with the two-stage procedures, in a 48-h period. Although the two types of procedures yielded slightly different baking results, the trends were the same for a diverse group of flour samples examined (Lee et al. 2002). Stronger doughs made crackers that were thicker, larger, and in harder texture than crackers made from weaker doughs.

1.5.5.2 Noodles

Asian noodles are another product, often made from blends of hard and soft wheat flours, which require stronger gluten. There are two basic kinds of Asian noodles: white salted (Udon) and alkaline noodles (Bettge 2004). Udon noodles are usually made from flour with 8 to 10% protein content and alkaline noodles 10.5 to 12% protein (Jun et al. 1998). The texture of Asian noodles is related to flour protein content and starch characteristics. The protein content of flour was positively correlated with noodle chewiness in a study by Baik et al. (1994b). Starch pasting properties have been shown to affect the overall texture of noodles, including softness and elasticity (Batey et al., 1997; Konik et al. 1992). Another important noodle quality determinant, especially with the higher pH of alkaline noodles, is discoloration from PPO activity. The Western Wheat Quality Laboratory of the USDA-ARS (Pullman, WA) has developed methods for testing alkaline and salted Asian noodles. Noodles are produced on a laboratory-scale machine and are evaluated based on color, texture, and yield.

1.6   EFFECTS OF FLOUR COMPONENTS ON COOKIES

Flour proteins, starches, pentosans, and lipids all affect the size or spread of cookies as well as their texture and appearance. With the wide variety of cookies produced, these components need to be taken into account when selecting flour.

1.6.1   PROTEINS

Soft wheat flour with low protein content is typically used in the production of cookies because of the deleterious effects on quality associated with the higher protein content in hard wheats. Sugar-snap cookies made from hard wheat flour are usually thicker, harder in texture, and have a smaller diameter (Miller and Hoseney 1997). Sugar-snap cookie diameter per unit of flour protein was negatively correlated with protein content in a study by Yamamoto et al. (1996) (Figure 1.3). Using a wire-cut cookie formulation, Gaines et al. (1996b) found a negative correlation between protein content and cookie diameter and a positive correlation with cookie height (Table 1.5). Harder texture was also positively correlated with increased protein content in this study. Higher flour protein content has been correlated with reduced cookie spread in other studies as well (Gaines 1985; Kaldy and Rubenthaler 1987). However, some studies have found a poor correlation between cookie quality and protein content (Abboud et al. 1985a; Yamazaki 1954).

Cookie spread is a function of the spread rate and the set time (Abboud et al. 1985b; Miller and Hoseney 1997). As cookie dough is heated, the decrease in viscosity

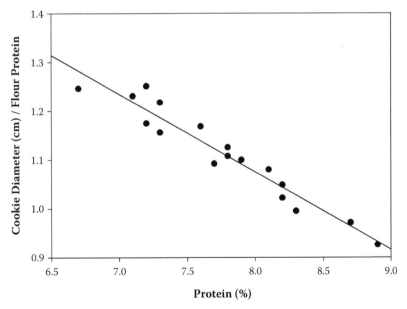

FIGURE 1.3   Relationship between protein content and sugar-snap cookie diameter per unit flour protein in cookies made from 17 soft wheat cultivars grown in the United States. (Adapted from Yamamoto, H., Worthington, S.T., Hou, G., and Ng, P.K.W., *Cereal Chemistry*, 73, 215–221, 1996.)

TABLE 1.5

Correlation Coefficients between Flour Protein Content and Wire-Cut Cookie Quality Characteristics

Parameter	Correlation Coefficient
Diameter	−0.57[a]
Height	0.64[a]
Hardness	0.79[a]

[a] Significant at the 5% level.

Source: Adapted from Gaines, C.S., Kassuba, A., and Finney, P.L., *Cereal Foods World*, 41, 155–160, 1996.

allows for the cookie to spread until it rises in viscosity and sets. Although gluten is not developed during mixing (Miller and Hoseney 1997), its glass transition temperature plays an important part in cookie set time. When the gluten reaches its glass transition temperature, the viscosity of the dough increases and spreading stops (Doescher et al. 1987; Miller et al. 1996). Miller and Hoseney (1997) examined the set time of different hard and soft wheat flours and found that within a group of hard or soft wheats, protein content affected the set time. However, the differences in protein content alone were not enough to fully explain the differences between the hard and soft wheat flour groups. Work has also been done to identify specific components of flour proteins that may affect cookie quality. Huebner et al. (1999) fractionated gliadins and glutenin subunits using size-exclusion high-performance liquid chromatography (HPLC). They found that flours with glutenin subunits 5+10 made better-quality cookies. Souza et al. (1994) found that the glutenin strength score, developed by Payne et al. (1987) to evaluate bread flours, was negatively correlated with cookie diameter. Hou et al. (1996b) separated the high (A subunits) and low (B and C subunits) molecular weight glutenin subunits and found that the ratio of the quantities of the B to C subunits was related to sugar-snap cookie diameter in flour from soft white winter wheat.

1.6.2 STARCH

The rate of spreading has been found to be faster in cookies made from soft wheat flour compared to cookies made from hard wheat flour (Abboud et al. 1985b; Miller et al. 1996; Miller and Hoseney 1997). A faster spread rate allows the cookie to spread to a larger diameter before setting occurs. Miller and Hoseney (1997) measured the spread rate of cookies made from soft wheat flour to be 7.8 mm/min compared to 4.6 mm/min in cookies made with hard wheat flour. The hard wheat flours were found to contain higher levels of soluble starch than the soft wheat flours. Removal of the soluble starch from hard wheat flours resulted in decreased dough viscosities and increased cookie spread rates. However, although the amount of soluble starch could explain the difference between the hard and soft wheat flour groups, it could not fully explain the difference in spread rates within the groups. Higher levels of damaged starch in milled hard wheat were also attributed to being part of the difference in spread rate between hard and soft wheat flours by Miller and Hoseney (1997).

During baking, minimal gelatinization of starch occurs due to the low water content of cookie dough, as shown by differential scanning calorimetry (Abboud and Hoseney 1984). However, damaged starch, with its greater water-holding capability, is known to negatively affect cookie diameter. Donelson and Gaines (1998) increased the damaged starch content of hard and soft wheat flours used to make sugar-snap cookies through the addition of ball-milled and pregelatinized starch. For both hard and soft wheat flours, the addition of damaged starch led to an increase in alkaline water retention capacity and a decrease in cookie diameter. They also made cookies with 100% of the flour replaced by combinations of prime and damaged starch. The soft wheat starch produced cookies with larger diameters than the hard wheat starch at all of the different levels of starch damage studied. Additionally, the hard wheat starch doughs had greater stiffness than those made from soft wheat starch. The authors concluded that there is a fundamental difference between hard and soft wheat starches that leads to their different performances in cookie baking.

1.6.3   PENTOSANS

With their ability to absorb large amounts of water, pentosans also affect cookie quality. Yamazaki (1955) found that the addition of purified starch tailings fractions, rich in pentosans, increased the hydration ability of soft wheat flour and reduced cookie spread. Bettge and Morris (2000) measured total, water-soluble, and grain membrane pentosans in 13 soft wheat flour samples. The amount of total pentosans had the largest negative correlation with sugar-snap cookie spread followed by the water-soluble and grain membrane pentosans. The grain membrane pentosans were also highly positively correlated with alkaline water retention capacity. Abboud et al. (1985a), on the other hand, reported a poor correlation between pentosan content and cookie diameter. Sucrose solvent retention capacity, which is associated with pentosans, was negatively correlated with sugar-snap cookie spread by Gaines (2004), even though in their study, alkaline water retention capacity was not. Using sucrose solvent retention capacity along with flour protein content and milling softness, Gaines (2004) was also able to generate a regression equation to predict cookie diameter.

1.6.4   LIPIDS

Studies involving the removal and reconstitution of flour lipids have shown that they are important to cookie spread, top grain (an "islanding" pattern formed on the surface of sugar-snap cookies), and structure. Cole et al. (1960) baked cookies with flour that had been extracted with water-saturated butanol and found that the cookies had decreased diameters. When the lipids were replaced, the cookie spread was returned to normal. Kissell et al. (1971) extracted free lipids from soft wheat flour and fractionated them into polar and nonpolar fractions. They were then reintroduced into the flour and baked into sugar-snap cookies. To achieve normal cookie spread and top grain, both the polar and nonpolar fractions were needed. Interchanging the lipids between different varieties of wheat flour did not affect the results, indicating that the presence of the mixed lipids is more important than the source. Fractionation studies by Clements and Donelson (1981), on the other hand, determined that the polar lipids (digalactosyl diglyceride and phosphytidyl choline along with glycolipids) were more important to sugar-snap cookie spread than the nonpolar lipids. The internal structure of cookies made from defatted flour is also negatively affected; these cookies have larger cells as opposed to the finer and more uniform cell structure found in good-quality cookies (Clements 1980).

1.7   EFFECTS OF FLOUR COMPONENTS ON CAKES

Cake batters are aerated emulsions of fat in water that expand during baking and set into a soft, porous gel (Mizukoshi et al. 1980; Shelke et al. 1990). During the initial phase of baking, there is a drop in batter viscosity as shortening melts and sugars become dissolved. This is followed by a rapid rise in viscosity when starch becomes gelatinized, absorbing free water and setting the cake (Howard et al. 1968; Shelke et al. 1990). Flour proteins and lipids along with the flour particle size also affect cake quality.

1.7.1   FLOUR PARTICLE SIZE

When measured by laser diffraction, soft wheat flour has been found to have a much higher percentage of particles below 41 μm in size than that of hard wheat flour (Hareland 1994). The particle sizes of various soft wheat flours have been negatively correlated to cake volume (Yamamoto et al. 1996; Yamazaki and Donelson 1972). In addition to varietal differences in particle size produced by normal milling, research has shown that further reduction of particle size through postmilling processing (pin-milling and air-classification) can improve the volume and quality of cakes (Chaudhary et al. 1981; Gaines and Donelson 1985a; Miller et al. 1967). Although reducing particle size is beneficial, it is important to limit starch damage, as damage levels greater than 5% have been negatively correlated with cake quality (Miller et al. 1967).

1.7.2   PROTEINS

Higher protein content in flour is generally associated with poorer quality for cake baking. According to Kaldy and Rubenthaler (1987), flour high in protein or with strong gluten results in cakes with lower volume and coarser texture due to protein disruption of the foam structure in cake batter. In their study of Canadian soft white winter and spring wheats, they found a significant negative correlation between flour protein content, Japanese sponge cake volume, and overall cake score. Yamamoto et al. (1996) also found a negative correlation between flour protein and Japanese sponge cake volume per unit protein (Figure 1.4). Gaines and Donelson (1985b) found that the volume and tenderness of white layer cakes were not significantly affected by protein content, although those of angel food cakes were. However, a difference of over 2% protein was needed to see an effect in the angel food cakes. Although an excess of protein may harm cake quality, soluble proteins (both from the flour and from other cake ingredients) are still needed for thermal stability of the cake foam structure (Howard et al. 1968). Protein composition in addition to content was shown to be important to Japanese sponge cake volume in work by Hou et al. (1996b). The presence of high-molecular-weight glutenin (HMW-GS) subunit

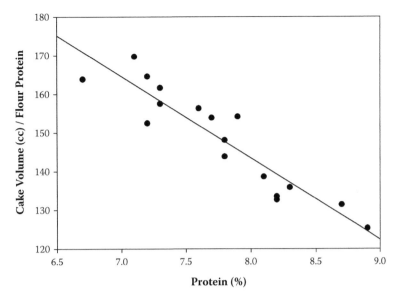

FIGURE 1.4 Relationship between protein content and Japanese sponge cake volume per unit flour protein in cakes made from 17 soft wheat cultivars grown in the United States. (Adapted from Yamamoto, H., Worthington, S.T., Hou, G., and Ng, P.K.W., *Cereal Chemistry*, 73, 215–221, 1996.)

1 in soft wheat flour resulted in larger cake volume, while the presence of HMW-GS subunit 2* resulted in smaller volume.

1.7.3 LIPIDS

Lipids make up only a small fraction of flour; however, they are important to cake volume and texture. Spies and Kirleis (1978) found that extraction of free flour lipids reduced volume and caused poorer texture in cakes made with a modified white layer cake formula. Reintroduction of the lipids restored most of the cake qualities. Interchanging the lipids between different varieties of wheat did not affect the results, indicating that the presence of lipids is more important than the source. Takeda (1994) extracted free lipids from flour, resulting in reduced sponge cake volume. The free lipids were also fractionated into polar and nonpolar fractions. Reintroduction of the polar lipids (monogalactosyl and digalactosyl diglycerides) returned the cake volume to its normal size, while the nonpolar fractions had only minor effects. Similar results were reported by Seguchi and Matsuki (1977a).

1.8 FLOUR CHLORINATION

To produce good-quality high-ratio cakes, chlorination of flour is necessary. Cakes made from nonchlorinated flour have poor volume, contour, crumb grain, and texture (Donelson et al. 2000; Montzheimer 1931; Smith 1932). Chlorine treatment also improves mouthfeel of cakes, making cakes drier and less sticky or gummy (Kissell and Yamazaki 1979; Seguchi and Matsuki 1977b). Chlorination is usually done with chlorine gas and can be monitored by a drop in pH of flour. Flour is typically chlorinated to a pH range of about 4.5 to 5.2 (Gough et al. 1978). Starch, lipids, and proteins are all affected by flour chlorination.

1.8.1 STARCH

Fractionation, interchange, and reconstitution studies of nonchlorinated and chlorinated flours have confirmed that the effects of chlorination on starch are important to cake quality. Cakes made from chlorinated flour with the starch interchanged with that from nonchlorinated flour had smaller volumes and poorer cake qualities (Johnson and Hoseney 1979a; Sollars 1958). The opposite was true when exchanging chlorinated starch into nonchlorinated flour. Gaines and Donelson (1982) used a modified Viscograph to examine the viscosity of cake batters made with chlorinated and nonchlorinated flours during heating. The apparent viscosity of heated batters increased faster in batters made from chlorinated flour compared to nonchlorinated flour. Chlorinated flour batters also showed greater expansion during baking. These results were in agreement with results from Kulp et al. (1972). Accelerated thickening of batters allows for improved setting and retention of larger cake volume (Donelson et al. 2000).

Donelson (1990) fractionated chlorinated and nonchlorinated flours and found that the chlorinated starch fraction had increased alkaline water retention capacity. These results were related to decreased sugar-snap cookie spread in his experiment. In addition to binding more water, chlorinated starch binds more oil as a result of increased starch granule hydrophobicity (Seguchi 1984). The oxidative depolymerization of starch that occurs during chlorination has been investigated as one of the reasons for these changes in starch properties (Huang et al. 1982; Johnson et al. 1980). Varriano-Marston (1985) hypothesized that the oxidative depolymerization increased the capillary size of starch granules, leading to the increased ability of chlorinated starch to bind water and oil.

1.8.2 LIPIDS

Various studies have determined that the effect of chlorination on lipids is important to cake quality. Kissell et al. (1979) chlorinated flours to pH 5.2, 4.8, and 4.0 and then extracted the free lipids with hexane. White layer cake volume was reduced in cakes baked without lipids; however, the normal volume was restored upon readdition of the extracted lipids. Flour chlorinated to pH 4.8 performed the best. By interchanging lipids from a chlorinated flour into a nonchlorinated one, Donelson et al. (1984) were able to increase high-ratio cake volume to that of the chlorinated flour. In contrast, Johnson et al. (1979), after conducting a lipid interchange study, came to the conclusion that although the presence of lipids is important, the effect of chlorination on them is not important to cake quality in cakes baked using Kissell's lean cake formulation (Kissell, 1959). In the study of Johnson et al. (1979), cakes baked from both chlorinated and nonchlorinated flours with their lipids extracted had poor grain. By adding either of the lipid fractions back to the chlorinated flour, they were able to restore the baking properties.

1.8.3 PROTEINS

The fractionation and interchange studies of Sollars (1958) found that chlorination of gluten and starch was of almost equal importance for the production of white layer cakes. Chlorination of gluten had an effect on yellow layer cakes as well, but to a lesser extent. Tsen et al. (1971) reported that chlorination of flour increases the extractability of proteins by water and acetic acid, and that this increased protein solubility may be part of the improving effects of chlorine treatment on flour. The changes in protein extractability were attributed to the actions of chlorine breaking hydrogen bonds, cleaving peptide bonds, degrading amino acids, and oxidizing sulfhydryl bonds.

The effect of chlorination on increasing the hydrophobicity of proteins in flours may also be important. Seguchi (1985) found that changes in the hydrophobicity of starch granules were due to conformational changes in surface proteins of the starch granules, and later, that chlorination also resulted in an increase in the amount of protein extracted (Seguchi 1990). Sinha et al. (1997) extracted gliadins from flour chlorinated to pH 4.8 and 4.3; gliadin protein hydrophobicity, as measured by fluorescence spectroscopy, increased with chlorination (Figure 1.5). Reversed-phase HPLC results suggested that the increases in hydrophobicity were due to conformational changes in the proteins (Sinha et al. 1997).

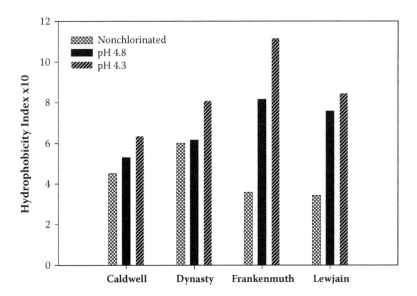

FIGURE 1.5 Relative hydrophobicities of gliadins extracted from chlorinated and non-chlorinated soft wheat flours, measured by fluorescence spectroscopy. (Adapted from Sinha, N.K., Yamamoto, H., and Ng, P.K.W., *Food Chemistry*, 59, 387–393, 1997.)

1.8.4 ALTERNATIVES TO CHLORINATION

Over the years, alternatives to chlorination have been explored as concern for the safety of chemically processed foods has grown. Russo and Doe (1970) improved cake volume by heating nonchlorinated flour; however, this also resulted in cakes with poor texture. The addition of ingredients such as starch, egg albumin, xanthan gum, L-cysteine, and hydrogen peroxide plus peroxidase have also been tested for their abilities to compensate for a lack of chlorination (Johnson and Hoseney 1979b; Russo and Doe 1970; Thomasson et al. 1995). In their cake formulation, Donelson et al. (2000) replaced nonchlorinated flour with either commercial hard wheat starch or a laboratory-produced soft wheat starch at a level equal to the area under an RVA pasting curve made from chlorinated flour. This was done to try and achieve the viscosity-modifying properties normally associated with chlorination. Egg albumin, soy lecithin, and xanthan gum were also added to their cake formula to improve texture and contour. The results after baking various types of cakes were equal to or better than those made with chlorinated flours. Most importantly, the cakes had crumbs with good texture rather than the gummy textures of cakes made with non-chlorinated flour. The ozone treatment of flour has also recently been investigated with promising results (Chittrakorn and MacRitchie 2006).

1.9 CONCLUSION

The different components of soft wheat flour collectively play a role in its quality. Softer kernel texture and lower protein content are typically favored for soft wheat products. Starches and lipids serve important functions in baked products such as cakes and cookies. Understanding wheat flour composition and how quality is measured provides a good base for further research and study of sweet goods.

REFERENCES

AACC International. 2000. Approved Methods of the American Association of Cereal Chemists, 10th ed. St. Paul, MN: American Association of Cereal Chemists.

Abboud, A.M. and R.C. Hoseney. 1984. Differential scanning calorimetry of sugar cookies and cookie doughs. *Cereal Chemistry* 61:34–37.

Abboud, A.M., G.L. Rubenthaler, and R.C. Hoseney. 1985a. Effect of fat and sugar in sugar-snap cookies and evaluation of tests to measure cookie flour quality. *Cereal Chemistry* 62:124–129.

Abboud, A.M., R.C. Hoseney, and G.L. Rubenthaler. 1985b. Factors affecting cookie flour quality. *Cereal Chemistry* 62:130–133.

Acker, L. and G. Becker. 1971. Recent studies on the lipids of cereal starches. Part 2. Lipids of various types of starch and their binding to amylose. *Die Stärke* 23:419–424.

Alexander, R.J. 1995. Potato starch: New prospects for an old product. *Cereal Foods World* 40:763–764.

Baik, B.K., Z. Czuchajowska, and Y. Pomeranz. 1994a. Comparison of polyphenol oxidase activities in wheats and flours from Australian and United States cultivars. *Journal of Cereal Science* 19:291–296.

Baik, B.K., Z. Czuchajowska, and Y. Pomeranz. 1994b. Role and contribution of starch and protein contents and quality to texture profile analysis of oriental noodles. *Cereal Chemistry* 71:315–320.

Batey, I.L., B.M. Curtin, and S.A. Moore. 1997. Optimization of Rapid-Visco Analyser test conditions for predicting Asian noodle quality. *Cereal Chemistry* 74:497–501.

Bettge, A., G.L. Rubenthaler, and Y. Pomeranz. 1989. Alveograph algorithms to predict functional properties of wheat in bread and cookie baking. *Cereal Chemistry* 66:81–86.

Bettge, A.D. and C.F. Morris. 2000. Relationships among grain hardness, pentosan fractions, and end-use quality of wheat. *Cereal Chemistry* 77:241–247.

Bettge, A.D., C.F. Morris, V.L. DeMacon, and K.K. Kidwell. 2002. Adaptation of AACC Method 56-11, solvent retention capacity, for use as an early generation selection tool for cultivar development. *Cereal Chemistry* 79:670–674.

Bettge, A.D. 2004. Collaborative study on L-DOPA—wheat polyphenol oxidase assay (AACC method 22-85). *Cereal Foods World* 49:338, 340–342.

Chaudhary, V.K., W.T. Yamazaki, and W.A. Gould. 1981. Relation of cultivar and flour particle size distribution to cake volume. *Cereal Chemistry* 58:314–317.

Chen, C.H. and W. Bushuk. 1970. Nature of proteins in triticale and its parental species. I. Solubility characteristics and amino acid composition of endosperm proteins. *Canadian Journal of Plant Science* 50:9–14.

Chittrakorn, S. and F. MacRitchie. 2006. Ozonation of cake flour as an alternative to chlorination. Poster: World Grains Summit: Foods and Beverages. San Francisco, CA.

Clements, R.L. 1980. Note on the effect of removal of free flour lipids on the internal structure of cookies as observed by a resin-embedding method. *Cereal Chemistry* 57:445–446.

Clements, R.L. and J.R. Donelson. 1981. Functionality of specific flour lipids in cookies. *Cereal Chemistry* 58:204–206.

Cole, E.W., D.K. Mecham, and J.W. Pence. 1960. Effect of flour lipids and some lipid derivatives on cookie-baking characteristics of lipid-free flours. *Cereal Chemistry* 37:109–121.

Cole, E.W. 1967. Isolation and chromatographic fractionation of hemicelluloses from wheat flour. *Cereal Chemistry* 44:411–416.

Creighton, D.W. and R.C. Hoseney. 1990. Use of a Kramer shear cell to measure cracker flour quality. *Cereal Chemistry* 67:111–114.

D'Appolonia, B.I. and S.K. Kim. 1976. Recent developments in wheat flour pentosans. *Baker's Digest* 50:45–49, 53–54.

Doescher, L.C. and R.C. Hoseney. 1985. Saltine crackers: Changes in cracker sponge rheology and modification of a cracker-baking procedure. *Cereal Chemistry* 62:158–162.

Doescher, L.C., R.C. Hoseney, and G.A. Milliken. 1987. A mechanism for cookie dough setting. *Cereal Chemistry* 64:158–163.

Donelson, J.R., W.T. Yamazaki, and L.T. Kissell. 1984. Functionality in white layer cake of lipids from untreated and chlorinated patent flours. II. Flour fraction interchange studies. *Cereal Chemistry* 61:88–91.

Donelson, J.R. 1990. Flour fraction interchange studies of effects of chlorination on cookie flours. *Cereal Chemistry* 67:99–100.

Donelson, J.R., and C.S. Gaines. 1998. Starch–water relationships in the sugar-snap cookie dough system. *Cereal Chemistry* 75:660–664.

Donelson, J.R., C.S. Gaines, and P.L. Finney. 2000. Baking formula innovation to eliminate chlorine treatment of cake flour. *Cereal Chemistry* 77:53–57.

Faridi, H. and V.F. Rasper. 1987. *The Alveograph Handbook*, 28-33. St. Paul, MN: The American Association of Cereal Chemists.

Finney, K.F. and W.T. Yamazaki. 1953. An alkaline viscosity test for soft wheat flours. *Cereal Chemistry* 30:153–159.

Finney, P.L. and L.C. Andrews. 1986. Revised microtesting for soft wheat quality evaluation. *Cereal Chemistry* 63:177–182.

Finney, P.L. 1989. Soft wheat: View from the eastern United States. *Cereal Foods World* 34:682, 684, 686–687.

Finney, P.L. 1994. Grain quality: Milling and baking requirements for soft wheat products. In *Cookie Chemistry and Technology*, Ed. K. Kulp, 51–87. Manhattan, KS: The American Institute of Baking.

Finney, P.L. and G.S. Bains. 1999. Protein functionality differences in eastern US soft wheat cultivars and interrelation with end-use quality tests. *Lebensmittel Wissenschaft und Technologie* 32:406–415.

Finnie, S.M., A.D. Bettge, and C.F. Morris. 2006. Influence of cultivar and environment on water-soluble and water-insoluble arabinoxylans in soft wheat. *Cereal Chemistry* 83:617–623.

Fuerst, E.P., J.V. Anderson, and C.F. Morris. 2006. Delineating the role of polyphenol oxidase in the darkening of alkaline wheat noodles. *Journal of Agricultural and Food Chemistry* 54:2378–2384.

Gaines, C.S. and J.R. Donelson. 1982. Cake batter viscosity and expansion upon heating. *Cereal Chemistry* 59:237–240.

Gaines, C.S. 1985. Associations among soft wheat flour particle size, protein content, chlorine response, kernel hardness, milling quality, white layer cake volume, and sugar-snap cookie spread. *Cereal Chemistry* 62:290–292.

Gaines, C.S. and J.R. Donelson. 1985a. Influence of certain flour quality parameters and postmilling treatments on size of angel food and high-ratio white layer cakes. *Cereal Chemistry* 62:60–63.

Gaines, C.S. and J.R. Donelson. 1985b. Effect of varying flour protein content on angel food and high-ratio white layer cake size and tenderness. *Cereal Chemistry* 62:63–66.

Gaines, C.S., P.F. Finney, L.M. Fleege, and L.C. Andrews. 1996a. Predicting a hardness measurement using the single-kernel characterizaiton system. *Cereal Chemistry* 73:278–283.

Gaines, C.S., A. Kassuba, and P.L. Finney. 1996b. Using wire-cut and sugar-snap formula cookie test baking methods to evaluate distinctive soft wheat flour sets: Implications for quality testing. *Cereal Foods World* 41:155–160.

Gaines, C.S., P.L. Finney, and L.C. Andrews. 1997. Influence of kernel size and shriveling on soft wheat milling and baking quality. *Cereal Chemistry* 74:700–704.

Gaines, C.S. 2000. Collaborative study of methods for solvent retention capacity profiles (AACC Method 56-11). *Cereal Foods World* 45:303–306.

Gaines, C.S., P.L. Finney, and L.C. Andrews. 2000. Developing agreement between very short flow and longer flow test wheat mills. *Cereal Chemistry* 77:187–192.

Gaines, C.S. 2004. Prediction of sugar-snap cookie diameter using sucrose solvent retention capacity, milling softness, and flour protein content. *Cereal Chemistry* 81:549–552.

Gajderowicz, L.J. 1979. Progress in the refrigerated dough industry. *Cereal Foods World* 24:44–45.

Gough, B.M., M.E. Whitehouse, and C.T. Greenwood. 1978. The role and function of chlorine in the preparation of high-ratio cake flour. *CRC Critical Reviews in Food Science and Human Nutrition* 10:91–113.

Greer, E.N. and B.A. Steward. 1959. The water absorption of wheat flour; relative effects of protein and starch. *Journal of the Science of Food and Agriculture* 10:248–252.

Greffeuille, V., J. Abecassis, C. Bar L'Helgouac'h, and V. Lullien-Pellerin. 2005. Differences in the aleurone layer fate between hard and soft common wheats at grain milling. *Cereal Chemistry* 82:138–143.

Gruppen, H., R.J. Hamer, and A.G.J. Voragen. 1992. Water-unextractable cell wall material from wheat flour. 2. Fractionation of alkali-extracted polymers and comparison with water-extractable arabinoxylans. *Journal of Cereal Science* 16:53–67.

Guttieri, M.J., C. Becker, and E.J. Souza. 2004. Application of wheat meal solvent retention capacity tests within soft wheat breeding populations. *Cereal Chemistry* 81:261–266.

Hareland, G. 1994. Evaluation of flour particle size distribution by laser diffraction, sieve analysis, and near-infrared reflectance spectroscopy. *Journal of Cereal Science* 20:183–190.

Hazen, S.P., P.K.W. Ng, and R.W. Ward. 1997. Variation in grain functional quality for soft winter wheat. *Crop Science* 37:1086–1093.

Hoseney, R.C. 1984. Functional properties of pentosans in baked foods. *Food Technology* 38:114–117.

Hoseney, R.C., P. Wade, and J.W. Finley. 1988. Soft wheat products. In *Wheat Chemistry and Technology Volume II*, Ed. Y. Pomeranz, 407–456. St. Paul, MN: American Association of Cereal Chemists.

Hou, G., H. Yamamoto, and P.K.W. Ng. 1996a. Relationships of quantity of gliadin subgroups of selected U.S. soft wheat flours to rheological and baking properties. *Cereal Chemistry* 73:352–357.

Hou, G., H. Yamamoto, and P.K.W. Ng. 1996b. Relationships of quantity of glutenin subunits of selected U.S. soft wheat flours to rheological and baking properties. *Cereal Chemistry* 73:358–363.

Howard, N.B., D.H. Hughes, and R.G.K. Strobel. 1968. Function of the starch granule in the formation of layer cake structure. *Cereal Chemistry* 45:329–338.

Huang, G., J.W. Finn, and E. Varriano-Marston. 1982. Flour chlorination. I. Chlorine location and quantitation in air-classified fractions and physiochemical effects on starch. *Cereal Chemistry* 59:496–500.

Huebner, F.R., J.A. Bietz, T. Nelsen, G.S. Bains, and P.L. Finney. 1999. Soft wheat quality as related to protein composition. *Cereal Chemistry* 76:650–655.

Ingelin, M.E. 1997. Comparison of two recording dough mixers: The Farinograph and Mixograph. In *The Mixograph Handbook*, Eds. C.E. Walker, J.L. Hazelton, and M.D. Shogren, 5–10. Lincoln, NE: National Manufacturing Division TMCO.

Johnson, A.C. and R.C. Hoseney. 1979a. Chlorine treatment of cake flours. III. Fractionation and reconstitution techniques for Cl_2-treated and untreated flours. *Cereal Chemistry* 56:443–445.

Johnson, A.C. and R.C. Hoseney. 1979b. Chlorine treatment of cake flour. II. Effect of certain ingredients in the cake formula. *Cereal Chemistry* 56:336–338.

Johnson, A.C., R.C. Hoseney, and E. Varriano-Marston. 1979. Chlorine treatment of cake flours. I. Effect of lipids. *Cereal Chemistry* 56:333–335.

Johnson, A.C., R.C. Hoseney, and K. Ghaisi. 1980. Chlorine treatment of cake flours. V. Oxidation of starch. *Cereal Chemistry* 57:94–96.

Jun, W.J., P.A. Seib, O.K. Chung. 1998. Characteristics of noodle flours from Japan. *Cereal Chemistry* 75:820–825.

Kaldy, M.S. and G.L. Rubenthaler. 1987. Milling, baking, and physical-chemical properties of selected soft white winter wheat and spring wheats. *Cereal Chemistry* 64:302–307.

Kim, Y.S. and R.A. Flores. 1999. Determination of bran contamination in wheat flours using ash content, color, and bran speck counts. *Cereal Chemistry* 76:957–961.

Kissell, L.T. 1959. A lean-formula cake method for varietal evaluation and research. *Cereal Chemistry* 36:168–175.

Kissell, L.T., Y. Pomeranz, and W.T. Yamazaki. 1971. Effects of flour lipids on cookie quality. *Cereal Chemistry* 48:655–662.

Kissell, L.T. and W.T. Yamazaki. 1979. Cake baking dynamics: Relation of flour-chlorination rate to batter expansion and layer volume. *Cereal Chemistry* 56:324–327.

Kissell, L.T., J.R. Donelson, and R.L. Clements. 1979. Functionality in white layer cake of lipids from untreated and chlorinated patent flours. I. Effects of free lipids. *Cereal Chemistry* 56:11–14.

Kitterman, J.S. and G.L. Rubenthaler. 1971. Assessing quality of early generation wheat selections with micro AWRC test. *Cereal Science Today* 16:313–314, 316, 328.

Konik, C.M., D.M. Miskelly, and P.W. Gras. 1992. Contribution of starch and non-starch parameters to the eating quality of Japanese white salted noodles. *Journal of the Science of Food and Agriculture* 58:403–406.

Kreis, M., P.R. Shewry, B.G. Forde, J. Forde, and B.J. Miflin. 1985. Structure and evolution of seed storage proteins and their genes with particular references to those of wheat, barley and rye. In *Oxford Surveys of Plant and Molecular Cell Biology*, Ed. B.J. Miflin, 253–317. London: Oxford University Press.

Kruger, J.E., D.W. Hatcher, and R. Depauw. 1992. A comparison of methods for the prediction of cantonese noodle colour. *Canadian Journal of Plant Science* 72:1021–1029.

Kruger, J.E., D.W. Hatcher, and R. Depauw. 1994. A whole seed assay for polyphenol oxidase in Canadian prairie spring wheats and its usefulness as a measure of noodle darkening. *Cereal Chemistry* 71:324–326.

Kulp, K. 1968. Penstosans of wheat endosperm. *Cereal Science Today.* 13:414–417, 426.

Kulp, K., C.C. Tsen, and C.J. Daly. 1972. Effect of chlorine on the starch component of soft wheat-flour. *Cereal Chemistry* 49:194–200.

Lee, L., P.K.W. Ng, and J.F. Steffe. 2002. A modified procedure (one-stage fermentation) for evaluating flour cracker-making potential. *Food Engineering Progress* 6:201–207.

Lin, W. and G. Vocke. 2004. Hard white wheat at a crossroads/WHS-04K-01, *Electronic Outlook Report from the Economic Research Service*, U.S. Department of Agriculture (December): www.ers.usda.gov/publications/whs/dec04/whs04K01/whs04K01.pdf.

Martin, C.R., R. Rousser, and D.L. Barbec. 1993. Development of a single-kernel wheat characterization system. *Transactions of the ASAE* 36:1399–1404.

Medcalf, D.G. and K.A. Gilles. 1968. Structural characterization of a pentosan from the water-insoluble portion of durum wheat endosperm. *Cereal Chemistry* 45:550–556.

Miller, B.S., H.B. Trimbo, and K.R. Powell. 1967. Effects of flour granulation and starch damage on the cake making quality of soft wheat flour. *Cereal Science Today* 12:245–247, 250–252.

Miller, R.A., R. Mathew, and R.C. Hoseney. 1996. Use of a thermo mechanical analyzer to study apparent glass transition in cookie dough. *Journal of Thermal Analysis and Calorimetry* 47:1329–1338.

Miller, R.A. and R.C. Hoseney. 1997. Factors in hard wheat flour responsible for reduced cookie spread. *Cereal Chemistry* 74:330–336.

Mizukoshi, M., H. Maeda, and H. Amano. 1980. Model studies of cake baking. II. Expansion and heat set of cake batter during baking. *Cereal Chemistry* 57:352–355.

Montzheimer, J.W. 1931. A study of methods for testing cake flour. *Cereal Chemistry* 8:510–517.

Morris, C.F., G.E. King, and G.L. Rubenthaler. 1997. Contribution of wheat flour fractions to peak hot paste viscosity. *Cereal Chemistry* 74:147–153.

Morrison, W.R. 1978a. Wheat lipid composition. *Cereal Chemistry* 55:548–558.

Morrison, W.R. 1978b. Stability of wheat starch lipids in untreated and chlorine-treated cake flours. *Journal of the Science of Food and Agriculture* 29:365–371.

Nemeth, L.J., P.C. Williams, and W. Bushuk. 1994. A comparative study of the quality of soft wheats from Canada, Australia, and the United States. *Cereal Foods World* 39:691–694, 696–698, 700.

Ng, P.K.W., L. Siler, and E. Tanhehco. 2007. *MSU wheat quality testing program report*, Department of Food Science and Human Nutrition, Michigan State University, East Lansing, MI.

Oda, M., Y. Yasuda, S. Okazaki, Y. Yamauchi, and Y. Yokoyama. 1980. A method of flour quality assessment for Japanese noodles. *Cereal Chemistry* 57:253–254.

Osborne, T.B. 1907. The protein of the wheat kernel. *Publication No. 84.* Carnegie Institute: Washington, DC.

Park, W.J., D.R. Shelton, C.J. Peterson, T.J. Martin, S.D. Kachman, and R.L. Wehling. 1997. Variation in polyphenol oxidase activity and quality characteristics among hard white wheat and hard red winter wheat samples. *Cereal Chemistry* 74:7–11.

Payne, P.I., M.A. Nightingale, A. Krattiger, and L.M. Holt. 1987. The relationship between HMW-glutenin subunit composition and the bread-making quality of British-grown wheat varieties. *Journal of the Science of Food and Agriculture* 40:51–56.

Perlin, A.S. 1951a. Isolation and composition of the soluble pentosans of wheat flour. *Cereal Chemistry* 28:370–381.

Perlin, A.S. 1951b. Structure of the soluble pentosans of wheat flours. *Cereal Chemistry* 28:382–393.

Pizzinatto, A. and R.C. Hoseney. 1980. A laboratory method for saltine crackers. *Cereal Chemistry* 57:249–252.

Pomeranz, Y. 1988. Soft wheat products. In *Wheat Chemistry and Technology Volume II*, Ed. Y. Pomeranz, 219–370. St. Paul, MN: American Association of Cereal Chemists.

Posner, E.S., and A.N. Hibbs. 1997. *Wheat Flour Milling*. St. Paul, MN: American Association of Cereal Chemists.

Ranhotra, G. and J. Gelroth. 1988. Soluble and insoluble fiber in soda crackers. *Cereal Chemistry* 65:159–160.

Russo, J.V. and C.A. Doe. 1970. Heat treatment of flours as an alternative to chlorination. *Journal of Food Technology* 5:363–374.

Seguchi, M. and J. Matsuki. 1977a. Studies on pan-cake baking. II. Effect of lipids on pancake qualities. *Cereal Chemistry* 54:918–926.

Seguchi, M. and J. Matsuki. 1977b. Studies on pan-cake baking. I. Effect of chlorination of flour on pan-cake qualities. *Cereal Chemistry* 54:287–299.

Seguchi, M. 1984. Oil-binding capacity of prime starch from chlorinated wheat flour. *Cereal Chemistry* 61:241–244.

Seguchi, M. 1985. Model experiments on hydrophobicity of chlorinated starch and hydrophobicity of chlorinated surface protein. *Cereal Chemistry* 62:166–169.

Seguchi, M. 1990. Study of wheat starch granule surface proteins from chlorinated wheat flours. *Cereal Chemistry* 67:258–260.

Shelke, K., J.M. Faubion, and R.C. Hoseney. 1990. The dynamics of cake baking as studied by a combination of viscometry and electrical resistance oven heating. *Cereal Chemistry* 67:575–580.

Sinha, N.K., H. Yamamoto, and P.K.W. Ng. 1997. Effects of flour chlorination on soft wheat gliadins analyzed by reversed-phase high-performance liquid chromatography, differential scanning calorimetry and fluorescence spectroscopy. *Food Chemistry* 59:387–393.

Slade, L. and H. Levine. 1994. Structure-function relationships of cookie and cracker ingredients. In *The Science of Cookie and Cracker Production*, Ed. H. Faridi, 23–141. New York: Chapman and Hall/AVI.

Smith, E.E. 1932. Report of the subcommittee on hydrogen-ion concentration with special reference to the effect of flour bleach. *Cereal Chemistry* 9:424–428.

Sollars, W.F. 1958. Cake and cookie fractions affected by chlorine bleaching. *Cereal Chemistry* 35:100–110.

Sollars, W.F. and G.L. Rubenthaler. 1971. Performance of wheat and other starches in reconstituted flours. *Cereal Chemistry* 48:397–410.

Souza, E., M. Kruk, and D.W. Sunderman. 1994. Association of sugar-snap cookie quality with high molecular weight glutenin alleles in soft white spring wheats. *Cereal Chemistry* 71:601–605.

Spies, R.D. and A.W. Kirleis. 1978. Effect of free flour lipids on cake-baking potential. *Cereal Chemistry* 55:699–704.

Takeda, K. 1994. Effects of various lipid fractions of wheat flour on expansion of sponge cake. *Cereal Chemistry* 71:6–9.

Thomasson, C.A., R.A. Miller, and R.C. Hoseney. 1995. Replacement of chlorine treatment for cake flour. *Cereal Chemistry* 72:616–620.

Tsen, C.C., K. Kulp, and C.J. Daly. 1971. Effects of chlorine on flour proteins, dough properties, and cake quality. *Cereal Chemistry* 48:247–255.

Uriyo, M.B., W.E. Barbeau, C.A. Griffey, and J. Rancourt. 2004. Examination of relationships between the rheological properties and baking performance of selected soft wheat flours. *Journal of Food Quality* 27:239–254.

USDA Economic Research Service. Updated July 21, 2006. Wheat: Background.www.ers. usda.gov/Briefing/Wheat/background.htm#classes.

USDA Foreign Agriculture Service. 2007. World Wheat Production, Consumption, and Stocks. Foreign Agricultural Service's Production, Supply and Distribution (PSD) online database. www.fas.usda.gov/psdonline/psdHome.aspx.

Varriano-Marston, E. 1985. Flour chlorination: New thoughts on an old topic. *Cereal Foods World* 30:339–343.

Wang, M., H.D. Sapirstein, A. Machet, and J.E. Dexter. 2006. Composition and distribution of pentosans in millstreams of different hard spring wheats. *Cereal Chemistry* 83:161–168.

Yamamoto, H., S.T. Worthington, G. Hou, and P.K.W. Ng. 1996. Rheological properties and baking qualities of selected soft wheats grown in the United States. *Cereal Chemistry* 73:215–221.

Yamazaki, W.T. 1953. An alkaline water retention capacity test for the evaluation of cookie baking potentialities of soft winter wheat flours. *Cereal Chemistry* 30:242–246.

Yamazaki, W.T. 1954. Interrelations among bread dough absorption, cookie diameter, protein content, and alkaline water retention capacity of soft winter wheat flour. *Cereal Chemistry* 31:135–142.

Yamazaki, W.T. 1955. The concentration of a factor in soft wheat flours affecting cookie quality. *Cereal Chemistry* 32:26–37.

Yamazaki, W.T. and D.H. Donelson. 1972. Relationship between flour particle size and cake-volume potential among eastern soft wheats. *Cereal Chemistry* 49:649–653.

Yamazaki, W.T. and L.C. Andrews. 1982. Experimental milling of soft wheat cultivars and breeding lines. *Cereal Chemistry* 59:41–45.

2 Functions of Ingredients in the Baking of Sweet Goods

Dasappa Indrani, Gandham Venkateswara Rao

CONTENTS

2.1   INTRODUCTION

Sweet goods are, as the name implies, sweet to taste, made from a formula high in sugar. The ingredients of sweet baked goods are flour, shortening, eggs, nonfat dry milk, yeast, salt, leavening agents, additives, flavors, water, and various other enriching ingredients, and so forth. The percentage levels of ingredients used in different sweet goods are presented in Table 2.1. Each one of these ingredients has its own role and function in the preparation of the product. The role of ingredients will vary from one type of product to another.

Among the various yeast-raised goods made in bakeries, sweet goods are the most common. Some of the examples of sweet goods are yeast-raised sweet breads and rolls, cakes, biscuits, cookies, and doughnuts. Yeast-raised sweet dough is similar to bread dough but contains high sugar and fat levels. Most sweet goods use white flour rather than whole wheat flour, because whole wheat flour can result in reduced

TABLE 2.1

Formulations, Products, and Ingredients

Ingredients	Yeasted Sweet Dough	Cakes	Biscuits	Cookies	Doughnuts
Wheat flour	100	100	100	100	100
Compressed yeast	2 –2.5	—	—	—	2–5
Sugar	10–30	60–120	15–35	50–60	5–12
Fat/shortening	10–20	50–100	10–25	50–60	10–15
Eggs	5–10	50–100	—	10	5–10
Salt	1.5–2	—	0.5	—	—
Nonfat dry milk	5–10	2–5	2–5	—	4–8
Baking powder	—	—	—	—	1–2
Sodium bicarbonate	—	—	0.5	—	—
Ammonium bicarbonate	—	—	1.0	—	—
Flavor	—	Variable	Variable	Variable	Variable
Water	Variable	Variable	Variable	—	Variable

volume. The sweetness in sweet baked goods often comes from sucrose. On a flour weight basis traditionally known as baker's percent, 15 to 25% sugar is typical for these products. Whole eggs and milk solids are constituents of most sweet dough formulations. These ingredients add richness, flavor, and tenderness to the product (Sharon, 2000).

Additives are substances intentionally added to bakery products in smaller quantities, with a view to improve the functional performance of the raw materials, processing characteristics, appeal, palatability, quality of products, and storage stability. The various additives commonly used in baking are oxidizing agents (potassium bromate or ascorbic acid), reducing agents (cysteine hydrochloride, potassium metabisulfite), vital wheat gluten, enzymes (fungal α-amylase, protease), surfactants and emulsifiers (glycerol monostearate, sodium stearoyl-2-lactylate, lecithin), and hydrocolloids (guar gum, xanthan, hydroxyl propyl methyl cellulose).

Discussed in this chapter are the functions of sugar, fat, eggs, leavening agents, water, salt, nonfat dry milk, and additives in yeasted doughs, cakes, cookies, and biscuits.

2.2   FUNCTIONS OF SUGAR

2.2.1   Yeasted Doughs

Sucrose is a disaccharide composed of a unit of dextrose plus a unit of fructose. The term *sugar* is commonly used to refer to sucrose.

Sugar is the most commonly used sweetener in sweet dough products. Sugar's main function is to provide food for yeast. When added to a dough, sugar is hydrolyzed, or inverted, almost instantly into glucose and fructose by the yeast enzyme invertase. Yeast ferments glucose and fructose into carbon dioxide and alcohol. In typical bread production, 2 to 3% sugar is adequate to sustain yeast activity. This food supply can come from added sugar or from the enzymatic conversion of the starch to sugar or from a combination of both. Even though adequate carbon dioxide gas production can be maintained with 2 to 3% sugar, higher levels are normally used in sweet dough formulations. Sugar that remains unfermented by yeast appears as residual sugar in the finished products. Residual sugar takes part in caramelization and the Maillard reaction (i.e., the reaction between reducing sugar and the proteins of flour to promote rapid color and taste formation). Sugar provides sweet taste in bread if used above 6% (Dubois, 1981b; Pyler, 1988a).

Sweet goods contain high levels of sugar, which affect the activity of yeast. Sugar has a strong inhibitory effect on the gassing power of yeast, caused by high osmotic pressure on the yeast cell. Sweet dough with 20% sugar requires two to three times more yeast to obtain the same gas production as that of typical lean dough.

The amount of protein, damaged starch, and pentosans in flour will influence the amount of water that it will hold. Ultimately, water makes up about 45% of bread dough. During mixing, it is known that a considerable amount of water becomes bound to many different ingredients, such as flour, sugar, and fat. Sugar will serve to decrease the strength of gluten development due to its competition for water. It inhibits the gliadin–glutenin–water complex, and gluten is thus weakened.

2.2.2 CAKES

Cake is a baked batter made from wheat flour, sugar, eggs, shortening, leavening agents, salt, nonfat dry milk, flavors, and water. High-ratio cakes, rich in sugar and fat, are extensively used in the baking industry. Cake batter is a complex fat-in-water emulsion composed of bubbles as the discontinuous phase and egg–sugar–water–fat mixtures as the continuous phase in which flour particles are dispersed (Mizukoshi, 1983; Ngo and Taranto, 1986; Shelke et al., 1990).

Sucrose is a principal ingredient in cakes. It provides energy and sweetness. It also facilitates air incorporation. It acts as a tenderizer by retarding and restricting gluten formation, increasing the temperatures of egg protein denaturation and starch gelatinization, and contributing to bulk and volume.

In high-ratio cake formulation, sugar results in a good air incorporation leading to a more viscous and stable foam (Paton et al., 1981). In addition, sugar affects the physical structure of baked products by regulating gelatinization of starch. Delay in starch gelatinization during baking allows air bubbles to expand properly due to vapor before the cake sets (Kim and Setser, 1992; Kim and Walker, 1992). At the concentration used in cakes (55 to 60%), sugar delays the gelatinization of starch from 57 to 92°C, which allows the formation of desired cake structure (Spies and Hoseney, 1982; Bean et al., 1978). Sugar's ability to limit the water available to the starch is thought to delay gelatinization (D'Appolonia, 1972; Derby et al., 1975; Hoseney et al., 1977). According to Spies and Hoseney (1982), sugar delays gelatinization through a combination of two independent mechanisms: lowering the water activity of the solution and interacting with starch chains to stabilize the amorphous regions of the granule.

Mizukoshi (1985) studied the effect of varying sugar content on shear modulus measured during cake baking, while keeping the proportion of other ingredients constant. He showed that below 20%, sugar has no effect on shear modulus, whereas 30 to 40% sugar reduces it appreciably, revealing the existence of a threshold value associated with the variation of sugar content in the formula.

2.2.3 COOKIES

The word *cookie* means "little cake." Cookies are made from soft and weak flours. They are characterized by a formula that is high in sugar and fat but low in water.

Sugar adds sweetness, acts as a tenderizing agent, and affects spread. Using a farinograph, Olewnik and Kulp (1984) observed that an increase in sugar concentration in a cookie dough reduces its consistency and cohesion. Sucrose acts as a hardening agent by crystallizing as the cookie cools and making the product crisp. However, at moderate amounts, it acts as a softener due to the ability of sucrose to retain water (Schanot, 1981). Sugar makes the cooked product fragile, because it controls hydration and tends to disperse the protein and starch molecules, thereby preventing the formation of a continuous mass (Bean and Setser, 1992).

2.2.4 Biscuits

The major ingredients of biscuit dough are flour, sugar, and fat. The quality of biscuits is governed by the nature and quality of the ingredients used. A variety of shapes and textures may be produced by varying the proportion of these ingredients. Rotary mold cookies are characterized by a formula fairly high in sugar and fat and very low in the amount of water.

Sugar is important in the taste and structure of most biscuits. The amount of sugar that goes into solution depends on the particle size of the sugar and influences the spread of biscuits and machining properties of dough to a great extent (Matz and Matz, 1978). Vetter (1984) studied the effect of sugar quality and its grain size on biscuit spreading. Vetter concluded that fine grain size and a high concentration of sugar contribute to significant spreading of the biscuit.

The limited amount of water used in biscuit formulation, and also its nonavailability to protein and starch, particularly contributes to the crispness of biscuits. The addition of sugar to the formula decreases dough viscosity and relaxation time. Sugar promotes biscuit length and reduces thickness and weight. Biscuits rich in sugar are characterized by high cohesive structure and crisp texture. Increasing the amount of sugar generally increases the spread and reduces the thickness of biscuits (Kissel et al., 1973; Vetter, 1984).

2.3 FUNCTIONS OF FAT

2.3.1 Yeasted Doughs

Fat or shortening is used at the level of 5 to 15% in bread making. It is thought that fat lubricates the gluten fibrils and makes the dough more extensible, thereby improving the gas retention capacity of the dough. The addition of fat facilitates dough handling and processing and improves loaf volume, crumb grain uniformity, tenderness, slicing properties, and shelf life.

2.3.2 Cakes

The major function of fat is to entrap air into the batter during mixing. In cake batter, the largest part of the fat crystals remains in the aqueous phase. When air starts expanding, fat crystals adsorbed to the air–water interface melt and thereby release the fat–water interface for bubble expansion. A large number of adsorbed crystals release sufficient interface to allow the bubbles to expand without rupturing (Brooker, 1993a, 1993b).

2.3.3 Cookies

Cookie dough characteristics depend on the quality and quantity of the ingredients used in the formulation. Cookie dough has high percentages of sugar or shortening and limited water. During baking, cookie diameter increases linearly and becomes suddenly fixed (Abboud et al., 1985; Miller et al., 1996; Yamazaki, 1959). The final cookie diameter depends upon the rate at which dough spreads and its setting time during baking. Cookie spread rate has been reported to be dependent on dough viscosity (Miller, 1989; Yamazaki, 1959). The viscosity of cookie dough depends upon the ratio of ingredients used in the cookie formula. The addition of fat influences the texture and taste of cookies, making the cookies crispier because this allows the dough to spread as it cooks on the hot cookie sheet.

2.3.4 BISCUITS

Fat is an essential ingredient in biscuit manufacture and is the largest component after flour and sugar. During mixing of the biscuit dough, fat acts as a lubricant; it also competes with the aqueous phase for the flour surface and prevents the formation of a gluten network in the dough (Wade, 1988). The addition of fat softens the biscuit dough and decreases the viscosity and relaxation time. Fat contributes to an increase in length and to a reduction in thickness and weight of biscuits, which are then characterized by a variable structure and are easy to break. Fat or shortening contributes to the plasticity of the dough as a lubricant. When present in large quantities, its lubricating effect is so pronounced that very little water is needed to achieve a soft consistency. When mixed with the flour before its hydration, the fat prevents the formation of a gluten network and produces less-elastic dough. High-elastic dough is not desirable in biscuit making, because it shrinks after lamination (Menjivar and Faridi, 1994). Starch swelling and its gelatinization are also reduced at high levels of fat, giving a crisp texture. Fat influences the dough machinability during processing, the dough spread after cutting out, and the textural and gustatory qualities of the biscuit after baking (Vetter, 1984).

2.4   FUNCTIONS OF EGGS

2.4.1 YEASTED DOUGHS

Eggs in yeasted doughs increase nutritive value, improve flavor and texture, produce color in crumb and crust, act as a binding agent to hold the ingredients together, aid in leavening, contribute to the emulsifying action due to the presence of the natural emulsifier lecithin, and produce a softer crumb because of the fat and other solids. Eggs contain 73 to 75% moisture, have the natural ability to bind and retain moisture, and hence improve quality. Eggs are an important source of iron, calcium, phosphorus, vitamin A, vitamin D, thiamine, and riboflavin, and they supply all essential amino acids (Pyler, 1988b).

The complex composition of eggs imparts numerous functional effects on baked products. The loaf volume advantage imparted by eggs documented for sweet baked goods may relate to increased dough water content or the emulsifying, coagulating, and leavening action of whole eggs (Forsythe, 1970).

2.4.2 CAKES

Eggs contribute structure to a baked product. They may serve to do this through their contribution of heat-denatured proteins, steam for leavening, or moisture for starch gelatinization. Egg white has the ability to form foams that are stable enough to support large quantities of flour or sugar. These foams must be capable of holding the other ingredients until heat coagulation can occur in the oven and a stable protein matrix develops. Globulins are primarily responsible for lowering surface tension and increasing viscosity where air gets incorporated. As foam develops, bubbles become smaller, the surface is greatly enlarged, ovomucin (protein) undergoes surface denaturation to form a solid foam, and the volume of foam increases. Ovalbumin, which is readily heat coagulable, sets up in heat and supports many times its weight of sugar and flour (MacDonnel et al., 1955).

Eggs may contribute liquid to a product and thus serve as a toughener. It is a toughener partially due to its contribution to gelatinization of starch and for development of gluten. The egg white portion appears to be particularly effective as a toughener. Actually, the yolk serves as a tenderizer probably due to its fat content. Egg also contributes to leavening action through the emulsification of fat and air incorporation, the foaming action, and the contribution of water to steam. Egg yolk is also a rich source of emulsifying agent and thus facilitates the incorporation of air, inhibits starch gelatinization, and contributes to a desirable golden color that gives rich appearance and flavor (Pyler, 1988b).

2.4.3 COOKIES

Eggs help in puffing, emulsifying the dough, and bringing the water and fat phases together to result in a creamier and smoother texture in cookies. Egg whites have a drying effect and contribute to the structure or shape.

2.5 FUNCTIONS OF LEAVENING AGENTS

Leavening is defined as a raising action that aerates dough or batter during mixing and baking so that the finished products are greater in volume and superior in texture. Leavening action in a baked product may be due to mechanical leavening, biological leavening, chemical leavening, and water vapor.

In mechanical leavening, air is incorporated by creaming, beating, or whisking by hand or machine the fat, sugar, and eggs. Cakes, sponge goods, and meringues are examples of mechanical aeration. In biological leavening, the baker's yeast (*Saccharomyces cereviseae*) converts simple sugars to carbon dioxide and alcohol. This carbon dioxide is responsible for the leavening of breads and other fermented bakery products.

Chemical leavening includes the use of chemicals such as baking soda, ammonium bicarbonate, and baking powder. Baking soda, also known as sodium bicarbonate, is used in recipes that contain an acidic ingredient, such as vinegar, buttermilk, chocolate, honey, or fruits. Baking soda liberates carbon dioxide when heated or when mixed with an acid, either hot or cool. Ammonium bicarbonate is commonly known as *vol*, derived from "volatile salt," because of its complete dissociation into carbon dioxide gas, ammonia gas, and water. It is important that all of the ammonia is driven off during baking or unpleasant tastes are encountered. Ammonium bicarbonate is therefore not suitable as a leavening agent in any product that leaves the oven with more than 5% moisture (Dubois, 1981a; Kichline and Conn, 1970).

Baking powder is a dry chemical leavening agent used in baking. There are several formulations—all contain an alkali, typically sodium bicarbonate, and an acid together with starch to keep it dry. When dissolved in water, the acid and alkali react and emit carbon dioxide gas, which expands existing bubbles to leaven the mixture. There are two types of baking powders—single acting and double acting. Baking powders that contain only the low-temperature acid salts, such as cream of tartar, calcium phosphate, and citrate, are called single acting. Double-acting baking powders contain two acid salts: one reacts at room temperature, producing a rise as soon as the dough or batter is prepared, and another reacts at a higher temperature, causing further rise during baking. Examples of high-temperature acid salts are aluminium salts, such as calcium aluminium phosphate.

Steam is a supplementary form of leavening in all products. It is formed when water is changed to water vapor as the temperature of cake batter or bread dough rises in the oven, thus exerting a greater pressure inside the cake batter or bread dough and resulting in an increase in volume of the finished products.

2.5.1 YEASTED DOUGHS

Yeast performs three important functions in yeasted dough: leavening, dough ripening, and flavor development. Yeast utilizes sugar to produce carbon dioxide and ethyl alcohol. Along with ethyl alcohol, yeast also produces several other organic compounds, including organic acids, aldehydes, and ketones, that impart typical flavor to the product.

2.5.2 CAKES

In cake making, three types of leavening action take place. By mechanical means, air is incorporated during the creaming of fat and sugar and the whipping of eggs. Chemically, air is incorporated by the use of baking powder which generates carbon dioxide during baking and by vapor pressure created by the water.

2.5.3 BISCUITS

Biscuits are mainly leavened by chemicals. The use of baking chemicals such as sodium bicarbonate and ammonium bicarbonate makes biscuits porous and crisp.

2.6   FUNCTIONS OF WATER

2.6.1 YEASTED DOUGHS

Water is an essential ingredient in dough formulation. It is necessary for solubilizing other ingredients, for hydrating proteins and carbohydrates, and for the development of a gluten network. It has been estimated that about 46% of the total water absorbed is associated with the starch, 31% with proteins, and 23% with the pentosans (Bushuk, 1966).

Water also acts as a solvent in the dough, and many of the reactions that take place during fermentation cannot occur if there is no solvent. For example, water acts as a solvent for some of the released carbon dioxide gas to form carbonic acid. Carbonic acid contributes to the acid pH of the dough during fermentation, providing a feasible atmosphere for the action of enzymes and yeast in the dough system.

2.6.2 CAKES

Water is present in sufficient quantity in cake batters to dissolve sugar, nonfat dry milk, salt, and other dry ingredients. Water adds moisture to the finished cakes and also regulates the consistency of the batter. It develops the protein in the flour to a very limited extent in order to retain the gas produced by baking powder.

2.6.3 BISCUITS

Water has a complex role in biscuits because it determines the conformational state of biopolymers, affects the nature of interactions between the various constituents of the formula, and contributes to dough structuring (Eliasson and Larsson, 1993). It is also an essential factor in the rheological behavior of flour doughs (Webb et al., 1970). Bloskma (1971) observed that adding water to the formula reduces the viscosity and increases dough extensibility. If the proportion of water is too low, the dough becomes brittle, not consistent, and exhibits a marked "crust" effect due to rapid dehydration of the surface. An increase in water results in the expansion of biscuits lengthwise with a smaller thickness.

2.7 FUNCTIONS OF SALT

2.7.1 YEASTED DOUGHS

Salt is used for flavor and taste, not necessarily its own, but to bring out or enhance the flavor of the other ingredients used in the dough. Usage levels are normally between 1.0 and 2.5%. Salt also inhibits fermentation due to osmotic pressure effect. Yeast cells will partially dehydrate due to the osmotic pressure. The fact that salt influences the fermentation can be used to control the fermentation rate. Salt toughens the gluten. According to Preston (1989), salt causes electrostatic shielding of charged amino acids on the surface of gluten proteins, resulting in increased interprotein hydrophobic and hydrophilic interaction, which results in an increase in dough strength.

2.7.2 CAKES

Salt is used as an adjustment of sweetness in cakes, brings out the flavor of other ingredients in cakes, lowers the caramelization temperature of the batter, and aids in obtaining crust color.

2.7.3 BISCUITS

Salt is used in all biscuit recipes for its flavor and flavor-enhancing properties. Its most effective concentration is around 0.5 to 1.0%. Salt also toughens the gluten and hence reduces stickiness.

2.8   FUNCTIONS OF NONFAT DRY MILK

2.8.1   Yeasted Doughs

Nonfat dry milk provides a richer color and a more tempting appearance to the finished product. It is recommended for use in all sweet dough formulas for the increased tolerances it gives to fermentation and makeup. Milk is high in lysine and calcium, and the overall quality of the milk protein is excellent. Milk improves the nutritional quality of the product. Milk has a buffer effect hence more stable pH, strengthening of the gluten if serum protein has been removed by heat treatment (Pyler, 1988c).

2.8.2   Cakes

Nonfat dry milk in cakes performs the function of structure formation and contributes to crust browning because of its protein and sugar content. Milk contains lactose sugar that regulates crust color.

2.8.3   Biscuits and Cookies

Nonfat dry milk is a minor dough ingredient in biscuits and cookies, used to give subtle flavor and textural improvements and to aid surface coloring.

2.9   FUNCTIONS OF ADDITIVES

2.9.1   Oxidizing Agents

The use of oxidizing agents improves the strength of sweet dough. As a result, they will improve dough handling for better machining and contribute to improved gas retention, giving better volume and a more uniform grain of the crumb. According to Cole (1973), oxidants promote the formation of disulfide bonds among extended molecules, thereby imparting gas-retaining and dough-strengthening properties to gluten film. Oxidizing agents are not beneficial in cakes, cookies, or biscuits.

2.9.2   Reducing Agents

The use of a reducing agent reduces disulfide bonds to the sulfhydryl group. The addition of L-cysteine hydrochloride and potassium metabisulfite weakens the dough; it normally reduces resistance to extension and extensograph area and increases extensibility (Mita and Bohlin, 1983). Reducing agents are useful in certain types of biscuit dough. A reducing agent such as sodium metabisulfite increases cookie spread by decreasing dough stability.

2.9.3   Surfactants and Emulsifiers

In yeasted doughs, a surfactant forms complexes with the protein and starch portions of the viscoelastic wheat flour dough and strengthens the extensible gluten–starch film and delays the setting of the dough during baking. The use of surfactants in yeast-raised products results in increased product volume, a more tender crust and crumb, a finer and more uniform cell structure with thin cell walls which causes a brighter crumb color, and a reduction in the rate of crumb firming due to the complex forming ability with an amylose moiety of a starch molecule and to prevention of its leaching out of the starch granule during storage.

In cake baking, emulsifiers aid the incorporation and subdivision of air into the liquid phase to promote foam formation and also promote uniform dispersion of fat that contains entrapped air cells, thereby providing more sites for the expansion of gas, resulting in greater volume and soft texture (Pyler, 1988d). Emulsifiers provide necessary aeration and gas bubble stability during the process until the cake structure is set (Sahi and Alava, 2003). Emulsifiers also coat the exterior of fat particles so that the surfaces of the particles are no longer disruptive to the protein film (Wootton et al., 1967). Kim and Walker (1992) reported that polysorbate-60 (PS-60) increased batter viscosity more than did sucrose esters. This may have been caused by more air bubbles being incorporated or a more air-viscous continuous phase interacting with water. The volume of high-ratio cake and overall quality increased and crumb firmness decreased with PS-60 addition. Jyotsna et al. (2004) reported that the use of emulsifier gels prepared using sodium stearoyl-2-lactylate (SSL), distilled glycerol monostearate (DGMS), propylene glycol monostearate (PGMS), polysorbate-60 (PS-60), and sorbitol monostearate (SMS) in cake making resulted in a decrease in batter density, an increase in the number of evenly distributed air cell bubbles, and an improvement in specific volume and texture of cake. Among different emulsifier gels, cakes with PS-60 showed a maximum increase in specific volume followed by cakes with SSL, DGMS, PGMS, and SMS gels.

Considerable information is available on the effect of different emulsifiers on the quality of cookies. Tsen et al. (1975) showed that emulsifiers improved the cookie spread and, more significantly, top grain score when they were creamed into a shortening and sugar mixture. A blend of surfactants in semisweet biscuits was shown to have several advantages, such as increased mixing time, greater mixing stability, reduced rate of dough breakdown, uniform fat distribution, prevention of moisture migration, and improvement in texture. Emulsifiers are used to reduce the shortening requirements and increase the shortening effect of fats by promoting the tendency of fat to cream and spread among slightly moist particles of sugars, fiber, and other ingredients. Emulsifiers tend to make cookies drier and improve their machinability (Kamel and Ponte, 1993).

Sai Manohar and Haridas Rao (1999) studied the effect of emulsifiers on rheological characteristics of biscuit dough and quality of biscuits. They reported that the addition of any of the emulsifiers glycerol monostearate, lecithin, or sodium stearoyl-2-lactylate lowered the elastic value, indicating their contribution to the shortening effect on gluten, and also resulted in a reduction in consistency and hardness and made the dough more cohesive. Glycerol monostearate and lecithin brought about a greater improvement in the quality of the biscuit when compared with sodium stearoyl-2-lactylate.

2.9.4   ENZYMES

The functions of enzymes in baked products include flour quality improvement, retardation of staling, dough improvement, and more efficient machinability. Fungal and bacterial enzymes available for use in bakery processing include α-amylase, protease, amyloglucosidases, pentosanase, glucanase, and phytase. The most important of these are α-amylase and protease.

α-amylase enzyme is beneficial in bread, buns, and rolls. Supplementation of α-amylase increases fermentable sugars, improves crust color, increases loaf volume, enhances flavor, improves gas retention through starch modification, increases moisture retention of crumb, and retards staling of yeasted dough products.

Supplementation of dough with protease enzyme helps to break down the gluten protein so that the dough is softer and more extensible (Mathewson, 2000). Protease enzyme is used to decrease the mixing requirement (in bread), improve the machinability of a dough (in saltine crackers), increase dough flow in the oven (cookies), increase pan flow (buns), and counteract the tendency to spring back when pizza dough is being sheeted (Dubois, 1980).

2.9.5   VITAL WHEAT GLUTEN

Vital wheat gluten is the natural wheat protein extracted from flour which still retains all of its gluten-forming characteristics. It is added to the dough to strengthen weak flour and to carry extra ingredients such as sugars, fibers, and grains. Vital wheat gluten is used in yeasted dough products formulation at levels varying from 2 to 4% based on flour. It increases the water absorption of the dough and imparts greater stability to the dough during fermentation. These functional properties of vital gluten are eventually reflected in increased loaf volume, improved grain, improved texture, and softness of crumb.

2.9.6   HYDROCOLLOIDS

Hydrocolloids or gums have been widely used in the food industry in order to improve food texture, slow the retrogradation of the starch, increase moisture retention, and extend the shelf life.

The addition of hydrocolloids increases the water absorption capacity of the flour. The highest increase in water absorption was observed by hydroxyl propyl methyl cellulose and alginate (Guarda et al., 2004). The increase in water absorption is due to the hydroxyl group in the hydrocolloid structure which allows more water interaction through hydrogen bonding (Friend et al., 1993). Use of guar gum up to a level of 1% greatly improved the overall bread-making quality of the flour with special reference to water absorption capacity of the dough, yield of bread, crumb softness, and crust appearance (Venkateswara Rao et al., 1985).

Hydrocolloids modify the pasting properties of the starch. These starch properties, including gelatinization temperature, paste viscosity, and retrogradation of starch, affect cake baking and the final quality of cakes (Christianson et al., 1981; Rojas et al., 1999; Rosell et al., 2001). Shelke et al. (1990) studied the effects of xanthan, guar gum, and carboxy methyl cellulose (CMC) on the quality of white layer cakes. The addition of hydrocolloids increased batter viscosity at ambient temperature over the control values, and xanthan gave higher batter viscosities than guar and CMC. High viscosity during heating would give the batter greater capacity to retain expanding air and nuclei and resist the setting of starch granules, thereby improving both cake volume and crumb grain.

High-quality cakes have various attributes, including high volume, uniform crumb structure, tenderness, long shelf life, and tolerance to staling (Gelinas et al., 1999). These attributes depend on the balanced formulas, aeration of cake batters, and stability of fluid batters in the early stages of baking and the thermal-setting stage. Then, the quality of a finished cake can be influenced by the addition of substances (such as hydrocolloids) that affect these properties. Gomez et al. (2007) studied the functionality of different hydrocolloids—sodium alginate, carrageenan, pectin, hydroxy-propyl-methylcellulose (HPMC), locust bean gum, guar gum, and xanthan gum—on the quality and shelf life of yellow layer cakes. The addition of hydrocolloids reduced the quantity of air retained on cake batter as demonstrated by the increase in its density. Cakes with hydrocolloids had higher volume than the control except when alginate was used. They concluded that the effect of hydrocolloids on yellow layer cake volume increase has to be related to the increase in batter viscosity and the chemical interaction between gums and starch that alters the setting temperature. The overall quality of the yellow layer cake depended notably on the type of gum added. The highest increase in the overall quality score was observed with HPMC followed by xanthan and alginate. Pectin showed a lower overall acceptance score than the control.

2.10 RECENT STUDIES ON THE EFFECTS OF INGREDIENTS ON QUALITY OF CAKES AND BISCUITS

In response to some population sectors with particular nutritional necessities, the food industry is being challenged to redesign traditional foods for optimal nutritional value and for taste that is as good as or better than that of the original. One way to achieve a healthy food product is to reduce or to omit some of the calorie-containing ingredients—especially sugar and fat—because, at present, obesity is frequently cited as a serious health problem (Ronda et al., 2005). There are a number of fat and sugar replacers in the market; however, it is important to consider the functionality of these fat and sugar replacers in a variety of high-sugar- and high-fat-containing products to obtain products with similar quality parameters (Kamel and Rasper, 1988).

Eggs play a multifunctional role in the cake system, affecting foaming, emulsification, texture, water binding, color, and flavor. However, in recent years, concerns over high cholesterol and high cost for people with specific dietary needs or restrictions (vegetarians, people with high cholesterol levels) and food safety issues have led to the replacement of eggs in cakes. The use of vegetable proteins for partial or total substitution of eggs in cake formulations was reported by Arozarena et al. (2001). They suggested the use of white lupine protein, emulsifiers, and xanthan gum to produce egg-free cake. Studies by several authors have suggested the use of bovine blood plasma to substitute for egg white (Lee et al., 1993) and soya flour to substitute for egg (Glibertson and Porter, 2001).

To take the next step toward calorie reduction, sugar and fat replacers are used in the baking industry. Polydextrose, a sugar and fat replacer, is a cross-linked, partially metabolized glucose polymer that adds body and texture to reduced-calorie foods. It provides 1 Kcal/g in comparison with 4 Kcal/g by sucrose and 9 Kcal/g by fat. Kocer et al. (2007) studied the effect of polydextrose as a sugar and fat replacer on the quality of high-ratio cake. They reported that the major outcome of polydextrose substitution was a decrease in the average pore size and pore size uniformity of the cake crust due to the combined effect of reduced batter stability and interference with the gelatinization mechanism. The high-ratio cake system with polydextrose allowed 25% fat and 22% sugar replacement, resulting in 22% reduction in calorific value based on total sugar and fat content.

Ronda et al. (2005) studied the effects of seven bulking agents—maltitol, mannitol, xylitol, sorbitol, isomaltose, oligofructose, and polydextrose—on the quality of sugar-free sponge cake. They obtained the best results with xylitol and maltitol, resulting in sponge cakes more similar to the control, manufactured with sucrose and with the highest acceptance level in sensory evaluations. They also concluded that mannitol proved to be the worst substitute of sucrose among all the bulking agents tested.

Gallagher et al. (2003a) studied the evaluation of raftilose, an oligosaccharide successfully used in food products as a sugar replacer in short dough biscuit production, where the sugar was replaced by 20 to 30%. They reported that at the lower and medium levels of sugar replacement, raftilose can be used successfully to reduce sugar in short dough biscuits.

Gallagher et al. (2003b) developed a formulation for a biscuit containing reduced fat and sugar levels as well as exhibiting functional properties using Novelose®, sodium caseinate, Raftilose®, and Simplesse®. They opined that a combination of the above four functional ingredients produced a biscuit of extremely high standards.

2.11 CONCLUSION

Ingredients play a major role in the development of a product. Sugar confers sweetness, fat lubricates dough and shortens the continuity of gluten, water is necessary for hydrating proteins, carbohydrates are necessary for the development of the gluten network and the solubilizing of other ingredients, nonfat dry milk provides a richer color and a more tempting appearance to the finished product, leavening agents aerate the dough or batter and contribute lightness to the product, and additives bring about improvements in the machinability of the dough and the quality characteristics of products. Replacers of fat, sugar, and egg are used in the bakery industry to improve the nutritional status of biscuits and cakes. Many of the functions of ingredients described in this chapter will be useful for a baking technologist in developing novel formulations for sweet goods, understanding the role of ingredients in the building up of the product, counteracting the problems that may arise during processing, improving the product quality, maintaining the consistency in the quality of the product, and handling bulk productions confidently.

REFERENCES

Abboud, A.M., R.C. Hoseney, and G.L. Rubenthaler. 1985. Factors affecting cookie flour quality. *Cereal Chemistry* 62:130–133.

Arozarena, I., H. Betholo, J. Empis, A. Bunger, and I. De Sousa. 2001. Study of the total replacement of egg by white lupeine protein, emulsifiers and xanthan gum in yellow cakes. *European Food Research Technology* 213:312–316.

Bean, M.M., W.T. Yamazaki, and D.H. Donelson. 1978. Wheat starch gelatinization in sugar solution. II. Fructose, glucose and sucrose: Cake performance. *Cereal Chemistry* 55:945–952.

Bean, M.M. and C.S. Setser. 1992. Polysaccharide sugars and sweeteners. In: Bowers, J. (Ed.). *Food Theory and Applications*, pp. 69–198. Macmillan, New York.

Bloskma, A.H. 1971. Rheology and chemistry of dough. In: Pomeranz, Y. (Ed.). *Wheat Chemistry and Technology*, 2nd ed., pp. 558–560. American Association of Cereal Chemistry, St. Paul, MN.

Brooker, B.E. 1993a. The stabilization of air in cake batters—The role of fat. *Food Structure* 12:285–296.

Brooker, B.E. 1993b. The stabilization of air in foods containing fat—A review. *Food Structure* 12:115–122.

Bushuk, W. 1966. Distribution of water in dough and bread. *Baker's Digest* 40:38–40.

Christianson, D.D., J.E. Hodge, D. Osborne, and R.W. Detroy. 1981. Gelatinization of wheat-starch as modified by xanthan gum, guar gum, and cellulose gum. *Cereal Chemistry* 58:513–517.

Cole, M.S. 1973. An overview of modern dough conditioners. *Baker's Digest* 47:21–23, 64.

D'Appolonia, B.L. 1972. Effect of bread ingredients on starch gelatinization properties as measured by the amylograph. *Cereal Chemistry* 49:532–543.

Derby, R.I., B.S. Miller, B.F. Miller, and H.B. Trimbo. 1975. Visual observations of wheat-starch gelatinization in limited water systems. *Cereal Chemistry* 52:702–713.

DeBois, D.K. 1980. Enzymes in baking. II. Applications. *Technical Bulletin*. American Institute of Baking 2(11): 1–6. Manhattan, KS.

Dubois, D.K. 1981a. Chemical leavening. *Technical Bulletin*. American Institute of Baking 3:1–6. Manhattan, KS.

Dubois, D.K. 1981b. Fermented doughs. *Cereal Foods World* 26:617–622.

Eliasson, A.C. and K. Larsson. 1993. *Cereal in Bread Making: A Molecular Colloidal Approach*, Chap. 6, pp. 261–266. Marcel Dekker, New York.

Forsythe, R.H. 1970. Eggs and egg products as functional ingredients. *Baker's Digest* 44:40.

Friend, C.P., R.D. Waniska, and L.W. Rooney. 1993. Effects of hydrocolloids on processing and qualities of wheat tortillas. *Cereal Chemistry* 70:252–256.

Gallagher, E., C.M. Obrien, A.G.M. Scannel, and E.K. Arendt. 2003a. Evaluation of sugar replacers in short dough biscuit production. *Journal of Food Engineering* 56:261–263.

Gallagher, E., C.M. Obrien, A.G.M. Scannel, and E.K. Arendt. 2003b. Use of response surface methodology to produce functional short dough biscuits. *Journal of Food Engineering* 56:269–271.

Gelinas, P., G. Roy, and M. Guillet. 1999. Relative effects of ingredients on cake staling based on an accelerated shelf-life test. *Journal of Food Science* 64:937–940.

Glibertson, D.B. and M.A. Porter. 2001. Replacing eggs in bakery goods with soy flour. *Cereal Foods World* 46:431–435.

Gomez, M., F. Ronda, P.A. Caballero, C.A. Blanco, and C.M. Rosell. 2007. Functionality of different hydrocolloids on the quality and shelf-life of yellow layer cakes. *Food Hydrocolloids* 21:167–173.

Guarda, A., C.M. Rosell, C. Benedito, and M.J. Galotto. 2004. Different hydrocolloids as bread improvers and antistaling agents. *Food Hydrocolloids* 18:241–247.

Hoseney, R.C., W.A. Atwell, and D.R. Lineback. 1977. Scanning electron microscopy of starch isolated from baked products. *Cereal Foods World* 22:56–60.

Jyotsna, R., P. Prabhasankar, D. Indrani, and G. Venkateswara Rao. 2004. Improvement of rheological and baking properties of cake batters with emulsifier gels. *Journal of Food Science* 69:SNQ 16–19.

Kamel, B.S. and V.F. Rasper. 1988. Effects of emulsifiers, sorbitol, polydextrose and crystalline cellulose on the texture of reduced calorie cakes. *Journal of Texture Studies* 19:307–320.

Kamel, B.S. and J.G. Ponte Jr. 1993. Emulsifiers in baking. In: Kamel, B.S., and Stauffer, C.E. (Eds.). *Advances in Baking Technology*, pp. 195–199. Blackie Academic and Professional, VCH, New York.

Kichline, T.P. and J.F. Conn. 1970. Some fundamental aspects of leavening agents. *Baker's Digest* 44:36–40.

Kim, S.S. and C.S. Setser. 1992. Wheat starch gelatinization in the presence of polydextrose or hydrolysed barley β-glucan. *Cereal Chemistry* 69:447–451.

Kim, C.S. and C.E. Walker. 1992. Interactions between starches, sugars and emulsifiers in high-ratio cake model systems. *Cereal Chemistry* 69:206–212.

Kissel, L.T., B.D. Marshall, and W.T. Yamazaki. 1973. Effect of variability in sugar granulation on the evaluation of flour cookie quality. *Cereal Chemistry* 50:225–264.

Kocer, D., Z. Hicsasmaz, A. Bayindirli, and S. Katnas. 2007. Bubble and pore formation of the high ratio cake formulation with polydextrose as a sugar and fat replacer. *Journal of Food Engineering* 78:953–964.

Lee, C.C., J.A. Love, and L.A. Johnson. 1993. Sensory and physical properties of cakes with bovine plasma products substituted for egg. *Cereal Chemistry* 70:18–21.

MacDonnel, L.R., R.E. Fenny, H.L. Hanson, A. Campbell, and T.F. Sugihara. 1955. The functional properties of the egg white proteins. *Food Technology* 9:49–53.

Mathewson, P.R. 2000. Enzymatic activity during bread baking. *Cereal Food World* 45:98–101.

Matz, S.A. and T.D. Matz. 1978. *Cookie and Cracker Technology*. AVI, Westport, CT.

Menjivar, J.A. and H. Faridi. 1994. Rheological properties of cookie and cracker doughs. In: Faridi, H. (Ed.). *The Science of Cookie and Cracker Production*, pp. 315–316. Chapman and Hall, New York

Miller, R.A. 1997. Factors in hard wheat flour responsible for reduced cookie spread. *Cereal Chemistry* 74: 330–336.

Miller, R.A., R. Mathew, and R.C. Hoseney. 1996. Use of a thermomechanical analyzer to study an apparent glass transition in cookie dough. *Journal of Thermal Analysis* 47:1329–1338.

Mita, T. and L. Bohlin. 1983. Shear stress relaxation of chemically modified gluten. *Cereal Chemistry* 60:93–97.

Mizukoshi, M. 1983. Model studies of cake baking. IV. Foam drainage in cake batter. *Cereal Chemistry* 60:399–402.

Mizukoshi, M. 1985. Model studies of cake baking. VI. Effects of cake ingredients and cake formula on shear modulus of cake. *Cereal Chemistry* 62:247–251.

Ngo, W.H. and M.V. Taranto. 1986. Effect of sucrose level on the rheological properties of cake batters. *Cereal Food World* 31:317–322.

Olewnik, M.C. and K. Kulp. 1984. The effect of mixing time and ingredient variation on farinogram of cookie doughs. *Cereal Chemistry* 61:532–537.

Paton, D., G.M. Larocque, and J. Holme. 1981. Development of cake structure: Influence of ingredients on the measurement of cohesive force during baking. *Cereal Chemistry* 58:527–529.

Preston, K.R. 1989. Effects of neutral salts of the lyotropic series on the physical dough properties of Canadian red spring wheat flour. *Cereal Chemistry* 66:144–148.

Pyler, E.J. 1988a. Sugars and syrups. In: Pyler, E.J. (Ed.). *Baking Science and Technology*, vol 1, pp. 412–413. Sosland, Kansas City, MO.

Pyler, E.J. 1988b. Eggs and egg products. In: Pyler, E.J. (Ed.). *Baking Science and Technology*, vol 1, pp. 540–541. Sosland, Kansas City, MO.

Pyler, E.J. 1988c. Dairy products and blends. In: Pyler, E.J. (Ed.). *Baking Science and Technology*, vol 1, pp. 513–519. Sosland, Kansas City, MO.

Pyler, E.J. 1988d. Bakery shortenings. In: Pyler, E.J. (Ed.). *Baking Science and Technology*, vol 1, pp. 474–477. Sosland, Kansas City, MO.

Rojas, F., C.M. Rosell, and C. Benedito de Barber. 1999. Pasting properties of different wheat flour–hydrocolloids systems. *Food Hydrocolloids* 13:27–33.

Ronda, F., M. Gomez, C.A. Blanco, and P.A. Caballero. 2005. Effects of polyols and nondigestible oligosaccharides on the quality of sugar-free sponge cakes. *Food Chemistry* 90:549–555.

Rosell, C.M., J.A. Rojas, and C. Benedito de Barber. 2001. Combined effect of different antistaking agents on the pasting properties of wheat flour. *Food Research Technology* 212:473–476.

Sahi, S.S. and J.M. Alava. 2003. Functionality of emulsifiers in sponge cake production. *Journal of the Science of Food and Agriculture* 83:1419–1429.

Sai Manohar, R. and P. Haridas Rao. 1999. Effect of emulsifiers, fat level and type on the rheological characteristics of biscuit dough and quality of biscuits. *Journal of the Science of Food and Agriculture* 79:1223–1231.

Schanot, M.A. 1981. Sweeteners: Functionality in Cookies and Crackers. *Technical Bulletin*. American Institute of Baking 3: 4. Manhattan, KS.

Sharon, G. 2000. Formulating irresistible sweet baked goods. *Food Product Design: Design Elements*. September—Snack Mixes, 1–9. www.foodproductdesign.com/articles/463/463-1000de.htm.

Shelke, K., J.M. Faubion, and R.C. Hoseney. 1990. The dynamics of cake baking as studied by a combination of viscometry and electrical resistance oven heating. *Cereal Chemistry* 67:575–580.

Spies, D. and R.C. Hoseney. 1982. Effect of sugars on starch gelatinization. *Cereal Chemistry* 59:128–131.

Tsen, C.C., L.J. Bauck, and W.J. Hoover. 1975. Using surfactants to improve the quality of cookies made from hard wheat flours. *Cereal Chemistry* 52:629–637.

Venkateswara Rao, G., D. Indrani, and S.R. Shurpalekar. 1985. Guar gum as an additive for improving the bread making quality of wheat flours. *Journal of Food Science and Technology* 22:101–104.

Vetter, J.L. 1984. *Technical Bulletin*. VI. American Institute of Baking, Manhattan, KS.

Wade, P. 1988. *Biscuits, Cookies and Crackers, the Principles of the Craft*, vol 1. Elsevier Applied Science, London.

Webb, T., P.W Heaps, P.W. Russell Eggitt, and J.B.M. Coppock. 1970. A rheological investigation of the role of water in wheat flour doughs. *Journal of Food Technology* 5:65–76.

Wootton, J.C., N.B. Howard, J.B. Martin, D.E. McOsker, and J. Holme. 1967. The role of emulsifiers in the incorporation of air into layer cake batter systems. *Cereal Chemistry* 44:333–343.

Yamazaki, W.T. 1959. The application of heat in the testing of flours for cookie quality. *Cereal Chemistry* 32:26–37.

3 Chemical Reactions in the Processing of Soft Wheat Products

Hamit Köksel, Vural Gökmen

CONTENTS

3.1   INTRODUCTION

Food chemistry traces the continuous series of chemical reactions in the chemical system that we call "food" through various stages such as harvesting, processing, storage, and consumption. A broad range of reactions occur particularly during food processing. These chemical reactions take place in lipids, proteins, carbohydrates, and other food constituents which primarily involve oxidation, degradation, denaturation, aggregation, hydrolysis, and polymerization. They have key importance in producing desirable and undesirable changes in food products and must be understood from the legal and toxicological points of view. A good understanding of the chemistry of a food-processing system is expected to have significant contributions to the successful control of the process as well as the quality of the final food product.

The baking quality of soft wheat products depends on the interactions of the various chemicals in flour and the other substances used in the recipe. Batter and dough systems, used in the production of soft wheat products, serve as chemical reaction pools with their wealthy composition. The wide range of ingredients in them affects the pH and results in characteristic flavors and colors, often desirable but sometimes undesirable. Baking results in various types of chemical reactions and physical changes due to simultaneous heat and mass transfer. The first stage of baking is characterized by a physical change in dough structure from "moist-soft" to "dry-hard." Evaporative cooling prevents the temperature of dough from exceeding 100°C in early stages. Hence, the types and progressions of chemical reactions are different in different stages of baking. It should be realized that an understanding of the chemistry of any food-processing system is necessary to attain successful control of the process. Therefore, the development of a well-controlled manufacturing process of soft wheat products makes it necessary for the food engineer or technologist to understand the main reactions in chemical leavening (i.e., foam generation and batter and dough stabilization), nonenzymatic browning, and various other reactions that affect different properties of soft wheat products, such as structure, texture, flavor, taste, and color. Many of these reactions initiated by processing are mediated by free radicals and involve site-specific reactions that lead to functional or nutritional changes in foods. However, this chapter is not intended to cover all of the chemical reactions that occur during the processing of soft wheat products. The main aim was to cover chemical leavening reactions and nonenzymatic browning reactions.

3.2   LEAVENING SYSTEMS: CHEMICAL LEAVENING VERSUS YEAST LEAVENING

Gas must be generated within the batter or dough during mixing and the early stages of baking in order to obtain a light-textured product with a characteristic porous cellular structure. The product will have a large number of gas cells and desired eating characteristics with suitable leavening action. The process starts with the incorporation of air (or other gases) into the batter of dough to form a nucleus of gas cells. These gas cells will expand with the effect of the leavening system and also during baking to create the structure of the final product (Hoseney et al., 1988).

Mainly, two types of leavening systems are used in baked goods: yeast leavening and chemical leavening. Anaerobic fermentation of sugars by yeast produces CO_2 and ethanol. Carbon dioxide produced by yeast is the major leavener for bread. Water–ethanol azeotrope, which has a boiling temperature of 78°C, might also contribute to the expansion of yeast-leavened products. It vaporizes and expands the dough while it is heated in the oven (Hoseney et al., 1988).

Yeast leavening is widely used in bread-type products but is hardly ever used in soft wheat products due to its undesirable influence on dough rheology and the texture of the final product. People also prefer chemical leavening to yeast leavening due to a couple of other reasons. Yeast generally requires 2 to 3 h of fermentation to produce CO_2 bubbles, and the action of chemical leavening agents is almost instantaneous. Chemically leavened cookies can be eaten 15 min after mixing a batch of dough. The doughs used in soft wheat products are usually rich in fat and sugar but usually low in water content. Their sugar content might exceed 50% (flour basis). Therefore, yeast is not suitable as a leavener in most of the soft wheat products.

3.2.1 GASES ACTING IN THE LEAVENING OF BAKERY PRODUCTS

Various gases are used for the leavening of baked goods. The main ones are air, carbon dioxide, water and ethanol vapor, and ammonia. Air is a mixture of gases (nitrogen 78%, carbon dioxide 0.03%, oxygen 21%, other gases <0.1%). It is present in all baked products and has some leavening action. Among these gases, carbon dioxide and oxygen do not contribute much to the formation of bubbles in a batter or dough because they are quite soluble in water. In contrast, nitrogen, the main constituent of air, is not very soluble in water and is incorporated into dough or batter during mixing to create bubbles. Water is also present in the batter and dough of all baked products. Evaporation of water in a batter or dough affects the expansion process to some extent. The expansion of gas cells by water depends on the excess vapor pressure in the product during baking. Thus, the temperature inside the batter or dough is important because water vapor pressure will increase gradually as temperature increases toward the boiling point of water. However, the leavening effect of water is rather low due to its relatively high boiling point. It may serve as an efficient leavener only if the batter or dough system is heated very rapidly (e.g., wafers).

3.2.2 CHEMICAL LEAVENING

Chemical leavening systems produce CO_2 by either chemical decomposition through the application of heat or a reaction of an acidic compound with a base. Gas can form during all phases of the processing (e.g., mixing, forming, and baking). The point at which gas formation occurs is controlled largely by the compounds used. The composition of the leavening system as well as the rate and stage at which the gas is released influence the quality (appearance, texture, color, and even the flavor) of the final product (Cauvain and Young, 2006).

Chemical leavening provides a high level of control of the reaction at the desirable stage (during the mixing, at the bench, or in the oven). The major development area in chemical leavening has been how to control the reaction and its speed. Common sources of carbon dioxide are sodium and ammonium bicarbonates. Although carbonates are also potential sources of carbon dioxide, they are not commonly used due to their high alkalinity (Hoseney et al., 1988).

Carbon dioxide can exist either in gaseous form (as free CO_2) or in two different ionic forms (bicarbonate: HCO_3^{-1} or carbonate: CO_3^{-2}). Their relative proportion depends on the pH and temperature of the system. The gaseous form (CO_2) is predominant at very low pH values, and the carbonate ion is predominant at very high pH values. If the pH is above 8.0, no leavening gas (CO_2) will be available. The pH values of most of the soft wheat products are around 7.0, and only part of the carbon dioxide is in the leavening gas form (CO_2). Because sodium bicarbonate is alkaline, the pH of the batter or dough increases to a value at which no leavening gas (CO_2) is released. The batter or dough must include an acid to release significant amounts of gas. Some of the ingredients used in soft wheat products are acidic (e.g., acidic fruits, buttermilk). The required acidity is usually provided using these ingredients in home baking. However, in commercial applications of chemical leavening systems, acids or acidic compounds are added to provide consistent and controlled gas production (Cauvain and Young, 2006; Hoseney, 1994).

3.2.2.1 Ammonium Bicarbonate

Ammonium bicarbonate (ABC), ammonium salt of carbonic acid, decomposes completely into ammonia, carbon dioxide, and water when heated to above 40°C.

$$NH_4HCO_3 \longrightarrow NH_3 + CO_2 + H_2O$$

It is also called "volatile salt" (*sal volatile* or "Vol" in short form) because of this complete dissociation.

One can easily smell the pungent odor escaping from the dough and product during processing. The pungent odor is due to ammonia gas (NH_3) that combines easily with water to form ammonium hydroxide (NH_4OH). After the decomposition, there is no residual salt. This is advantageous because residual salts might influence the dough rheology and the taste of the final product. Products including ammonium bicarbonate as a chemical leavener must be baked thoroughly. Ammonium bicarbonate dissipates in low-moisture products, such as cookies and crackers. In higher-moisture products, on the other hand, water retains the ammonia and ammonium hydroxide is formed (especially in the moist crumb of baked goods), giving the product an undesirable flavor and odor. The crumb takes on a greenish color and produces an alkaline taste.

Some gas is released from ammonium bicarbonate at room temperature during mixing and forming, but most is generated during baking. The dissociation is faster during baking as the temperatures reach to around 60°C and above. Acidic conditions accelerate the reaction at lower temperatures. The rapid release of gas results in relatively large cells. Ammonium bicarbonate not only increases the volume or height, but also is likely to increase the spread in cookies.

Ammonium bicarbonate is marketed as a white crystalline powder that has a tendency to lumping. Therefore, it is not suitable for storage even at dry conditions. It is better to dissolve ammonium bicarbonate in water before using to avoid problems that might be caused due to severe lumping.

3.2.2.2 Sodium Bicarbonate

Sodium bicarbonate (baking soda; $NaHCO_3$) is the most popular leavening agent. It is a white crystalline alkaline compound that reacts by effervescing (fizzing) when it comes into contact with acids, thus producing carbon dioxide. This chemical reaction facilitates the rising action in baked goods. There are various advantages of sodium bicarbonate (Hoseney et al., 1988; Hoseney, 1994):

- Commercial baking soda is relatively inexpensive, and it is readily obtainable at high purity and various size grades.
- It is stable during storage and easy to handle.
- It does not affect the taste of the final product to a large extent.
- It is nontoxic.

When water and acidic compounds are not available, sodium bicarbonate will release some of its CO_2 and will be converted into sodium carbonate upon heating. This occurs at high temperatures ($\geq 90°C$) and cannot be used for chemical leavening. It can be represented by the following equation:

$$2NaHCO_3 \xrightarrow{\text{heat}} Na_2CO_3 + CO_2 + H_2O$$

When an aqueous solution of $NaHCO_3$ is heated, just a fraction of CO_2 will be released. If this solution is cooled, complex salt crystals ($Na_2CO_3 \cdot 2NaHCO_3 \cdot 2H_2O$) containing one part Na_2CO_3 and two parts $NaHCO_3$ will be formed which can be represented as follows:

$$4NaHCO_3 \xrightarrow{\text{heat}} Na_2CO_3.2NaHCO_3 + CO_2 + H_2O$$

On the other hand, if water and acidic compounds are available, sodium bicarbonate will react with acidic compounds to liberate CO_2 and decompose to give water and the sodium salt of the acidic compound:

$$NaHCO_3 + H^+ \longrightarrow Na^+ + CO_2 + H_2O$$

This reaction is more desirable because reaction with an acid converts all of the sodium bicarbonate to CO_2 and water, while the first two reactions convert sodium bicarbonate to sodium carbonate, CO_2, and water. Sodium carbonate residue is not desirable in the final product due to its bitter and soapy taste (Hoseney, 1994).

The CO_2 released is usually required as a raising agent and an early reaction is not desirable. Therefore, sodium bicarbonate should be kept away from other ingredients as long as possible. For example, sodium bicarbonate is added at the last stage of multistage mixing procedures. If sodium bicarbonate is added at the last stage, it must be sieved to evenly distribute it into the dough or batter system and to eliminate lumps.

There are various particle size grades of sodium bicarbonate suitable for bakery products. Use of the coarser grades can reduce the tendency of sodium bicarbonate with acid for prereaction and increase the system's stability. However, sodium bicarbonate with coarser granulation may not dissolve quickly during the preparation of the batter or dough and may cause darker specks on the surface of cookies. An excessive amount of sodium bicarbonate will result in cookies with an alkaline characteristic and yellowish crumb and crust coloration. This is accompanied by an unpleasant taste (known as soda-bite). If the level of soda is excessive, high pH values may also cause soapy flavors produced by reaction with the fats in formula. If the level is too low, it will allow the acidic flavors to come through. Therefore, it is usually aimed to achieve a pH value of around 7.0 ± 0.5 in cookies by the use of an appropriate amount of sodium bicarbonate (Cauvain and Young, 2006; Manley, 1991). However, a few special types of products deflect from this pH value (soda crackers are alkaline, cheese crackers are acidic). Sodium bicarbonate can be encapsulated by using fat-based coatings to increase its stability, particularly for refrigerated doughs. The encapsulated forms are quite successful but costly, so they are usually used in value-added products.

3.2.2.3 Potassium Bicarbonate

Potassium bicarbonate ($KHCO_3$) is a potential source of carbon dioxide, especially in reduced-sodium products. Because it has a greater molecular weight than sodium bicarbonate, it requires approximately 20% more for the equivalent leavening action (for neutralization of the acids used in the formula). Potassium bicarbonate results in a higher crumb pH than sodium bicarbonate and may contribute a sharp aftertaste to products. It is hygroscopic. It is not commonly used because it has a tendency to impart slight bitterness to the products. The reaction of potassium bicarbonate with acid is similar to that of sodium bicarbonate:

$$KHCO_3 + H^+ \longrightarrow K^+ + CO_2 + H_2O$$

3.2.2.4 Sodium Carbonate

Sodium carbonate (Na_2CO_3) is also a potential source of carbon dioxide. It is not used due to its high alkalinity, which increases the risk of getting a localized region of very high pH in dough. Such high pH regions might be detrimental to the quality characteristics of the final product.

3.2.3 BAKING POWDERS

A number of leavening acids (or acid salts) can be incorporated into the leavening system to obtain a larger yield of gas and to control the rate of CO_2 release. Baking powders consist of mixtures of sodium bicarbonate, one or more acidic compounds (acidic salts or acids), and an inert diluent such as starch or corn flour. The diluent physically separates the acid and base and prevents their early reaction during storage of the baking powder. The salt dissociates to give an acidic reaction in solution at different stages of processing. The purpose of preparing such a mixture of chemicals is to produce CO_2 bubbles during mixing (at room temperature) or as the dough or batter is heated in the oven. Cream of tartar, calcium phosphate, and citrate are common low-temperature acid salts. High-temperature acid salts are usually aluminum salts, such as calcium aluminum phosphate. They can be found in many baking powders. Although dietary aluminum is not definitely known to be harmful to human health, baking powders are available without aluminum salts for people who are afraid of consuming aluminum, and also for those sensitive to the taste (Conn, 1981; Hoseney, 1994).

Strong acids such as sulfuric acid or hydrochloric acid dissociate completely and give high levels of hydrogen ion in solution. They react very rapidly with sodium bicarbonate. On the other hand, weak acids such as tartaric acid and lactic acid dissociate incompletely and generate lower levels of hydrogen ion as compared to strong acids. As a consequence, they react slowly with sodium bicarbonate. Other acidic compounds, such as partially neutralized acids, have an intermediate position. Neutral salts such as sodium aluminum sulfate can function as acids through the hydrolysis reaction that creates hydroxides and hydrogen ions:

$$NaAl(SO_4)_2 + 3H_2O \longrightarrow Al(OH)_3 + Na^+ + 2(SO_4)^{-2} + 3H^+$$

$$3H^+ + 3NaHCO_3 \longrightarrow 3Na^+ + 3CO_2 + 3H_2O$$

The amount of acid needed in a formulation is determined by the amount of sodium bicarbonate and the neutralizing value of acid or acid salt used in the formula. Because the acidity is usually provided by acidic salts, in some cases, stoichiometry of the reaction is not clear. Therefore, the concept of neutralizing value (NV) was introduced in order to determine the amount of acid needed in a formulation, to compare the CO_2-releasing power of various leaveners, and to allow the correct level of usage. NV is a measure of the acid required within a specific bakery formulation (Conn, 1981; Hoseney, 1994).

The neutralizing value of a leavening acid is the amount (g) of sodium bicarbonate needed to completely neutralize 100 g of that acid (Equation 3.1). The neutralizing values of various leavening agents are presented in Table 3.1 (Conn, 1981; Thacker, 1997). In most applications, the goal is to have little or no sodium bicarbonate or leavening acid remaining in the finished product. However, sometimes an additional amount of sodium bicarbonate or acid is used to provide a specific pH-related effect, such as color or flavor modification or adjustment.

$$\text{Neutralization value} = \frac{\text{g of NaHCO}_3}{100\,\text{g acidic salt}} \times 100$$

$$(3.1)$$

Other ingredients included in the formulation also affect the pH of a product, so the addition of leavening acids and bases in neutral proportions does not assure a neutral pH in the final product. Changing the pH may influence the speed and reactivity of the leavening system. The degree of flour bleaching and adding ingredients like acidic fruits, buttermilk, cocoa powder, or high-fructose corn syrup can significantly alter the pH of the final product and affect its volume. Many ingredients contain organic acids that will react rapidly with the sodium bicarbonate. Therefore,

TABLE 3.1

Neutralizing Values of Leavening Agents

Leavening Acids	Reaction Rate	Neutralizing Value[a]
Monocalcium phosphate	Very fast	80
Sodium aluminum phosphate	Slow	100
Sodium aluminum sulfate	Medium	100
Glucono-δ-lactone	Continuous	45
Sodium acid pyrophosphate	Medium	72
Dicalcium phosphate dihydrate	Slow	33
Cream of tartar	Very fast	45
Calcium acid pyrophosphate	Medium	67
Adipic acid	Very fast	115
Fumaric acid	Medium	145
Citric acid	Very fast	159

[a] Grams of sodium bicarbonate needed to neutralize 100 g of leavening acid.

Source: From Conn, J.F., *Cereal Foods World*, 26, 119–123, 1981; Thacker, D., in *The Technology of Cake Making*, 6th ed., Chapman & Hall, London; New York, 1997, 100–106.

it is important to make sure if the pH of the batter or dough is correct after formulation. It may be necessary to regulate the pH to get the right result in the final product (Hoseney, 1994).

3.2.3.1 Single-Acting and Double-Acting Baking Powders

Baking powders might be either single acting or double acting. Combinations of acids can be used to create double-acting baking powders. In this type of leavening system, one acid reacts at room temperature, usually releasing a small amount of gas, and another acid reacts when batter or dough is heated in the oven, releasing the rest of the gas. The initial gas release (during mixing) provides small gas cells (nucleation) that promote homogeneous expansion of the product during baking. This gas cell nucleation can also occur by incorporating air during mixing. The better the dispersion of these nucleating gas cells, the finer the grain in the final product. During baking, the rate at which CO_2 is formed and the continuity of CO_2 production are both important. If an excessive amount of CO_2 is produced at the beginning and the reaction finishes, the cake will collapse when it is taken from the oven. Therefore, some baking powders have two different acids to guarantee fast initial reaction and continuity of the CO_2 production during baking (Cauvain and Young, 2006; Hoseney, 1994).

3.2.3.2 Leavening Acids

The original leavening acids for baking were the lactic acid in soured milk and cream of tartar. The technological developments led to the utilization of other compounds that are less expensive or less reactive. Hence, CO_2 is released when the product is placed in the oven rather than during mixing. Several commercial products are used as leavening acids. They vary in neutralizing value, reaction rate at various temperatures, and effect on the finished product (Table 3.1).

3.2.3.2.1 **Cream of Tartar** *($KHC_4H_4O_6$; Monopotassium Salt of Tartaric Acid, Potassium Acid Tartrate, or Potassium Bitartrate)*

$$HOOC-\underset{\underset{OH}{|}}{CH}-\underset{\underset{OH}{|}}{CH}-COO^-\,K^+$$

Cream of tartar has traditionally been used in baking applications, but its use in commercial applications is limited due to its high cost and very fast reaction rate. It is obtained as a by-product of wine production. It reacts quickly at room temperature.

$$KHC_4H_4O_6 + NaHCO_3 \longrightarrow KNaC_4H_4O_6 + CO_2 + H_2O$$

| *Cream of tarter* | *Sodium bicarbonate* | *Potassium sodium tartrate* |

In reduced-temperature batters, cream of tartar has limited solubility resulting in limited gas development during the initial stages of mixing at lower temperatures. The rate of reaction increases at higher temperatures (room temperature and above). Because of these characteristics, and its pleasant taste, cream of tartar is used in some baking powders and in the leavening systems of a number of baked goods and dry mixes.

3.2.3.2.2 **Tartaric Acid** *($H_2C_4H_4O_6$)*

$$HOOC-\underset{\underset{OH}{|}}{CH}-\underset{\underset{OH}{|}}{CH}-COOH$$

Tartaric acid is soluble in both cold and hot water. In cold water, it reacts with sodium bicarbonate instantly. The salt remaining after the reaction (sodium tartrate) has a laxative effect. Hence, some people do not choose to use it in baking applications. However, the quantities used in baking are very small and should not have a significant laxative effect.

$$H_2C_4H_4O_6 + 2NaHCO_3 \longrightarrow Na_2C_4H_4O_6 + CO_2 + 2H_2O$$

Tartaric	Sodium	Sodium
acid	bicarbonate	tartrate

3.2.3.2.3   Monocalcium Phosphate *(MCP)*

MCP also reacts quickly at room temperature with the sodium bicarbonate when dissolved, releasing approximately two thirds of the CO_2 in the first 2 min:

$$3Ca(H_2PO_4)_2 + 8NaHCO_3 \longrightarrow Ca_3(PO_4)_2 + 4Na_2HPO_4 + 8CO_2 + H_2O$$

Hence, MCP often acts as a nucleating acid. It is often combined with a slower-acting acid in products requiring a double-acting leavener, such as pancake batter. It is also called *calcium biphosphate* or *acid calcium phosphate* (ACP), and the commercial product is usually in the monohydrate form, monocalcium orthophosphate monohydrate, with the chemical formula $Ca(H_2PO_4)_2 \cdot H_2O$. It is necessary to use a finely powdered form of ACP to eliminate the risk of black speck formation on the surface of the end product.

The monohydrate form of ACP reacts as fast as cream of tartar, but the salt obtained by drying the monohydrate form (anhydrous salt) reacts at only about 80% of this rate. Coated anhydrous monocalcium phosphate may be used in applications where the initial gas release must be slowed. The initial reaction releases about 20% of the CO_2, with approximately 50% released after 10 to 15 min. This type of product is used in cake mixes and self-rising flour or cornmeal.

3.2.3.2.4   Sodium Aluminum Sulfate *(SAS)*

SAS was used as the second acid in double-acting baking powders in combination with MCP. It reacts too slowly to be used extensively in commercial applications, although sometimes it is used in formulations such as retail cake mixes. The main disadvantages of SAS are its slight astringent taste and its weakening effect on crumb texture:

$$NaAl(SO_4)_2 + 3NaHCO_3 \longrightarrow Al(OH)_3 + 2Na_2SO_4 + 3CO_2$$

3.2.3.2.5   Sodium Acid Pyrophosphate *(SAPP; $Na_2H_2P_2O_7 \cdot H_2O$)*

SAPP has a slow rate of reaction, especially under cold conditions. It releases most of the CO_2 during baking:

$$Na_2H_2P_2O_7 + 2NaHCO_3 \xrightarrow{\text{heat}} Na_4P_2O_7 + 2H_2O + 2CO_2$$

| Sodium | Sodium | Sodium |
| pyrophosphate | bicarbonate | phosphate |

There are several types of SAPPs. They show various rates of reaction, depending on how they are made. They differ in terms of processing conditions, porosity, and granulation, but they are chemically the same ($Na_2H_2P_2O_7$). SAPP typically releases 20 to 40% of the CO_2 gas in the first 2 min of mixing, little or none during holding, and the rest during baking. High levels of SAPP can result in a slightly bitter aftertaste (pyro taste) in the finished product. The pyro taste does not come from pyrophosphates. The pyrophosphates are broken down to sodium phosphate by enzymes in the dough. It appears that the aftertaste is due to the exchange of calcium with the sodium from the surface of the teeth. The problem has been partially solved by adding various forms of calcium (e.g., calcium lactate) to the baking powder in small amounts. SAPPs might also cause a scratchy sensation at the back of the throat and a peppery aftertaste, which is noticeable in products with lower sugar content and may be masked by high levels of sweetness. SAPPs are often used in refrigerated and frozen dough and cake doughnuts.

3.2.3.2.6 Sodium Aluminum Phosphate
($NaH_{14}Al_3(PO_4)_8 \cdot 4H_2O$,
Sodium Acid Aluminum Phosphate, SALP)

SALP is a neutral salt. It has high neutralizing value (Table 3.1) and is economical to use. The production of acid from SALP can be represented by the following:

$$NaH_{14}Al_3(PO_4)_8 \cdot 4H_2O + 5H_2O \longrightarrow 3Al(OH)_3 + Na^+ + 4H_2PO_4^- + 4HPO_4^{-2} + 11H^+$$

SALP reacts slowly and does not result in off-flavors or aftertaste in the finished product. It is suitable for use in doughs or batters that might stand for long periods before baking. SALP is commonly used as the second acid in double-acting baking powders. It is used in place of SAPP in products like cake and muffin mixes and gives whiter crumb than SAPP. SALP has a tenderizing effect on baked products. It initially releases about 20% of the carbon dioxide, with the rest generated during baking.

SALP is most commonly used for producing "self-raising" flour due to its greater resistance to reacting with sodium bicarbonate when mixed in flour that usually has 14% (or lower) moisture content.

3.2.3.2.7 Dicalcium Phosphate Dihydrate (DPD; $CaHPO_4 \cdot 2H_2O$)

Dicalcium phosphate is never used by itself as a leavening acid, but it can be used in combination with other leavening acids to release CO_2 later in baking, usually after 20 min baking time. It is commonly used in dehydrated form, and it is not an acidic salt. However, at higher temperatures, DPD disproportionates and gives an acidic reaction, resulting in a leavening effect. It is used in special baking applications that require very slow gas release. It has no activity during mixing or on the bench, and it reacts only when the temperature exceeds 57°C. This means that it can be used only in products that have extended baking times. It is useful for adjusting the pH of the final product.

3.2.3.2.8 Glucono-δ-Lactone *(GDL; $C_6O_6H_{10}$)*

Glucono-δ-lactone is an intramolecular ester of gluconic acid. Glucose oxidase enzyme oxidizes D-glucose into D-glucono-δ-lactone.

D-Glucose   α-D-Glucopyranose   D-Glucono-1,5-lactone   D-Gluconate

It is produced commercially by fermentation involving *Aspergillus niger* or *A. suboxydans*. Although it is not an acid, in aqueous solution it slowly hydrolyzes to an equilibrium mixture of gluconic acid and its δ- and γ-lactones. At room temperature, this hydrolysis reaction takes around 2 h and yields about 60% gluconic acid. The hydrolysis rate increases at higher temperatures.

D-Glucono-1,5-lactone   D-Gluconic acid   D-Glucono-1,4-lactone
(delta lactone)   (gamma lactone)

It has a slow but continual reaction rate. The main gas release occurs during baking as the ingredient is slowly hydrolyzed. It also has an advantage in that there is no aftertaste. Although GDL is relatively expensive, there are certain specialized types of products such as pizza dough, cake doughnuts, and refrigerated and frozen doughs for which it is very suitable as an acid component of the leavening system.

3.2.3.2.9 Calcium Acid Pyrophosphate *(CAPP)*

CAPP is utilized in a few specific applications such as with rye flour doughs, crackers, and frozen (yeast) doughs, and for dough strengthening.

3.2.3.2.10 Dimagnesium Phosphate *(DMP)*

This is a slow-acting acid salt that is heat activated (40 to 44°C). It may need to be used in combination with a faster leavening agent.

3.2.3.2.11   **Other Acids**

Various organic acids, such as adipic, fumaric, citric, and lactic acids, can be used as leavening acids. They might also be constituents of formula ingredients (e.g., soured milk and cream, fruit juices). They often act as nucleating acids for baking powders, and they can lower the pH for special applications.

3.2.3.3 Dough Reaction Rate and Batter Reaction Rate

Dough rate of reaction (DRR) is used to measure the CO_2 generated during mixing and holding stages in a dough system. In order to better understand the selection of leavening acid for specific product applications, it is important to understand the DRR. The test quantifies the percent of CO_2 released from dough versus time from the start of mixing. It measures the reactivity of leavening acids (CO_2 generated) with sodium bicarbonate during mixing and the subsequent bench time of the dough (Conn, 1981; Parks et al., 1960). Leavening acids with the same neutralizing values might have different DRR values. Hence, they may result in different CO_2 releasing rates during mixing and holding stages. The test is useful in the development of leavening acids with different reaction rates in the actual dough system.

Different formula ingredients may influence the DRR value due to their differences in acidity. The flour is the main constituent providing the acidity. The level of gas produced by the flour depends on factors such as extraction rate and age of the flour. Other formula ingredients may also affect the DRR value. Sugar might decrease the CO_2 reaction rates up to 10% depending upon the amount of sugar in the formula. Because of its acidity, milk has a small effect on DRR value. However, calcium ions contributed by milk slow the solution rate of most leaveners, and this might also affect the DRR value (Conn, 1981; Parks et al., 1960).

Batter reaction rate (BRR) is defined as the time required for a given percent of reaction to take place at a given temperature between a leavening acid and sodium bicarbonate. Usually the time needed for 60% of the reaction to take place is measured. BRR can be used to estimate when a certain percent of reaction will occur. This can easily be done by determining the product temperature at different stages of the process, especially during baking (Conn, 1981; Parks et al., 1960). The test is useful to investigate the behavior and suitability of different leavening acids in actual batter systems with various formulations.

3.2.3.4 Nutritional Values of Baking Powders

Although most batter and dough formulations include only a small quantity of baking powder (around 1 to 2%), chemical leavening agents supply considerable amounts of calcium, phosphorus, sodium, and potassium to soft wheat products. Calcium, phosphorus, and sodium contents of a baking powder containing sodium bicarbonate, SAS, MCP, and calcium sulfate are around 60 mg/g, 15 mg/g, and 100 mg/g, respectively (no potassium). On the other hand, calcium, phosphorus, and potassium contents of a low-sodium baking powder containing potassium bicarbonate and acidic potassium salts are around 50 mg/g, 70 mg/g, and 110 mg/g, respectively (no sodium). Hence, a piece of soft wheat product containing 3 g baking powder may contribute around one quarter of the Recommended Daily Allowances (RDAs) of calcium and phosphorus. (For adults, the RDA for calcium and phosphorus is 800 mg.) This contribution might be important, but consumers should read labels on products, because there is a wide variation in their calcium contents. For example, baking powders with tartrate contain no calcium. Some people may have to use a low-sodium baking powder in order to restrict dietary sodium intake. Fortunately, low-sodium baking powders are available in which the sodium salts have been replaced by potassium salts. These low-sodium baking powders have similar leavening properties (Ensminger et al., 1995).

3.2.3.5 Utilization of Blends of Chemical Leavening Agents and Premixes

Blends of acidulants are also marketed under various trademarks. These commercial blends are usually mixed with a diluent such as starch. Therefore, when purchasing an acidic salt, the information should be clear as to whether it is pure chemical or in diluted form. Lack of knowledge might cause complications due to the incorrect proportion of sodium bicarbonate acid mixtures in dough or batter.

Some cookie formulations include a lot of ingredients, some of which are used in very small quantities. Therefore, the weighing operations might cause problems. In most industrial processing units, the small quantities of ingredients are manually added, and various systems have been developed to streamline operations. Inevitable errors might be encountered because these small quantities of ingredients must be weighed repeatedly, usually for each batch every day. The errors might occur both in accuracy and in omission. The operation can be simplified by preparing blends of ingredients suitable for each recipe. However, the preparation of a homogenous and stable blend is a prerequisite (Manley, 1991).

Some of the ingredients and leavening agents do not disperse completely during dough mixing due to various reasons. First, their quantities are very small and the mixing action of the mixer might be insufficient to disperse them completely. Chemical leavening agents such as sodium bicarbonate and ammonium bicarbonate have a tendency to lump during storage or when wetted. Hence, it is often necessary to grind, sieve, or dissolve them in water before adding them to the mixer. Therefore, there are some advantages of preparing premixes (Manley, 1991):

- Reduce the number of separate weighing operations for each batch of dough
- Reduce the incidence of metering errors
- Reduce the incidence of ingredient omissions
- Reduce the mix-cycle time for each batch of dough by allowing a shorter mixer loading time
- Improve the means of metering leavening agents (e.g., pumping rather than metering) and the potential for automatic metering

Because it is easier to meter liquids than solids, there is an interest in using the dough water as a carrier in premixes. However, the solubility of the chemical leavening agents must be considered because most of these soluble chemicals form saturated solutions at quite low concentrations (e.g., solubility of sodium bicarbonate is around 10 g per 100 ml of water at room temperature). Mixtures of chemical leavening agents and other ingredients might be incompatible due to the pH of the premix or chemical reactions resulting in loss of CO_2 or other gases. Furthermore, in some cases, the amount of water in the recipe might not be sufficient to dissolve the items to be incorporated in the premix (Manley, 1991).

3.2.3.6 Effect of Batter Viscosity and CO_2 Production Rate on the Quality of Soft Wheat Products

A cake batter is a complex mixture of ingredients. It is principally an aqueous system, but it has a number of dispersed phases such as starch granules, fat, and air. If the viscosity of the batter is very low, the phases separate easily, even before entering the oven, resulting in cakes with inferior quality. Suitable batter viscosity slows the separation of phases so that the separation seems to be stopped during the time required to make a cake. During mixing, air is incorporated into the batter. After it is mixed, chemical leavening agents cannot generate new gas bubbles in batter or dough. During mixing, air must be included in batter or dough in order to supply preexisting bubbles. Air bubbles within the batter or dough are essential to provide nuclei. Other gases, produced by yeast or chemical leaveners, can diffuse into these bubbles and expand them (Hoseney, 1994). However, air cells can be lost from the batter by rising to the surface and also by the coalescing of two cells into one.

As the batter is heated in the oven, at the beginning its viscosity decreases. Starch is dense and may settle, forming a tough layer at the bottom of the pan and a light foam-like structure at the top. A batter with a lower viscosity cannot keep the gas bubbles colliding with sufficient force to join together to form larger bubbles (coalesce). Although small gas bubbles can be trapped in the batter, large bubbles will have sufficient buoyancy to rise to the surface of the batter and they are lost. The heating of the batter is continued in the oven, and its viscosity increases in later stages. This is caused mainly by starch gelatinization. As starch gelatinizes, its water-binding capacity increases to a large extent. Hence, the viscosity of the batter increases tremendously, resulting in a solid-like appearance. This is called setting. Most of the available CO_2 must be released during expansion and up to a certain stage of baking before the dough or batter reaches its set-point temperature. This is more important for cakes than cookies. If the gas-releasing rate is too fast, a coarse structure may result. The structure may collapse if the gas release is completed before the structure sets. If the gas-releasing rate is extremely slow, the volume of the final product is usually smaller and structural rupturing or cracking may occur due to the effects of gases released after setting (Thacker, 1997). Of course, it should be kept in mind that carbon dioxide gas expands more quickly at higher altitudes and therefore has greater leavening action. Therefore, the quantity of baking powder should be decreased when baking at higher altitudes.

The balance of leavening acids and sodium bicarbonate is also important and reached by using the neutralizing value to match the amount of leavening acid to the amount of sodium bicarbonate so that the highest level of CO_2 is produced. If an insufficient amount of leavening acid is added, a smaller amount of CO_2 is produced, and the remaining sodium bicarbonate increases the pH of the final product. If an excessive amount of leavening acid is added, gas production is maintained at the same level, but some of the leavening acids such as SAPP and GDL will leave a bitter aftertaste in the final product. For example, the NV of pure SAPP is 72, but it is also available in standardized form with an NV of 50. For an application that utilizes sodium bicarbonate at 2% of flour weight, standardized SAPP with an NV of 50 should be added at 4% of flour weight to provide the right leavening balance:

$$\frac{2\% \, NaHCO_3}{50} \times 100 = 4\% \, SAPP$$

(3.2)

The type of leavening acid used in the formulation of baking powder also affects the rate of CO_2 production, which in turn affects the product. For example, dough-

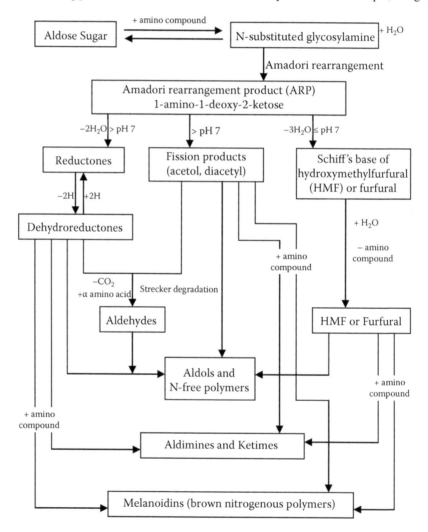

FIGURE 3.1 The major steps in the Maillard reaction.

nuts require little leavening during mixing, and then a fast rate of CO_2 production is required so that the batter is leavened quickly and will have buoyancy to float on the hot oil during frying, ensuring a crisp product. SAPP is generally used as the leavening acid, sometimes together with additional MCP for early activity or GDL for delayed activity.

3.3   MAILLARD REACTION

A Maillard reaction is a nonenzymatic chemical reaction involving condensation of an amino group and a reducing group, resulting in the formation of intermediates that ultimately polymerize to form brown pigments. The reaction was first described by Maillard (1912). Hodge (1953) presented a quite complete scheme of the several stages of the reaction, which was fully applied, and it is still valid to interpret many characteristics of the products and of the Maillard reaction kinetics (Figure 3.1).

The chemistry of the Maillard reaction includes a complex series of reactions leading to the formation of a variety of compounds, including flavors and colors, which are of fundamental importance to define the quality of food products. There are three major stages of the reaction. The first stage consists of glycosylamine formation and rearrangement of N-substituted-1-amino-1-deoxy-2-ketose (Amadori compound). The second stage involves loss of the amine to form carbonyl intermediates, which upon dehydration or fission form highly reactive carbonyl compounds through several pathways. The third stage occurring upon subsequent heating involves the interaction of the carbonyl flavor compounds with other constituents to form brown nitrogen-containing pigments, called melanoidins.

In addition to the desired consequences of the Maillard reaction, such as the formations of color and flavor, there are undesirable consequences in thermally processed foods including bakery foods. The changes commonly encountered in thermally processed foods as a result of the Maillard reaction are summarized in Figure 3.2.

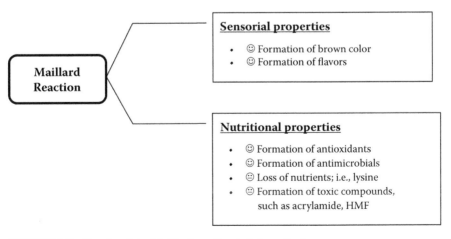

FIGURE 3.2   Results of the Maillard reaction in bakery foods.

10 min	15 min	20 min	25 min	30 min

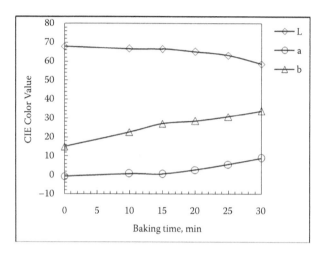

FIGURE 3.3   Change of color in cookies during baking at 160°C. (See color insert after p. 158.)

3.3.1   FORMATION OF COLOR

The colored products of the Maillard reaction are of two types: the high-molecular-weight macromolecule materials commonly referred to as melanoidins and the low-molecular-weight colored compounds containing two or three heterocyclic rings (Ames et al., 1998). Color development increases with increasing temperature, time of heating, increasing pH, and intermediate moisture content (a_w 0.3 to 0.7). In general, browning occurs slowly in dry systems at low temperatures and is relatively slow in high-moisture foods. Color generation is enhanced at pH 7. Of the two starting reactants, the concentration of reducing sugars has the greatest impact on color development. Of all the amino acids, lysine makes the largest contribution to color formation, and cysteine has the least effect on color formation.

Change of color in bakery foods during baking is a dynamic process in which certain color transitions occur as the baking proceeds. Color is usually measured in *Lab* units, which is an international standard for color measurements adopted by the Commission Internationale d'Eclairage (CIE) in 1976 (Papadakis et al., 2000). In cookies, the lightness parameter (*L*) tends to decrease while the chromatic parameters (*a* and *b*) increase as the baking proceeds, as illustrated in Figure 3.3.

3.3.2 Formation of Flavor and Aroma Compounds

Flavor compound formation in the Maillard reaction (MR) depends on the following:

- Type of sugars and amino acids involved
- Reaction temperature, time, pH, and water content (Jousse et al., 2002)

In general, the first factor mentioned determines the type of flavor compounds formed, and the second factor influences the kinetics.

The most common route for formation of flavors via the Maillard reaction includes the interaction of α-dicarbonyl compounds (intermediate products in the MR, stage 2) with amino acids through the Strecker degradation reactions. Alkyl pyrazines and Strecker aldehydes belong to commonly found flavor compounds from the MR. For example, low levels of pyrazines are formed during the processing of potato flakes when the temperature is less than 130°C, but increase tenfold when the temperature is increased to 160°C, and decrease at 190°C, probably owing to evaporation or binding to macromolecules. The aroma profile varies with the temperature and the time of heating. At any given temperature–time combination, a unique aroma, which is not likely to be produced at any other combination of heating conditions, is produced (van Boekel, 2006).

The total amount of amino acids and reducing sugars and their relative proportions on the surface of dough during baking are the limiting factors in crust aroma quality. Both depend on their formation from enzymatic activities and their involvement in metabolic processes. Intensity of reaction depends on temperature and moisture in the oven (Kaminski et al., 1981). The type of aroma is influenced by the amino acid structure (El-Dash, 1971), so glucose plus leucine, arginine, or histidine gives a fresh bread aroma, while dihydroxyacetone plus proline gives a cracker-type aroma (Coffman, 1965). Sugar type affects reaction rate more than aroma type (El-Dash, 1971).

Many compounds have been detected in doughs and breads (crust and crumb), and they were summarized (Coffman, 1965; Hansen and Hansen, 1994; Lund et al., 1989; Maga, 1974; Schieberle and Grosch, 1985, 1987). Wheat and rye flour doughs and breads contain many common components, but rye flour leads to additional components not identified in wheat products (Maga, 1974; Schieberle and Grosch, 1985, 1987). The compounds identified in preferments, doughs, and breads are based on acids, alcohols, aldehydes, esters, ethers, furan derivatives, ketones, pyrrole derivatives, pyrazines, and sulfur compounds (Maga, 1974; Schieberle and Grosch, 1985, 1987).

Carbonyls (aldehydes and ketones) can result from many different reactions in breadmaking (Maga, 1974). Some proceed from fermentation, such as acetoin or diacetyl (Lawrence et al., 1976), but the majority are produced during nonenzymatic browning reactions (Johnson et al., 1966). Also, some carbonyl compounds generated during fermentation can volatilize during baking and appear again during browning reactions (Johnson et al., 1966). Bread crust usually contains a larger number of carbonyl compounds than bread crumb (El-Dash, 1971). Lipoxygenase is a source of carbonyls by decomposition of hydroperoxides. Thus, linoleic acid produces hexanal, in addition to pentanol and other products from the partial oxidation of hexanal. Like aldehydes, ketones are mainly formed during Maillard reactions. All amino acids promote acetone formation in the crust, which is also formed during fermentation, as well as 2-butanone (Maga, 1974).

Furan derivatives result from thermal degradation of sugars in bread crust. Some are intermediate compounds in Maillard reactions undergoing condensation with amino acids. Their characteristics are more influenced by amino acid than by sugar type (Maga, 1974).

Pyrrole derivatives, pyrazines, and sulfur compounds also proceed from Maillard reactions; some pyrazines can be provided by flour or milk solids included in formulation (Maga, 1974).

Pyrazines resulting from thermal reactions also have typical aroma features able to be imparted to bread (Maga, 1974). Finally, the least volatile fractions, such as melanoidins, dihydroxyacetone, ethyl succinate, and succinic and lactic acids, contribute more to the taste than to the aroma of bread (Baker et al., 1953).

3.3.3 FORMATION OF ANTIOXIDANTS

There are several reports on the formation of antioxidative Maillard reaction products in food processing. Pronyl-lysine, a Maillard reaction product, present in bread crust has been demonstrated to have beneficial effects on human health (Lindenmeier et al., 2002). The addition of amino acids or glucose to cookie dough has been shown to improve oxidative stability during the storage of the cookies (Summa et al., 2006).

Although the antioxidant effect of the Maillard reaction products has been extensively investigated, the exact nature of the antioxidants formed is not yet well known. It was reported that the intermediate reductone compounds of Maillard reaction products could break the radical chain by donation of a hydrogen atom. Maillard reaction products were also observed to have metal-chelating properties and retard lipid peroxidation. Melanoidines were reported to be powerful scavengers of reactive oxygen species (Hayase et al., 1989). Recently, it was suggested that the antioxidant activity of xylose–lysine MRPs may be attributed to the combined effect of reducing power, hydrogen atom donation, and scavenging of reactive oxygen species (Yen and Hsieh, 1995). In the Maillard reaction, high antioxidant capacity was generally associated to the formation of brown melanoidins (Aeschbacher, 1990; Anese et al., 1993, 1994; Eichner, 1981; Gomyo and Horikoshi, 1976; Hayase et al., 1989; Kirigaya et al., 1968; Yamaguchi et al., 1981; Yen and Hsieh, 1995; Yen and Tsai, 1993). Although in its early stages the Maillard reaction leads to the formation of well-known Amadori and Heyn's products (Ames, 1988; Rizzi, 1994), little information is available on the chemical structure of the hundreds of brown products formed by a series of consecutive and parallel reactions including oxidations, reductions, and aldol condensations among others (Eriksson, 1981; Yaylayan, 1997).

A number of parameters can help in selecting the prevalent mechanism of the overall reaction and its rate, leading to the formation of different chemical species that are expected to exert different antioxidant properties. In particular, the antioxidant properties of Maillard reaction products have been reported to be strongly affected by the physicochemical properties of the system and by the processing conditions (Eichner and Ciner-Doruk, 1981; Homma et al., 1997; Huyghebaert et al., 1982; Kim and Harris, 1989; Lingnert and Eriksson, 1981; Manzocco et al., 1999; Obretenov et al., 1986; Stamp and Labuza, 1983; Waller et al., 1983). Moreover, it must be kept in mind that polyphenols, ascorbic acid, and other carbonyl compounds—even if formed during oxidative reactions—can take part in the Maillard reaction (Eriksson, 1981; Rizzi, 1994; Yaylayan, 1997). The contribution of these compounds to the formation of heat-induced antioxidants is still unknown.

3.3.4 LOSS OF NUTRITIONAL QUALITY

The loss of available lysine is the most significant consequence of the Maillard reaction, and it is of the greatest importance in those foods where this amino acid is limiting, such as in cereals (Henle et al., 1991). Evaluation of the early stages of the Maillard reaction can be achieved by determination of the furosine (ε-N-(furoylmethyl)-L-lysine) amino acid formed during acid hydrolysis of the Amadori compound, fructosyl-lysine, lactulosyl-lysine, and maltulosyl-lysine produced by the reaction of e-amino groups of lysine with glucose, lactose, and maltose (Erbersdobler and Hupe, 1991). For this reason, the estimation of the extent of protein damage caused by heating in the first stages of that reaction are often based on determinations of the amount of furosine that is formed during the acid hydrolysis of foods. Furosine determination has been used in cereals to control the processing of pasta (Resmini and Pellegrino, 1994), bakery products (Henle et al., 1995), baby cereals (Guerra-Hernández and Corzo, 1996; Guerra-Hernández et al., 1999), and bread (Ramirez-Jimenez et al., 2000; Ruiz et al., 2004).

3.3.5 FORMATION OF TOXIC COMPOUNDS

3.3.5.1 Acrylamide

One of the most significant consequences of the Maillard reaction is the formation of acrylamide (Mottram et al., 2002; Stadler et al., 2002), which was first discovered in thermally processed foods in April 2002 by Swedish researchers (Tareke et al., 2002). Acrylamide is classified as a probable human carcinogen by the International Agency for Research on Cancer (IARC, 1994).

Studies to date clearly show that the amino acid asparagine is mainly responsible for acrylamide formation in cooked foods after condensation with reducing sugars or a carbonyl source (Figure 3.4). Moreover, the sugar–asparagine adduct, N-glycosylasparagine, generates high amounts of acrylamide, suggesting the early Maillard reaction as a major source of acrylamide (Stadler et al., 2002). In addition, decarboxylated asparagine, when heated, can generate acrylamide in the absence of reducing sugars (Zyzak et al., 2003). A recent study revealed that, besides acrylamide, 3-aminopropionamide, which may be a transient intermediate in acrylamide formation, was also formed during heating when asparagine was reacted in the presence of glucose (Granvogl and Scieberle, 2006).

In certain bakery products, acrylamide contents up to 1000 mg/kg have been observed (Croft et al., 2004). The highest contents have been found in products pre-

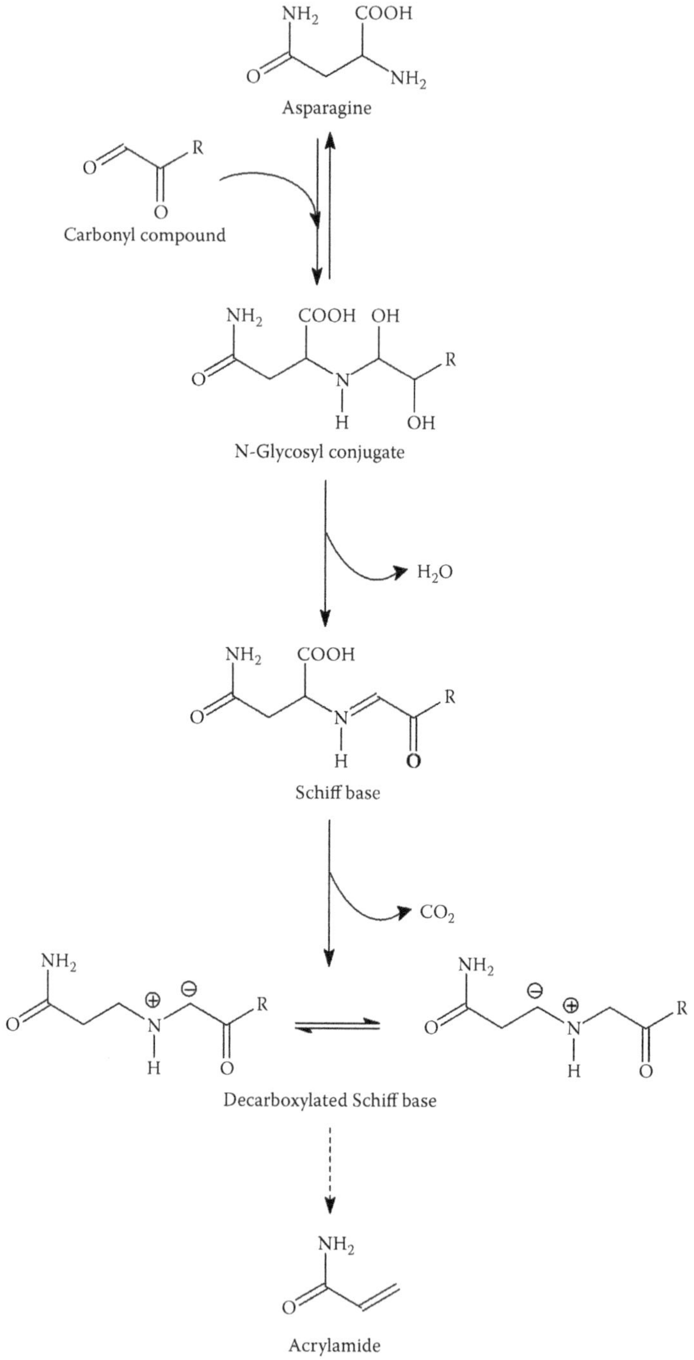

FIGURE 3.4 Formation of acrylamide during the pyrolysis of asparagine with glucose. (Adapted from Gökmen, V., and Senyuva, H.Z., *European Food Research and Technology* (225, 815–820), 2007. With permission.)

pared with the baking agent ammonium hydrogen carbonate, such as gingerbread products (Amrein et al., 2004; Konings et al., 2003). Model experiments showed that ammonium bicarbonate strongly promotes acrylamide formation in sweet bakery (Biedermann and Grob, 2003; Weisshaar, 2004). Replacing this baking agent by sodium hydrogen carbonate presents a very effective way to limit the acrylamide content of bakery goods (Amrein et al., 2004; Vass et al., 2004). In addition to the baking agent, the content of reducing sugars and free asparagine as well as the process conditions influence the acrylamide formation in bakery goods (Amrein et al., 2004; Gökmen et al., 2007; Surdyk et al., 2004; Vass et al., 2004).

Even though the baking temperature is high enough to produce acrylamide in the crust, the total baking time is not long enough to increase the center temperature above 120°C, at which point acrylamide begins to form. However, it was shown that acrylamide is present in both zones of the biscuit, but apparently lower amounts are found in the center of the biscuit (Taeymans et al., 2004). Similar to biscuits, the crust and crumb of bread contain significantly different amounts of acrylamide (Surdyk et al. 2004). The crust layers have been shown to contain up to 718 µg/kg of acrylamide while the crumb was free of acrylamide (Şenyuva and Gökmen, 2005). These results suggest that acrylamide formation in baked cereals mostly occurs by a surface phenomenon.

When biscuits are toasted to a near burnt state, the acrylamide concentration is decreased by up to 50% (Taeymans et al., 2004). Similar results have been shown for several other forms of cereals. Acrylamide levels in biscuits are in contrast with those for the potato but are consistent with the suggestion made elsewhere that the acrylamide content results from a balance between formation and elimination, with the latter being more rapid at higher temperature (Taeymans et al., 2004).

Ingredients play an important role in acrylamide formation in bakery foods, as different ingredients have various amounts of free asparagine and reducing sugars

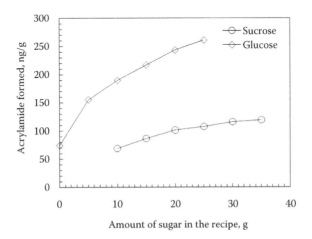

FIGURE 3.5 Effect of sugar type and amount on acrylamide formation in cookies during baking. (From Gökmen, V., Açar, Ö.Ç., Köksel, H., and Acar, J., *Food Chemistry*, 104, 1136–1142, 2007. Reprinted with permission from Elsevier.)

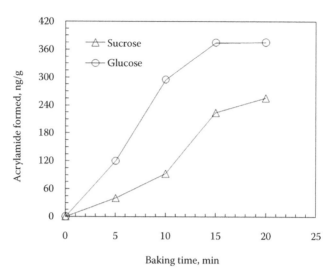

FIGURE 3.6   Changes in acrylamide concentration in cookies during baking at 200°C with time. (Adapted from Gökmen, V., Açar, Ö.Ç., Köksel, H., and Acar, J., *Food Chemistry*, 104, 1136–1142, 2007. Reprinted with permission from Elsevier.)

precursors. Sugars seem to be the most important ingredient in the dough formula from the viewpoint of acrylamide formation, because the free asparagine level of wheat flour is relatively low. Noti et al. (2003) reported levels of 150 to 400 mg/kg of asparagine in ten samples of wheat flour. Surdyk et al. (2004) measured asparagine levels of 170 mg/kg in white wheat flour. Figure 3.5 shows the formation of acrylamide in cookies during baking as influenced by the type and amount of sugars in the recipe.

Baking temperature and time are also closely related with acrylamide formation in the baking process. There is no acrylamide present in uncooked dough, but the acrylamide level rises with time. In cookies containing sucrose, acrylamide concentrations showed a rapid increase after an initial lower rate period, reaching a plateau within a baking time of 15 to 20 min at 200°C (Figure 3.6). When sucrose was replaced with glucose, the initial lower rate period disappears and the acrylamide concentration of cookies increases rapidly, attaining the plateau values earlier as shown in Figure 3.6 (Gökmen et al., 2007).

The two major approaches to reducing acrylamide levels in cookies are replacing NH_4HCO_3 by $NaHCO_3$, and using sucrose instead of reducing sugars (Graf et al., 2006). In crackers, a combination of the first two approaches has been even more effective (Vass et al., 2004), which could also apply to other products. The measure focusing on the replacement of reducing sugars by sucrose is limited to products where browning is not of primary importance. The addition of glycine during dough making has been shown to reduce acrylamide in flat breads and bread crusts up to 90% (Bråthen et al., 2005). A moderate addition of organic acids may also be considered for mitigation of acrylamide in cookies if the recipe is formulated with the invert syrup instead of sucrose. Lowering pH may result in excessive hydrolysis of

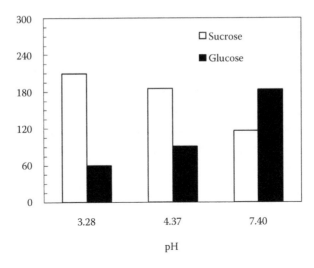

FIGURE 3.7   The effect of dough pH on the amount of acrylamide formed in cookies having different sugars during baking at 205°C for 11 min. (From Gökmen, V., Açar, Ö.Ç., Köksel, H., and Acar, J., *Food Chemistry*, 104, 1136–1142, 2007. Reprinted with permission from
Elsevier.)

sucrose, which will make the composition more favorable to form acrylamide during
baking (Gökmen et al., 2007). For cookie dough that incorporates sucrose, lowering the pH value from 7.40 to 3.28 by the addition of citric acid has been shown to
almost double the amount of acrylamide formed in cookies during baking. However,
the addition of citric acid has limited the formation of acrylamide for cookies incorporating glucose (Figure 3.7).

3.3.5.2 Hydroxymethylfurfural (HMF)

Among the many products formed, HMF, a possible mutagen (Surh et al., 1994),
seems particularly interesting because of its accumulation during the baking process. Although the toxicological relevance of HMF is not clear, as *in vitro* studies
on genotoxicity and mutagenicity have given controversial results (Cuzzoni et al.,
1988; Janzowski et al., 2000; Lee et al., 1995), its presence is undesired in thermally
processed foods.

HMF is naturally formed as an intermediate in the Maillard reaction (Ames
et al., 1998), and from dehydration of hexoses under mild acidic conditions (Kroh,
1994) during thermal treatment applied to foods (Figure 3.8). HMF levels found in
cereal products are highly variable.

During the baking of bread, the water content on the surface becomes lower than
that in the center; this combined with the high temperature is one of the factors that
makes the crust different from the crumb (Thorvaldsson and Skjoldebrand, 1998).
HMF levels in crumb have been found between 0.6 and 2.2 mg/kg and those in crust
were notably greater, from 18.3 to the 176.1 mg/kg in white bread (Ramírez-Jiménez
et al., 2000).

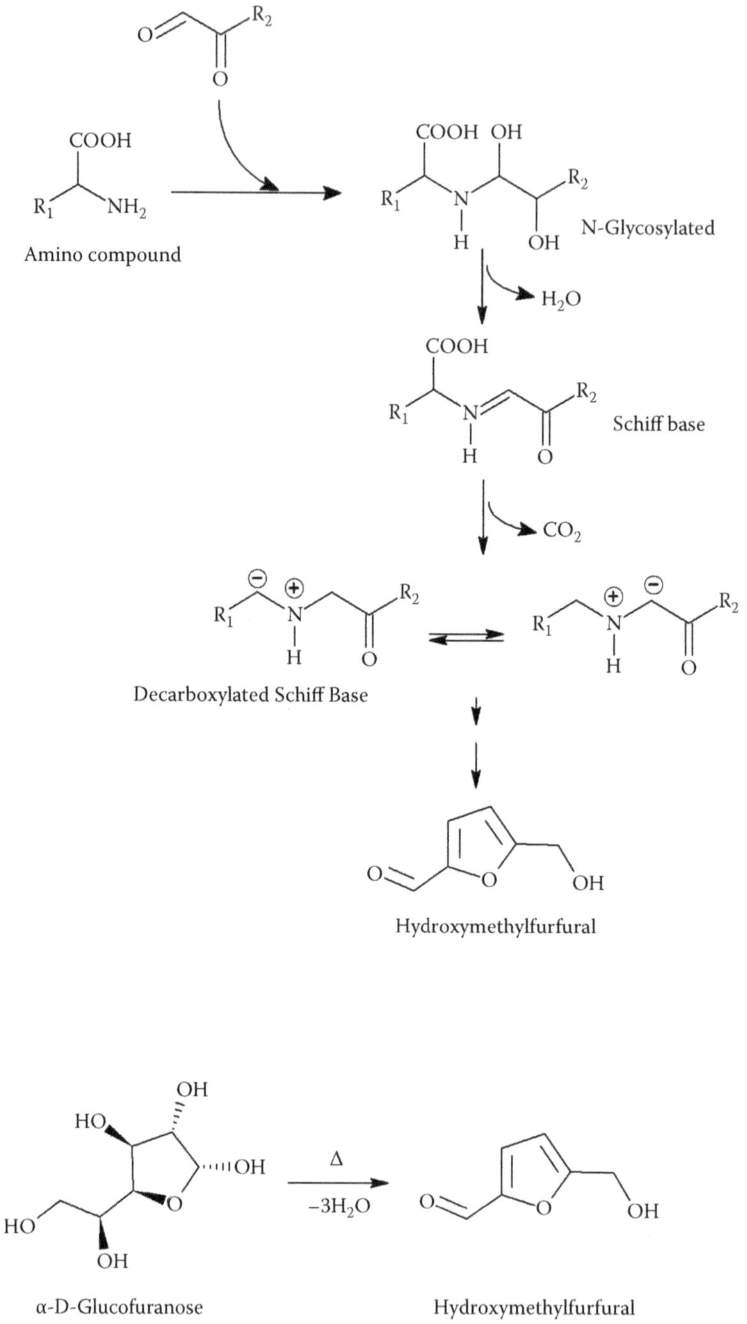

FIGURE 3.8 Formation of hydroxymethylfurfural (HMF) as a consequence of (a) the Maillard reaction and (b) the pyrolysis of sugar.

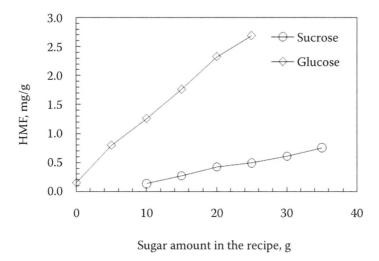

FIGURE 3.9   Effect of sugar type and amount on hydroxymethylfurfural (HMF) formation in cookies during baking. (From Gökmen, V., Açar, Ö.Ç., Köksel, H., and Acar, J., *Food
Chemistry*, 104, 1136–1142, 2007. Reprinted with permission from Elsevier.)

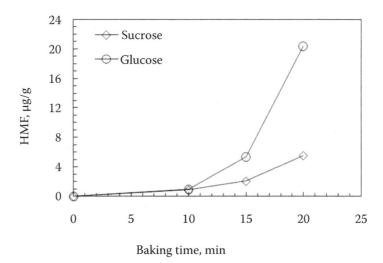

FIGURE 3.10   Changes in hydroxymethylfurfural (HMF) concentration in cookies during
baking at 210°C with time. (Adapted from Gökmen, V., Açar, Ö.Ç., Köksel, H., and Acar, J.,
Food Chemistry, 104, 1136–1142, 2007. With permission.)

The presence of a high amount of reducing sugars in the recipe makes the dough more susceptible to form HMF during baking (Figure 3.9). Increasing the baking temperatures also increases the rate of HMF formation during baking (Figure 3.10).

The dough composition and its changes during baking are the most critical factors affecting the formation of HMF in bakery foods. As the water activity of cookies decreases during baking, the conditions become more favorable to form HMF (Ameur et al., 2006). The removal of water to a level corresponding to water activity of 0.4 is most probably reflecting a stage in baking where the temperature of cookies begins to rise above 100°C, which accelerates HMF formation thermodynamically. Keeping the temperature below 200°C may prevent an excessive decomposition of sucrose, and thus an excessive formation of HMF during baking.

REFERENCES

Aeschbacher, H.U. 1990. Anticarcinogenic effect of browning reaction products. In *The Maillard Reaction in Food Processing, Human Nutrition and Physiology*, Ed. P.A. Finot, H.U. Aeschbacher, R.F. Hurrel, and R. Liardon, 335–348. Birkhäuser Verlag, Basel.

Ames, J., R.J. Bailey, and J. Mann. 1998. Recent advances in the analysis of colored Maillard reaction products. In *The Maillard Reaction in Foods and Medicine*, Eds. J. O'Brien, H.E. Nursten, M. James, C. Crabbe, and J. Ames, 76–82. London: Royal Society of Chemistry.

Ames, J.M. 1988. The Maillard Browning Reaction—An Update. *Chemiistry & Industry* 5: 558–561. London.

Ameur, L.A., G. Trystram, and I. Birlouez-Aragon. 2006. Accumulation of 5-hydroxymethyl-2-furfural in cookies during the baking process: Validation of an extraction method. *Food Chemistry* 98: 790–796.

Amrein, T.M., B. Schönbächler, F. Escher, and R. Amado. 2004. Acrylamide in gingerbread: Critical factors for formation and possible ways for reduction. *Journal of Agricultural and Food Chemistry* 52: 4282–4288.

Anese, M., M.C. Nicoli, and C.R. Lerici. 1994. Influence of pH on the oxygen scavenging properties of heat-treated glucose/glycine systems. *Italian Journal of Food Science* 3: 339–343.

Anese, M., P. Pittia, and M.C. Nicoli. 1993. Oxygen consuming properties of heated glucose/glycine aqueous systems. *Italian Journal of Food Science* 1: 75–79.

Baker, J.C., H.K. Parker, and K.L. Fortmann. 1953. Flavor of bread. *Cereal Chemistry* 30: 22–30.

Becalski, A., B.P.Y. Lau, D. Lewis, and S.W. Seaman. 2003. Acrylamide in foods: Occurrence, sources and modeling. *Journal of Agricultural and Food Chemistry* 51: 802–808.

Biedermann, M. and K. Grob. 2003. Model studies on acrylamide formation in potato, wheat flour and corn starch; ways to reduce acrylamide contents in bakery ware. *Mitteilungen aus Lebensmitteluntersuchung und Hygiene* 94: 406–422.

Bråthen, E., A. Kita, S.H. Knutsen, and T. Wicklund. 2005. Addition of glycine reduces the content of acrylamide in cereal and potato products. *Journal of Agricultural and Food Chemistry* 53: 3259–3264.

Cauvain, S. and L. Young. 2006. *Baked Products: Science, Technology and Practice*. Blackwell Scientific, Oxford.

Coffman, J.R. 1965. Bread flavor. In *The Symposium on Foods: The Chemistry and Physiology of Flavors*. Westport. CT: The AVI Publishing Co. 185–202.

Conn, J.F. 1981. Chemical leavening systems in flour products. *Cereal Foods World* 26: 119–123.

Croft, M., P. Tong, D. Fuentes, and T. Hambridge. 2004. Australian survey of acrylamide in carbohydrate-based foods. *Food Additives and Contaminants* 21(8): 721–736.

Cuzzoni, M.T., G. Stoppini, G. Gazzani, and P. Mazza. 1988. Influence of water activity and reaction temperature of ribose-lysine and glucose-lysine Maillard systems on mutagenicity, absorbance and content of furfurals. *Food and Chemical Toxicology* 26: 815–822.

Eichner, K. 1981. Antioxidative effect of Maillard reaction intermediates. In *Progress in Food Nutrition and Science 5,* Ed. C. Ericksson, 441–451. Pergamon Press, Oxford.

Eichner, K. and M. Ciner-Doruk. 1981. Early indication of the Maillard reaction by analysis of reaction intermediates and volatile decomposition products. In *Progress in Food Nutrition and Science 5*, Ed. C. Eriksson, 115–135, Pergamon Press, Oxford.

El-Dash, A.A. 1971. The precursors of bread flavor: Effect of fermentation and proteolytic activity. *Bakers' Digest* 45(6): 26–31.

Ensminger, A.H., M.E. Ensminger, J.L. Konlande, J.R.K. Robson. 1995. *Concise Encyclopedia of Foods and Nutrition*, 2nd ed., CRC Press, Boca Raton, FL.

Erbersdobler, H.F., and A. Hupe. 1991. Determination of lysine damage and calculation of lysine bio-availability in several processed foods. *Z. Ernaehrungswiss.* 30: 46–49.

Eriksson, C. 1981. Maillard reaction in food: Chemical, physiological and technological aspects. In *Progress in Food Nutrition and Science 5*, Ed. C. Eriksson, 441–451, Pergamon Press, Oxford.

Gökmen, V. and H.Z. Senyuva. 2007. Effects of some cations on the formation of acrylamide and furfurals in glucose-asparagine model system. *European Food Research and Technology.* 225: 815–820.

Gökmen, V., Ö.Ç. Açar, H. Köksel, and J. Acar. 2007. Effects of dough formula and baking conditions on acrylamide and hydroxymethylfurfural formation in cookies. *Food Chemistry* 104: 1136–1142.

Gomyo, T. and M. Horikoshi. 1976. On the interaction of melanoidins with metal ions. *Agricultural Biology and Chemistry* 40: 33–40.

Graf, M., T.M. Amrein, S. Graf, R. Szalay, F. Escher, and R. Amadò. 2006. Reducing the acrylamide content of a semi-finished biscuit on industrial scale. *LWT* 39: 724–728.

Granvogl, M. and P. Scieberle. 2006. Thermally generated 3-aminopropionamide as a transient intermediate in the formation of acrylamide. *Journal of Agricultural and Food Chemistry* 54: 5933–5938.

Guerra-Hernández, E. and N. Corzo. 1996. Furosine determination in baby cereal by ion-pair reversed phase liquid chromatography. *Cereal Chemistry* 73: 729–731.

Guerra-Hernández, E., N. Corzo, and B. García-Villanova. 1999. Maillard reaction evaluation by furosine determination during infant cereal processing. *Journal of Cereal Science* 29: 171–176.

Hansen, A. and B. Hansen. 1994. Influence of wheat flour type on the production of flavour compounds in wheat sourdoughs. *Journal of Cereal Science* 19: 185–190.

Hayase, F., S. Hirashima, G. Okamoto, and H. Kato. 1989. Scavenging of active oxygens by melanoidines. *Agricultural Biology and Chemistry* 53: 3383–3385.

Henle, T., H. Walter, I. Krause, and H. Klostermeyer. 1991. Efficient determination of individual Maillard compounds in heat-treated milk products by amino acid analysis. *International Dairy Journal* 1: 125–135.

Henle, T., G. Zehetner, and H. Klostermeyer. 1995. Fast and sensitive determination of furosine. *Z Lebensm Untersuch Forsch* 200: 235–237.

Hodge, J.E. 1953. Dehydrated foods, chemistry of browning reactions in model systems. *Journal of Agricultural and Food Chemistry.* 1: 928–943.

Homma, S., N. Terasawa, T. Kubo, N. Yoneyama-Ishii, K. Aida, and M. Fujimaki. 1997. Changes in chemical properties of melanoidin by oxidation and reduction. *Bioscience, Biotechnology, and Biochemistry* 61: 533–535.

Hoseney, R.C. 1994. *Principles of Cereal Science and Technology*, 2nd ed. American Association of Cereal Chemists, St. Paul, MN.

Hoseney, R.C., P. Wade, and J.W. Finley. 1988. Soft wheat products. In *Wheat Chemistry and Technology*, 3rd ed., Ed. Y. Pomeranz, 407–456. American Association of Cereal Chemists, St. Paul, MN.

Huyghebaert, A., L. Vandewalle, and G. Van Landshoot. 1982. Comparison of the antioxidative activity of Maillard and caramelization reaction products. In *Recent Developments in Food Analysis*, Ed. W. Baltes, P.B. Czedik-Eysenberg, and W. Pfannhauser, 409, Weinheim Verlag, Chemie.

International Agency for Research on Cancer (IARC). 1994. Acrylamide. IARC monographs on the evaluation of the carcinogenic risk of chemicals to humans, vol. 60, Lyon, France, 389–433.

Janzowski, C., V. Glaab, E. Samimi, J. Schlatter, and G. Eisenbrand. 2000. 5-Hydroxymethylfurfural: assessment of mutagenicity, DNA-damaging potential and reactivity towards cellular glutathione. *Food and Chemical Toxicology* 38: 801–809.

Johnson, J.A., L. Rooney, and A. Salem. 1966. Chemistry of bread flavor. *Flavor Chemistry* 56: 153.

Jousse, F., W. Jongen, W. Agterof, S. Russell, and P. Braat. 2002. Simplified kinetic scheme of flavour formation by the Maillard reaction. *Journal of Food Science* 67: 2534–2342.

Kaminski, E., R. Przybilsky, and L. Gruchala. 1981. Thermal degradation of precursors and formation of flavour compounds during heating of cereal products. Changes of amino acids and sugars. *Nahrung* 25: 507–518.

Kim, N.K., and N.D. Harris. 1989. Antioxidant effect of non-enzymatic browning reaction products on linoleic acid. In *Trends in Food Science*, Ed. A.H. Ghee, 19–23. Singapore Institute of Food Science and Technology, Singapore.

Kirigaya, N., H. Kato, and M. Fujimaki. 1968. Studies on antioxidant activity of non-enzymatic browning reaction products. Part I. Reaction of colour intensity and reductones with antioxidant activity of browning reaction products. *Agricultural Biology and Chemistry* 32: 287–290.

Konings, E.J.M., A.J. Baars, J.D. van Klaveren, M.C. Spanjer, P.M. Rensen, M. Hiemstra, J.A. van Kooij, and P.W.J. Peters. 2003. Acrylamide exposure from foods of the Dutch population and an assessment of the consequent risk. *Food and Chemical Toxicology* 41(11): 1569–1579.

Kroh, L.W. 1994. Caramelisation in food and beverages. *Food Chemistry* 51: 373–379.

Lawrence, R.C., T.D. Thomas, and B.E. Terzagui. 1976. Reviews of the progress of dairy science: Cheese starters. *Journal of Dairy Science* 43: 141–193.

Lee, Y.C., M. Shlyankevich, H.K. Jeong, J.S. Douglas, and Y. Surh. 1995. Bioactivation of 5-hydroxymethyl-2-furaldehyde to an electrophilic and mutagenic allylic sulfuric-acid ester. *Biochemistry and Biophysics Research Communications* 209: 996–1002.

Lindenmeier, M., V. Faist, and T. Hofmann. 2002. Structural and functional characterization of pronyl-lysine, a novel protein modification in bread crust melanoidins showing in vitro antioxidative and phase I/II enzyme. *Journal of Agricultural and Food Chemistry* 50: 6997–7006.

Lingnert, H. and C.E. Eriksson. 1981. Antioxidative effect of Maillard reaction products. In *Progress in Food Nutrition and Science 5*, Ed. C. Eriksson, 453–466. Pergamon Press, Oxford.

Lund, B., A. Hansen, and M.J. Lewis. 1989. The influence of dough yield on acidification and production of volatiles in sourdough. *Food Science and Technology* 22: 150–153.

Maga, J.A. 1974. Bread flavor, 5:55–142. *Critical Reviews in Food Science and Technology, 5:* 55–142.

Maillard, L.C. 1912. Action des acides amines sur les sucres formation des melanoidines par voie methodique. *Council of Royal Academy Science Series 2* 154: 66–68.

Manley, D. 1991. *Technology of Biscuits, Crackers and Cookies*, 2nd ed. Ellis Horwood Ltd., New York.

Manzocco, L., M.C. Nicoli, and E. Maltini. 1999. DSC analysis of Maillard browning and procedural effects. *Journal of Food Processing and Preservation* 23: 317–328.

Mottram, D.S., B.L. Wedzicha, and A.T. Dodson. 2002. Acrylamide is formed in the Maillard reaction. *Nature* 419: 448–449.

Noti, A., S. Biedermann-Brem, M. Biedermann, K. Grob, P. Albisser, and P. Realini. 2003. Storage of potatoes at low temperature should be avoided to prevent increased acrylamide during frying or roasting. *Mitteilungen aus Lebensmitteluntersuchung und Hygiene* 94: 167–180.

Obretenov, T., S. Ivanov, and D. Peeva. 1986. Antoxidative activity of Maillard reaction products obtained from hydrolysates. In *Amino-Carbonyl Reactions in Food and Biological Systems*, Eds. M. Fujimaki, M. Namiki, E. Kato, 281–299. Elsevier, Tokyo.

Papadakis, S.E., S. Abdul-Malek, R.E. Kamdem, and K.L. Yam. 2000. A versatile and inexpensive technique for measuring color of foods. *Food Technology* 54(12): 48–51.

Parks, J.R., A.R. Handleman, J.C. Barnett, and F.H. Wright. 1960. Method for measuring reactivity of chemical leavening systems I. Dough rate of reaction. *Cereal Chemistry* 34: 503–518.

Ramírez-Jiménez, A., E. Guerra-Hernández, and B. García-Villanova. 2000. Browning indicators in bread. *Journal of Agricultural and Food Chemistry* 48: 4176–4181.

Resmini, P. and L. Pellegrino. 1994. Occurrence of protein-bound lysilpyrrolaldehyde in dried pasta. *Cereal Chemistry* 7: 254–262.

Rizzi, G.P. 1994. The Maillard reaction in food. In *Maillard Reaction in Chemistry Food and Health*. Eds. T.P. Labuza, G.A. Reineccius, V.M. Morrier, J. O'Brien, and J.W. Baynes, 11–19. The Royal Society of Chemistry, Cambridge.

Ruiz, J.C., E. Guerra-Hernández, and B. García-Villanova. 2004. Furosine is a useful indicator in pre-baked breads. *Journal of the Science of Food and Agriculture* 84: 366–370.

Schieberle, P., and W. Grosch. 1985. Identification of volatile flavour compounds of wheat bread crust. Comparison with rye bread crust. *Zeitschrift für Lebensmittel Untersuchun und Forschung* 180: 474–478.

Schieberle, P. and W. Grosch. 1987. Evaluation of the flavor of wheat and rye bread crusts by aroma extract dilution analysis. *Zeitschrift für Lebensmittel Untersuchun und Forschung* 185: 111–113.

Şenyuva, H.Z. and V. Gökmen. 2005. Survey of acrylamide in Turkish foods by an in-house validated LC-MS method. *Food Additives and Contaminants* 22(3): 204–209.

Stadler, R.H., I. Blank, N. Varga, F. Robert, J. Hau, P.A. Guy, M.C. Robert, and S. Riediker. 2002. Acrylamide from Maillard reaction products. *Nature* 419: 449–450.

Stamp, J.A. and T.P. Labuza. 1983. Kinetics of Maillard reaction between aspartame and glucose in solutions at high temperatures. *Journal of Food Science* 48: 543–544.

Summa, C., T. Wenzl, M. Brohee, B. De La Calle, and E. Anklam. 2006. Investigation of the correlation of the acrylamide content and the antioxidant activity of model cookies. *Journal of Agricultural and Food Chemistry* 54: 853–859.

Surdyk, N., J. Rosén, R. Andersson, and P. Åman. 2004. Effects of asparagine, fructose, and baking conditions on acrylamide content in yeast-leavened wheat bread. *Journal of Agricultural and Food Chemistry* 52(7): 2047–2051.

Surh, Y.J., A. Liem, J.A. Miller, and S.R. Tannenbaum. 1994. 5-Sulfooxymethylfurfural as a possible ultimate mutagenic and carcinogenic metabolite of the Maillard reaction product, 5-hydroxymethylfurfural. *Carcinogenesis* 15: 2375–2377.

Taeymans, D., J. Wood, P. Ashby, I. Blank, A. Studer, R.H. Stadler, P. Gonde, P. Van Eijck, S. Lalljie, H. Lingnert, M. Lindblom, R. Matissek, D. Müller, D. Tallmadge, J. O'Brien, S. Thompson, D. Silvani, and T. Whitmore. 2004. A review of acrylamide: An industry perspective on research, analysis, formation and control. *Critical Reviews in Food Science and Nutrition* 44(5): 323–347.

Tareke, E., P. Rydberg, P. Karlsson, S. Eriksson, and M. Törnqvist. 2002. Analysis of acrylamide, a carcinogen formed in heated foodstuffs. *Journal of Agricultural and Food Chemistry* 50: 4998–5006.

Thacker, D. 1997. Chemical aeration. In *The Technology of Cake Making*, 6th ed., Ed. A.J. Bent, 100–106. Chapman & Hall, London; New York.

Thorvaldsson, K. and C. Skjoldebrand. 1998. Water diffusion in bread during baking. *LWT* 31: 658–663.

Van Boekel, M.A.J.S. 2006. Formation of flavour compounds in the Maillard reaction. *Biotechnology Advances* 24: 230–233.

Vass, M., T.M. Amrein, B. Schönbächler, F. Escher, and R. Amadó. 2004. Ways to reduce the acrylamide formation in cracker products. *Czech Journal of Food Science* 22: 19–21.

Waller, G.R., R.W. Beckel, and B.O. Adeleye. 1983. Conditions for the synthesis of antioxidative arginine-xylose Maillard reaction products. In *The Maillard Reaction in Foods and Nutrition*, Eds. G.R. Waller and M.G. Feather, 127–140. ACS Symposium Series 215, American Chemical Society, Washington, DC.

Weisshaar, R. 2004. Acrylamid in Backwaren-Ergebnisse von Modellversuchen. *Deutsche Lebensmittel-Rundschau* 100(3): 92–97.

Yamaguchi, N., Y. Koyama, and M. Fujimaki. 1981. Fractionation and anti-oxidative activity of browning reaction products between D-xylose and glycine. In *Progress in Food Nutrition and Science 5*, Ed. C. Ericksson, 429–439. Pergamon Press, Oxford.

Yaylayan, V.A. 1997. Classification of the Maillard reaction: A conceptual approach. *Trends in Food Science and Technology* 8: 13–18.

Yen, G.C. and P. Hsieh. 1995. Antioxidative activity and scavenging effects on active oxygen of xylose-lysine Maillard reaction products. *Journal of the Science of Food and Agriculture* 67: 415–420.

Yen, G.C. and L.C. Tsai. 1993. Antimutagenity of a partially fractionated Maillard reaction product. *Food Chemistry* 47: 11–15.

Zyzak, D.V., R.A. Sanders, M. Stojanovic, D.H. Tallmadge, B.L. Eberhart, D.K. Ewald, D.C. Gruber, T.R. Morsch, M.A. Strothers, G.P. Rizzi, and M.D. Villagran. 2003. Acrylamide formation mechanism in heated foods. *Journal of Agricultural and Food Chemistry* 51: 4782–4787.

4 Cake Emulsions

Sarabjit S. Sahi

CONTENTS

4.1 INTRODUCTION

Cake batters are a complex mixture of numerous air bubbles and dispersed fat particles in a continuous aqueous phase. The batters can be divided into two categories, those with high and low fat content. In batters with high levels of fat, the air is predominantly beaten into the fat phase, thus creating an unusual system where the foam is trapped inside the emulsified fat phase which is then mixed into an aqueous phase. In low-fat or fatless batters, the air is occluded directly into the aqueous phase and thus forms a conventional liquid foam. The dispersion of air and fat generates considerable area in terms of gas–liquid and liquid–liquid interfaces, and these newly created surfaces have to be stabilized to prevent the dispersed phases from reuniting and separating out of the aqueous phase. During baking, a further increase in interfacial area occurs as a result of bubble expansion, and this area also has to be stabilized to prevent bubbles from coalescing and impacting on cake quality. Interfacial properties therefore play a crucial role in both the batter and the baking stages of cake production. This chapter reviews in brief a number of the chemical and physical properties of surface-active materials and explains some of the functional roles that are relevant to cake emulsion systems. The type of improvements required from surface-active materials are discussed, and the ways these improvements are brought about are explained. The production of surface-active materials *in situ* by the use of enzymes is specifically covered.

4.2   WHAT ARE EMULSIONS?

By definition an emulsion is a dispersion of droplets of one immiscible liquid within another. The emulsion may be oil-in-water such as milk in which the oil droplets are dispersed in a continuous aqueous phase. It may also be a water-in-oil emulsion in which the aqueous phase is dispersed in a continuous oil phase (e.g., a margarine). Even a simple foam where air is dispersed in a continuous liquid phase can be classed as an emulsion. With these definitions in mind, a cake batter can be considered as a complex emulsion system in which fat and air are mechanically dispersed in a continuous aqueous sugar phase.

Formation of an emulsion requires two processes: First, the immiscible phase needs to be dispersed into small uniform droplets. This is normally achieved by expenditure of mechanical energy. Such breakdown of the immiscible phase results in considerable increase in the interfacial area. This is a thermodynamically unstable situation, and there is a strong tendency for the droplets to coalesce, eventually leading to complete phase separation. The newly created interfacial area is stabilized by surface-active materials, commonly referred to as emulsifiers in the food industry. Surfactants or emulsifiers are materials that consist of molecules that possess dual solubility within the same molecule. This is possible as the molecules consist of both hydrophilic and hydrophobic parts. The hydrophobic part of the molecule may consist of a fatty acid, the length of which can range from 12 to 18 carbon atoms. The hydrophilic part of the molecule may consist of glycerol, sucrose, or other chemical groupings. Such materials are surface active—that is, they have a strong tendency to diffuse from the bulk phase (usually the continuous phase), in which they are dispersed, to accumulate at interfaces between the immiscible liquids. The concentration of the emulsifier molecules at the interface results in the formation of interfacial films. It is the composition and the physical properties of these films that play a crucial role in stabilizing dispersed phases once they are formed by mechanical action.

In the same way as an emulsion cannot be created without the application of mechanical energy, foam creation requires the expenditure of energy, although foam can also be created by the vigorous introduction of air into a liquid or when gas trapped in a liquid is subjected to a sudden drop in pressure. Emulsions and foams are thermodynamically unstable, and in the absence of adequate stabilization, it is only a matter of time before the dispersed phase is separated from the continuous phase. The only option a food technologist has is to slow the rate at which the various processes destabilize foam and emulsions. This is achieved, to a degree, by the use of surfactants or emulsifiers.

The key physical events that are important in relation to emulsion stability are creaming (or sedimentation), coagulation, and coalescence of the dispersed phase droplets. Creaming results from differences in density between the oil and the aqueous phases. The inequality between the two phases should be kept as small as possible to assist in maintaining the uniform dispersion of the dispersed fraction. The process of coalescence is somewhat more complex. Coalescence is initiated by droplet coagulation, which is followed by displacement of interfacial material from the region of droplet contact. Material displacement would be expected to be easier with an expanded film than with a close-packed film. The stability of an emulsion can therefore be influenced by the correct selection of emulsifiers that form a close-packed or a condensed film at an interface. Optimum film properties are usually achieved by a mixture of surfactants, typically by combining oil-soluble and water-soluble surfactants which gives the required balance of hydrophilic and hydrophobic properties in the emulsifier system for a specific recipe formulation.

Foams consist of liquid lamellae filled with gas and are destabilized and eventually destroyed by drainage of liquid from the lamellae region which leads to film rupture. To slow the rate at which a foam collapses, the drainage has to be opposed. Surfactants can do this by creating surface tension gradients. The role of surface tension in creating and stabilizing foams is unclear. It is widely known that surfactants that are highly effective in reducing interfacial tension, which would be expected to stabilize, actually do not, whereas others that have similar surface activity have good foaming properties and stabilize foams effectively. It would appear that it is the rate at which the surface tension changes with surfactant concentration, rather than how much it changes, that plays a key role in distinguishing foaming and emulsification behavior of one type of surfactant from another.

4.3   EMULSIFIER TYPES AND FORMS

There are two main types of surface-active materials that are used in cake manufacture: those derived from lipid-based materials and those based on proteins. Lipid-based emulsifiers have become increasingly important ingredients as cake manufacturing has become a more and more mechanized process. In the early days of cake production, the key emulsifiers were eggs and lecithin. The functionality of these materials is based on the presence of surface-active lipoproteins and phospholipids. However, these natural emulsifiers lacked the effectiveness and the degree of functionality required to withstand the severity of commercial manufacturing operations to produce products with sufficiently long shelf life.

The first widely used chemical emulsifiers were mono- and diglycerides, formed by the reaction between the fatty acids of triglyceride molecules and the alcohol glycerol. These early emulsifiers contained approximately 40% of the active ingredient, the monoglyceride, the rest being the nonsurface-active diglyceride which has little or no emulsifying properties. For this reason, these early emulsifiers were used at a relatively high level of about 3%. The development of the distilled monoglycerides in which the active component was increased to 90% led to considerable benefits to the cake industry, allowing increased use of fat and sugar in the recipe formulation resulting in the production of high-ratio cakes. High-ratio cakes are rich and moist eating and with much improved keeping qualities. Continued improvements in emulsifier products have produced further benefits to the manufacturing processes as well as to the quality of the cakes. A significant step was the emulsifier glycerol monostearate (GMS), produced by reacting stearic acid (C18) with glycerol. It was found that esters formed with the saturated fatty acids possessed good complexing properties with the amylose fraction of starch, resulting in improved antistaling properties.

Other derivatives of monoglycerides that have been found to give useful advantages to cake quality have also been produced. Replacing glycerol with propylene glycol and reacting it with stearic acid generates the emulsifier propylene glycol monostearate (PGMS). This emulsifier has found specific uses in the production of enriched sponges—that is, sponges made with oil. The PGMS was demonstrated to form a strong film around the dispersed oil droplets and isolate them from the aqueous phase, hence preventing the oil from interfering with the protein-stabilized foam. Another derivative of the monoglycerides is made by reacting polyglycerol with selected fatty acid to produce polyglycerol esters (PGE) of fatty acids.

The physical form of emulsifiers such as the monoglycerides is important in order to achieve the optimum functionality. It is generally recognized that emulsifiers like GMS give the best performance when hydrated to the most active and water-dispersible α-crystalline form. This is most conveniently achieved by dispersing the GMS powder in about three times its weight of water heated to about 60 to 65°C, when a transparent homogenous liquid crystalline lamellar phase is formed (Krog and Larsson, 1968). In this form, the emulsifier consists of bilayers of monoglyceride molecules separated by water layers between the polar groups. On cooling, the lamellar phase sets to a gel phase, the α-crystalline phase that has the most efficient whipping and batter-stabilizing properties. This phase is known to be unstable and on storage begins to convert to the β-crystalline form in which the lipid bilayers are stacked on top of each other and the crystals are relatively large, giving this crystalline phase poor aeration and creaming properties. It is therefore not useful in cake production. The active GMS is available commercially in convenient ready-to-use paste form in which the active form is maintained by including other emulsifiers such as polyglycerol esters or polypropylene glycol esters. The paste contains water, and this has to be accounted for when adding the correct amount of the active ingredient to a particular recipe. In this form, they tend to be most effective in foaming properties and emulsification of the oil into the aqueous batter phase. Emulsifiers that form the α-phase when hydrated are referred to as α-tending—examples include monoglycerides, polyglycerol esters, and polyglycol monostearate. It has been suggested that the films consisting of these emulsifiers, after exceeding the solubility limit, solidify as a result of crystallization of the aligned molecules. When this happens to the film around oil droplets, the film essentially forms a solid physical barrier between the oil and water phase, preventing the two from coming together. This process effectively prevents the oil from interfering with the stability of foam stabilized by proteins in the batter stage.

Protein material with emulsifying properties can be derived from egg, cereals, or dairy sources, but the key requisite property is solubility. Proteins must be soluble in the aqueous phase in order for the surface properties to be expressed. The emulsifying properties of egg yolk have been attributed mainly to lipoproteins because of their ability to interact at the surfaces of oil droplets to form protein layers, but other proteins are likely to be involved. The presence of lipovitellin, lipovitellenin, and livetin contributes to a reduction in the drainage of the emulsifier film in oil-in-water emulsions. Lipovitellenin provides the best overall emulsion stability. Other proteins that have been used in oil-in-water emulsion stability are caseins, albumins, and globulins. More information regarding the role of protein in foam and emulsions will be discussed later in the chapter.

It must be pointed out that individually both the lipid-based emulsifiers and proteins are capable of forming foams and emulsions. However, foam instability results when the two types of surface-active materials are present together, particularly when the lipid emulsifiers are present at low concentration. This is demonstrated in Figure 4.1 where GMS was added at 0.25 and 0.75% to sponge cake batter mixed to a density of 0.65 g/ml. The egg is considered to be the foaming agent in sponge batters (Figure 4.1a), and the presence of small amounts of oil or fat reduces foam stability. Figure 4.1b shows signs of foam instability with bubbles coalescing to form larger bubbles. Cakes baked with 0.25% GMS displayed severe cores or collapse, indicating breaking of the foam created during mixing. Cakes made with no GMS were of good volume but with fairly open crumb structure. This would suggest that the protein and GMS do not combine together to stabilize interfacial films around the gas bubbles. At 0.75% addition of GMS, Figure 4.1c, there were greater numbers of gas bubbles with uniform size distribution compared with 0.25% addition. The sponge cakes produced good volume and crumb structure with no coring present. The results would indicate that the GMS at 0.25% is not present in sufficient quantity to stabilize gas bubbles at the batter or the baking stage but is able to interfere with the protein–protein interactions stabilizing the gas films. At the 0.75% level, the GMS completely replaces the protein films and is able to saturate the air–liquid interface created by the mixing process and maintain stability during the baking stage.

4.4   CONCEPT OF HYDROPHILIC–LIPOPHILIC BALANCE (HLB)

In chemical terms, the molecular structure of emulsifiers consists of polar (water-loving) and nonpolar (oil-loving) groups. Emulsifiers can be classed by a system called hydrophilic–lipophilic balance (HLB). It is a way of indicating the overall attraction of an emulsifier to either water or oil. The HLB scale ranges from 0 to 20, with 1 indicating a totally oil-soluble material and 20 indicating a highly water-soluble material. GMS, for example, has a HLB value of 3.8 and is oil soluble, whereas high-monoester-content sucrose ester possesses a HLB of 15 and is soluble in water. Incidentally, diglyceride esters have a HLB value of 0 and therefore have poor emulsification properties. In fact, if present in sufficient amounts with monoglycerides, diglycerides can actually be detrimental to the functionality of monoglycerides,

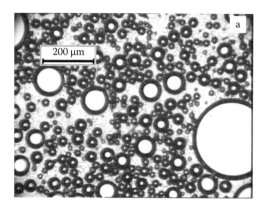

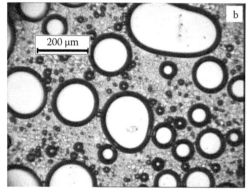

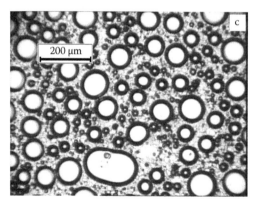

FIGURE 4.1 Influence of glycerol monostearate (GMS) concentration on gas bubbles in sponge batters: (a) control batter—protein-stabilized aqueous foam; (b) 0.25% GMS—disruption of protein-stabilized bubble; (c) 0.75% GMS—foam stabilization by emulsifier alone. (Courtesy of Campden & Chorleywood Food RA, Chipping Campden, UK.)

and hence the widespread use of distilled monoglycerides that contain low levels of diglycerides. HLB numbers are also indicative of which type of emulsion a particular emulsifier or a mixture of emulsifiers would be most likely to form. For example, the oil-soluble GMS would stabilize a water-in-oil emulsion such as margarine. On the other hand, the water-soluble sucrose monostearate would be suitable as a stabilizer of an oil-in-water emulsion such as a cake batter or a mayonnaise. Emulsifier types of intermediate HLB are not good emulsifiers but have good wetting properties. The HLB system works well with basic emulsion systems where the reduction of interfacial tension and the stabilization of the dispersed phase are the key roles required of the emulsifier system. However, cake systems are more complex, and other factors in addition to the interfacial properties are involved. This is demonstrated by the fact that the best results are usually obtained by the use of a blend of emulsifiers with different hydrophilic and lipophilic properties. Also, the HLB system cannot predict which HLB value will produce optimum emulsifier stability. This and other important factors, such as concentration of the emulsifiers to be used, the relative ratio of the two immiscible phases, and the means by which the emulsification process is to be achieved, have to be determined experimentally.

4.5 FACTORS AFFECTING THE STABILITY OF FOAMS AND EMULSIONS IN CAKE BATTERS

Cake batters made with fat or oil represent a classical emulsion system of fat dispersed in a continuous water phase. Coupled with that is a foam system, initially in the fat phase but being transferred to the aqueous phase when the fat melts during baking, resulting in a complex emulsion and foam system. It has been demonstrated that the air cells in a fat particle will coalesce within the fat particle rather than transfer as discrete air cells to the aqueous phase (Shepherd and Yoell, 1976). In the same publication, it was observed that air bubbles released from the fat phase on melting remained attached to the fat surface, suggesting that the fat surface may be coated with egg proteins or added emulsifiers during mixing which are transferred to the gas bubbles when they transfer from the fat to the aqueous phase. Continued heating of the batter during baking causes rapid expansion of the bubbles. At this point, the viscoelastic properties of the films surrounding the gas cells become important in the maintenance of the integrity of the bubbles. Expansion ceases when the cake structure is set and the discrete foam systems break to form an open network sponge. At present, there is no established mechanism to explain how complex systems like cake emulsions are stabilized, but some progress has been made to help explain the essential roles played by key ingredients. However, there are a number of fundamental factors that govern emulsion stability, and these are also likely to aid the stability of the dispersed fractions in a cake batter to some degree.

4.5.1 INTERFACIAL TENSION

It is accepted that the interfacial tension between the dispersed phase and the continuous phase is an important factor in controlling stability in such systems. A high tension between the two phases is likely to be detrimental to emulsion and foam stability, whereas a low tension, as a result of surface-active molecules accumulating at the interface, is likely to be beneficial in stabilizing such systems. The interfacial tension between oil and water in the presence of emulsifiers can range from 1 to 10 mN/m. Surface-active materials naturally present in key ingredients used in cake systems and added emulsifiers can therefore stabilize emulsions. Interfacial measurements are sensitive to the composition of the interface and can be used to identify the nature of the predominant material, for example lipid-based surfactant or proteins. Methods to measure surface or interfacial tensions include the Ring method, the Wilhelmy plate method (Shaw, 1980), as well as a number of methods based on imaging a pendant drop hanging from the tip of a capillary tube (Ambwani and Fort, 1979).

4.5.2   INTERFACIAL RHEOLOGY

The rheological properties of the interfacial films play a key role in stabilizing dispersed phases. The rheology is chiefly governed by the composition of the interfacial region. Protein adsorption at an interface usually results in the formation of a two-dimensional viscoelastic film, whereas lipid-like materials form relatively thin mobile films. Film rheology can be measured using either shear or dilatational methods. Surface shear measurements are usually performed with a ring or a bicone located at the interface, and the film is deformed with an oscillatory stress or strain. The analysis of the applied stress or strain and the resulting response characterizes the viscoelastic properties of the interface. This technique was used to demonstrate the disruptive influence of flour lipids on protein-stabilized interfacial films, giving an insight into the negative effect that such materials can have on foam stability (Sahi, 1994). The dilatational approach again involves oscillatory methods, but here the surface area of the interface is changed and the change in surface or interfacial tension is monitored. This method is useful for studying mixed film systems such as protein–lipid mixtures. Whereas dispersed phases stabilized by lipid-based materials (emulsifiers) influence emulsion and foam stability through their effects on interfacial tension, protein-based stabilization arises from the physical presence of the adsorbed film rather than from a reduction in the tension of the dispersed phase interfaces. Such films can be many layers thick and viscoelastic and are able to dampen perturbations experienced in the interface.

4.5.3   STABILIZATION BY SOLID PARTICLES

Emulsions can also be stabilized by the presence of solid particles at the interface. Finely dispersed solids that have a contact angle between 0 and 180° have a natural tendency to collect at an oil–water interface. When dispersed particles come close to one another, the physical presence of the solid particles prevents one droplet from touching another. Such a mechanism may be significant in cake emulsions where the presence of egg and flour particles could contribute to emulsion stability.

4.5.4   SURFACE CHARGES

Electrical double-layer repulsion is an important stabilizing mechanism in oil-in-water emulsions. The surface charges on the droplets can repel droplets when they come close together by the mutual repulsion of their electrical double layers (Bergenstahl and Claesson, 1990). The emulsifiers in a cake recipe formulation, present both naturally and as additives, on adsorption to an oil–water interface will have an overall negative electrical charge. In this respect, flour can contribute lipoproteins as well as polar lipids; for example, phospholipids and similar types of materials are also present in egg. When such materials are adsorbed onto oil or fat droplets, an overall negative charge results, which will prevent close contact of the droplet and contribute toward emulsion stability. Chemically synthesized emulsifiers are able to repel oil droplets in a similar fashion.

4.5.5 DISPROPORTIONATION AND OSTWALD RIPENING

Disproportionation is the process by which large gas bubbles grow at the expense of small bubbles. This is predicted by Laplace's equation that states that the pressure in a bubble (P_{bubble}) will be

$$P_{bubble} = P_a + \frac{4\gamma}{R} \qquad (4.1)$$

where P_a is the atmospheric pressure, γ is the surface tension of the liquid, and R is the radius of the bubble.

Gas will diffuse from a smaller bubble to a large bubble in contact with each other driven by the pressure differential between the bubbles. This mechanism suggests that in order to obtain a uniform distribution of bubbles in a cake batter, the interfacial tension must be as low as possible in order to decrease the value of the excess pressure that would increase as bubble size becomes small.

Ostwald ripening occurs in emulsions and is caused by the diffusion of liquid from smaller droplets into larger ones. Larger bubbles thus grow larger and then are more susceptible to coalescence. A uniform distribution of droplet size helps to stabilize an emulsion with respect to Ostwald ripening.

4.5.6 FILM THINNING

Film thinning behavior provides insight into the factors that promote foam and emulsion stability. Thinning occurs as a result of gravitational forces, and the viscosity of the aqueous phase plays an important role: the higher the viscosity of the aqueous phase, the slower is the rate of film thinning. The composition of the film plays an important role in controlling and maintaining thickness. For example, lipid-stabilized films exhibit rapid drainage (Clark et al., 1990), whereas protein-stabilized films display slow drainage (Clark et al., 1994).

4.5.7 FILM RUPTURE

Film thinning eventually results in film rupture. As mentioned previously, proteins form strong viscoelastic films that resist rupture. However, lipid films tend to possess low lateral cohesion, and localized stresses of the film cause it to expand, resulting in a decrease of surfactant concentration in the region of stress and an increase in surface tension (the Gibbs effect). Because a finite time is required for surfactant molecules to diffuse to this region to restore the original surface tension (the Marangoni effect), the increased surface tension may persist long enough to recover the original thickness. The absence of the Gibbs–Marangoni effect is the main reason why pure liquids do not foam.

4.6   APPLICATION OF EMULSIFIERS IN CAKE PRODUCTION

There are numerous reasons why emulsifiers are used in the manufacture of cakes. Primarily, these include foaming, emulsification, and batter stability. Further benefits include the homogeneous dispersion of fat in the batter, greater tolerance to recipe formulation changes, and the possibility of changing from multistage mixing to an all-in mixing method. Emulsifiers also give cakes a tender eating quality as well as help to improve shelf life. The latter is important for industrially produced cakes, as there is considerable length of time before they reach the consumer. The use of emulsifiers has also led to key changes in the recipe formulations from the earlier pound cake to the high-ratio cakes made today. Some of the key functions of emulsifiers are described in more detail below.

4.6.1   Dispersion of Shortening

Using shortening alone in a high-ratio-type recipe can result in the batter curdling during mixing. This can be avoided by inclusion of an emulsifier during production of the shortening, such a shortening being referred to as high-ratio shortening. Emulsifiers such as GMS and mono- and diglycerides are used in soft fats at levels up to 10% to manufacture high-ratio shortenings. Because of the functionality of the emulsifier, these shortenings can be easily dispersed in high-ratio batters without curdling. Cake margarines also contain emulsifiers added by the manufacturer, both to maintain a smooth product and to facilitate easy dispersion in the batter. In addition, emulsifier inclusion can improve creaming properties in both shortenings and margarines.

4.6.2   Reduction in Mixing Time

Specific emulsifiers such as GMS, polyglycerol esters, and lactic acid esters of monoglycerides possess marked foam-promoting properties. When added to sponge batters (0.5 to 1.0% of batter weight), whisking time can be greatly reduced. The emulsifiers do this by reducing the interfacial tension of the aqueous phase, thus allowing the mixing action to break the interface more easily and to incorporate air into the batter. The lowering of the interfacial tension also aids the breakup of the air bubbles to produce a smaller and more uniform cellular structure of the cake. An additional advantage is a more stable batter, resistant to mechanical depositing stresses and more tolerant to variations in holding time after mixing.

4.6.3   To Reduce Fat and Egg Content

Foam-promoting emulsifiers such as GMS, polyglycerol esters, propylene glycol esters, or blends of these can help to reduce the fat content of a cake or substitution of the fat by a smaller quantity of vegetable oil. Oil addition was originally found to give low volume, open grain, and poor structure of the cake. However, the use of the correct emulsifier or blends of emulsifiers can produce cakes of good quality, using oil instead of hard fat. Emulsifiers can also be used to replace part of the egg in sponge cakes. In such a case, the aeration properties of the egg are performed by the emulsifier, and egg content can be reduced sufficiently to support the foam structure and contribute to the eating quality of the cake.

4.6.4 TO PREPARE CAKE MIXES

Emulsifiers are included in dry cake-mix formulations to improve batter aeration. Special cold-water-dispersible types are available consisting of a mixture of emulsifiers similar to those found in paste emulsifiers but dried onto a carrier such as skimmed milk powder. Spray-dried emulsions of cake emulsifiers such as distilled monoglycerides, propylene glycol monostearate, and lactylated monoglycerides containing a nonfat milk powder as a carrier are used in sponge cakes with the same benefit as aqueous gels.

Practical experience shows that combinations of emulsifiers are usually more effective than any single emulsifier used alone, suggesting synergistic effects between blends of emulsifiers. A blend of two or even three emulsifiers can be used; an example of a blend widely used is mono-diglyceride esters and polyglycerol esters of monoglycerides.

4.7 METHODS TO CREATE CAKE BATTERS AND THE ROLES PLAYED BY SPECIFIC INGREDIENTS

A cake batter is formed from a basic formula consisting of flour, fat, sugar, and egg. The ingredients can be mixed in stages using the creaming method (sugar and shortening), the flour batter method, or the all-in method where all the ingredients are added simultaneously into the mixing bowl. In the creaming method, the fat and sugar are first mixed together to aerate the mixture. Small air bubbles are introduced into the dispersed particles of fat. Optimum aeration of the batter depends on the size of the fat crystals, and small crystals (β′-form) give the best aeration properties (Hoerr, 1960). Such crystals are able to orientate themselves around the gas cells and provide a protective shell. The amount of air trapped in the creaming process is important to the structure development of the cake as no new gas cells are created during the remaining stages of the cake-making process (Carlin, 1944). In the flour batter method, the shortening and the flour are mixed to form an aerated mass, and in a separate container the egg and sugar are whipped into a foam. The two are combined with the addition of the other ingredients at the same time. In the all-in method, all the key ingredients are mixed into a smooth batter. The mixing of a fat-containing batter results in the formation of a fat emulsion with air bubbles also being trapped in the emulsified fat. Aeration of fatless recipes occurs in the aqueous phase, with egg proteins playing an important role as whipping agents and foam stabilizers. The important processes are the homogenous mixing of the ingredients to create a silky smooth batter containing a large volume of air in the form of small cells in the fat phase or the aqueous phase depending on the recipe. In the respective mixing methods, the major ingredients play a key role at each stage. The contribution of each ingredient is examined with respect to foam formation and emulsification.

4.7.1 FATS AND SHORTENING

The functional properties of fats play a key role in bakery products in general, and this is especially important in the production of cakes. The functionality of plastic fats is influenced by the ratio of liquid oil to crystalline solids as well as the crystalline form achieved during the processing. The two fundamental functional properties of fats are creaming and emulsification. The process of creaming fat with sugar results in the breakup of the fat as well as the trapping of air into the dispersed particles of fat. This stage of the mixing therefore has a crucial bearing on the volume of cake that can be achieved at the end of baking. The plasticity of the fat and the particle size of the sugar are both important in achieving optimum fat particle size and air incorporation. Caster sugar with particle size specification of <10% above 425 μm and <22% below 212 μm is typically used as it possesses the optimum surface area to break down the fat.

In addition to the aeration capacity of the cake shortening, the crystallinity of the fat has been suggested to contribute to stabilization of the foam in the batter and during baking. The fat crystals with the best functionality are small and needle shaped and can align themselves at the interface of air bubbles trapped into the fat droplets (Brooker, 1993), thus stabilizing the air bubbles trapped during the creaming process. The same researcher also hypothesized that the fat crystals come out of the fat droplets, and as they emerge into the aqueous phase of the batter, become coated with surface-active water-soluble proteins. This then confers surface properties to the fat crystals, which then align at the air–water interface in the batter. During the baking process, the air bubbles expand, generating new interfacial area that needs to be stabilized to prevent rupture of the foam before sufficient expansion has been achieved. The timely melting of the fat crystals releases the surface-active proteins adsorbed on the fat crystal. This protein then becomes available to stabilize the expanding air bubbles. The fat crystals therefore act as a reservoir of surface-active material that is in close proximity to the gas bubbles as they expand.

Recent work on the role of fat in the stabilization of air in "all-in-one" cake batters has shown that fat crystals are ejected from the emulsified shortening during mixing, become enveloped by a fat (crystal)–water interface, and are able to stabilize large numbers of small air bubbles by adsorbing to their surface. During baking, air bubbles can expand without rupturing because of extra interfacial material provided by the adsorbed fat crystals when they melt. The outcome of this mechanism is that a batter can expand during baking without collapse to produce a high-volume cake of fine crumb structure.

The use of fluid shortenings is on the increase in cake manufacturing and other bakery systems. In the past, they were found to be detrimental to cake quality, resulting in poor crumb quality and low cake volume (Knightly, 1988). This was largely overcome by the use of a reduced amount of total fat in the recipe and a careful combination of hydrophilic and hydrophobic emulsifiers dispersed in fluid shortenings. Fluid shortenings are mixtures of hard fat and surfactants in vegetable oil. Typical surfactants used are a combination of monoglycerides, lactylated monoglycerides, and propylene glycol esters. The level of emulsifier used can be quite high at 10 to 15% of the oil weight. The addition of the emulsifier in the α-crystalline form is particularly effective. In this form, the emulsifier system is able to emulsify the oil into fine particles in the aqueous phase and, as a result, shield the protein-stabilized foam from the destabilizing effect of the oil. Other benefits of these shortenings have also been realized in recent times. Health concerns regarding the intake of saturated fats and trans fats have led to replacement of the plastic shortenings and the challenge to the industry has been to achieve this without losing the benefits derived from them. Fluid shortenings that produce finished products identical to those made with plastic shortenings have been developed by using α-stable emulsifier gels.

4.7.2   WATER-SOLUBLE PROTEINS

Protein naturally present in egg and flour are good foam and emulsion promoters and stabilizers. In order for the protein to be surface active, it must possess good solubility in an aqueous system. Examples of such proteins include albumins and globulins. The mechanism by which proteins function at interfaces is different than that of the lipid-based emulsifiers. Protein molecules diffuse from the bulk of the solution to the air–water or oil–water interface. On reaching the interface, some segments of the molecules attach themselves and some unfolding of the protein chain may also occur. Intermolecular interactions, for example H-bonding between sections of the protein chains, link the neighboring polypeptide chains together. Further adsorption of proteins underneath the primary interfacial layer can lead to the formation of a thick, viscoelastic interfacial layer that is able to stretch without breaking and hence absorb fluctuations in the interfacial region. The physical presence of the thick protein film also prevents bubbles coalescing on contact with each other. It must be emphasized that proteins are not as effective as lipid-based surfactants in reducing interfacial tension. For example, commonly used emulsifiers can lower air–water tension to 30 mN/m or lower, whereas the large protein molecules with limited number of hydrophobic sites in their polypeptide chains can only attain values of about 45 mN/m. However, the lowering of the interfacial tension is not the only requirement in foam and emulsion systems.

4.7.3   WHEAT FLOUR

Wheat flour used in cake production contains numerous components that can play various roles in foam and emulsion stability. The proteins and lipids are a source of surface-active materials that have the capacity to accumulate at the air–water and oil–water interfaces. Wheat flour also contains three major types of polysaccharides—starches, hemicelluloses, and β-glucans. Starches from heat-treated and chlorinated flours allow better absorption of liquid in the batter; hence, batters made from treated flours have a higher viscosity compared with batters made with untreated flour. This has a positive benefit on aeration of the batter, as air can be trapped and subdivided more easily and the bubbles have a lesser tendency to float out of the batter compared with a less-viscous system. Batter viscosity is also influenced by the hemicellulose material that is present at 2 to 3% of the flour mass (Montgomery and Smith, 1956). The hemicelluloses have two forms, denoted by their solubility in water: insoluble materials and soluble materials. Both are composed of mainly arabinoxylans and are commonly known as pentosans. In line with other hydrocolloids, they have the ability to bind eight to ten times their own weight in water. This is beneficial at the batter stage where water absorption by these materials increases batter viscosity. The β-glucans also contribute to batter viscosity in a similar fashion. The hemicelluloses have also been shown to possess surface-active properties. Izydorczyk et al. (1991) reported that arabinoxylans and arabinogalactans reduced the surface tension of water from 72 to 52 and 50 mN/m, respectively. These values are similar to the surface tension values reported for some proteins, suggesting that these materials in wheat flour could potentially act as foam and emulsion stabilizing materials. Hemicellulose types of materials are believed to stabilize foam films by retarding gas diffusion through the gas–film interface or acting as steric stabilizers of the films surrounding the gas bubbles (Prins, 1988). Although the foaming properties of the arabinoxylans and arabinogalactans were inferior compared to the protein bovine serum albumin, the viscosity and elasticity of the interfacial film surrounding the gas bubbles is likely to be important for foam stability.

The starch fraction in the wheat flour does not appear to play a direct role in the stabilization of foam or emulsion systems in cake batters. However, starch does absorb water in the recipe, and the amount of water absorbed is increased by chlorination or heat treatment of the cake flour. This gives increased viscosity that helps to trap and subdivide air in the batter during mixing. It might be expected that, because starch granules have a large surface-area-to-volume ratio, interactions between starch and emulsifiers may influence gas bubble stability. However, there is no reported evidence to suggest that the starch granules interact with air bubbles trapped in the batter. The reason for this may be that the starch granules have a relatively small surface area compared with that of the gas bubbles, especially when the bubbles begin to expand. However, with the mechanism of stabilization by solid particles in mind, the potential contribution of starch granules toward interfacial stability cannot be ruled out.

4.8   APPLICATION OF ENZYMES TO GENERATE SURFACE-ACTIVE MATERIALS

In addition to natural or synthetic surface-active agents, highly active materials can also be created *in situ* by the action of specific enzymes on substrates that are present in raw materials, such as flour, egg, or shortenings. Such enzymes have the potential to replace added emulsifiers. This is of considerable importance in terms of the health image of a product, because enzymes have no function in the final product and do not have to be declared on the packaging. This has led to the application of enzymes to produce the required surface-active materials *in situ*. There are a number of esterhydrolases that can modify the chemical structure of phospholipids—namely, phospholipases A1, A2, C, and D. Phopholipase A has been used by the food industry for many years to improve the functionality of egg yolk by breaking down the lipid complexes. Lipase enzymes are present in cereals in small concentrations and are known to produce fatty acids in flours stored for long periods. The action of lipase is to cleave the bond between the fatty acid esters and glycerol. The net result is the production of a mixture of fatty acids and the surface-active monoglycerides, which it is claimed, may be used to replace monoglycerides usually added as an ingredient.

The functionality of the monoglycerides produced would be dependent on the source of the triglyceride acting as the substrate. Saturated fatty acids with 14 to 18 carbon atoms in their chains are the most effective antistaling monoglycerides.

However, it is not clear if the required concentration levels of emulsifiers could be produced by added lipase, or whether the large amounts of fatty acids produced at the same time would be detrimental or beneficial to the baking process.

Continuing progress in biotechnology has produced a new generation of lipases that act on the nonpolar triglycerides and polar lipids naturally present in wheat flour. Wheat flour contains 2 to 3% lipid material consisting of both polar and nonpolar fractions. However, a large proportion of this lipid is found in the starch granular structure and hence is not available as substrate for lipase action. The nonstarch lipid forms about 1.6% of the flour weight, and out of this about 0.6% is composed of the functionally important polar fraction. Examples of the polar lipids found in flour include the phospholipid lecithin and the galactolipid digalactosyl diglyceride. Action of the new lipase on triglyceride molecules would be expected to produce mono- and diglycerides as well as free fatty acids. These by-products of enzyme action are similar to those generated by the traditional 1,3 specific lipase. However, it is the action on the polar lipids which is of interest in terms of increasing the surface activity of these materials. The action of the new lipase on lecithin cleaves one of the fatty acyl chains to produce lysolecithin, a more surface-active material compared to the original substrate. Likewise, the cleavage of an acyl chain from a diglyceride digalactosyl molecule produces the more surface-active digalactosyl monoglyceride. The utility of this enzyme to produce surface-active materials *in situ* was investigated in the manufacture of high-ratio cakes using a range of flours. Initially, the performance of the enzyme was assessed by its effect on the batter density with mixing time. Mixing time to achieve the target batter density (0.72 g/ml) was reduced by a minimum of 24% at 90 ppm addition of the enzyme for one flour and up to 35% for another flour based on the mixing time with no enzyme added. Two other flours examined had mixing time reduced to intermediate values between the two extremes. The use of emulsifiers to reduce mixing time of batters is well recognized in the cake-making industry, and the same effect observed with the use of the new lipase would suggest the production of surface-active materials by the enzyme. Fundamental studies with cake batter suggested the presence of surface-active materials. These materials lowered the surface tension and the surface viscosity (Table 4.1), probably by displacing the protein complexes with mono-acyl lipids such as fatty acids, lysolecithin, and digalactosyl monoglycerides (Guy and Sahi, 2006). The presence of these materials was confirmed by analysis of free fatty acids and an increase in amylose–lipid complexes found in the baked product.

Enzyme addition also increased batter viscosity in proportion to the amount of enzyme used, and this would benefit the stabilization of the bubbles in the batter after mixing and during their expansion in the baking process. It is thought that the surface-active materials may have wetted-out the proteins and helped to increase their hydration volumes, because the increases in viscosity were lost at high shear levels in the batter. Initial baking trials with three heat-treated flours used to manufacture high-ratio cakes showed immediate benefits in increased cake volume (Figure 4.2). In total, this improvement was found with six different flours, from three different commercial milling companies (Guy and Sahi, 2006). Analysis of the cake

TABLE 4.1

Effect of New Lipase on Surface Properties of the Aqueous Phases of High-Ratio Cake Batters at 21°C

Sample	Surface Tension (mN/m)	Surface Viscosity[a] (uNm.s/m)
Water	72.5 ± 0.3	
Control batter	31.7 ± 0.2	118 ± 10
Batter with 60 ppm lipase	31.1 ± 0.1	80 ± 5
Batter with 90 ppm lipase	30.1 ± 0.2	38 ± 5

[a] Surface viscosity achieved after 30 min.

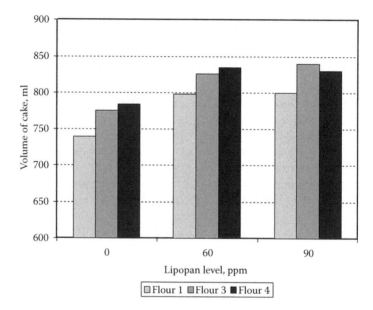

FIGURE 4.2   Effect of new lipase addition on the volume of high-ratio cakes.

crumb structure suggested that the cells were more stable, because they were generally similar in number in the cakes, but had increased in size to give extra volume (Figure 4.3).

Enzymatic modification of proteins has also been attempted in order to improve their emulsifying and foaming properties. This usually involves incorporation of substituent groups into the protein structure in order to improve the solubility. For example, the sulfhydral groups of proteins have been modified to improve solubility and other functional properties (Regenstein and Regenstein, 1984). Enzymes have been used to modify gelatin proteins producing a mixture of polypeptides known as enzymatically modified gelatin-6 (EMG-6) and EMG-12. Bread baking trials demonstrated that EMG-12 could improve loaf volume (Arai and Watanabe, 1988), but no trials were performed on cake systems.

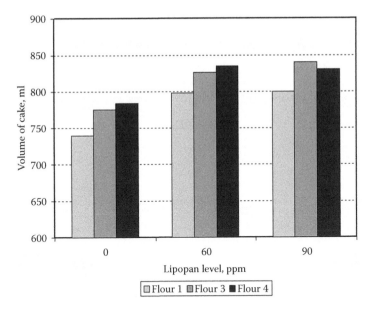

FIGURE 4.3 Functionality of new lipase in high-ratio cakes. (Courtesy of Campden & Chorleywood Food RA, Chipping Campden, UK.)

4.9 CONCLUSION

This chapter presented an overview of the interfacial properties relevant to the formation and the stability of emulsions and foams in relation to cake emulsions. The role of emulsifiers and other key recipe ingredients and their components in relation to influencing interfacial properties was discussed. The interactions of emulsifier molecules at air–water and oil–water interfaces are crucial to stabilizing the dispersed air and fat phases in cake batters. Emulsifier interactions at a molecular level with other components such as fat and proteins have been shown to be important to foam and emulsion stability. Chemically synthesized emulsifiers have played a key role in the manufacture of cakes for many years. However, the trend of consumer demand for natural ingredients has led to alternative ways of producing emulsifiers. The use of methods to produce surface-active materials *in situ* with the application of enzymes is likely to increase. Initially, this may serve to partially replace chemically synthesized emulsifiers, with full replacement possible only in recipe formulations that depend on relatively low levels of added emulsifiers. However, developments in biotechnology may lead to enzymes that can generate greater amounts of surface-active materials *in situ*, which along with changes in recipe formulation or processing methods may lead to full replacement of chemically synthesized emulsifiers.

SUGGESTED FURTHER READING

Bennion, E.B. and G.S.T. Bamford. 1997. In: *Technology of Cake Making*, Ed. A.J. Bent, Blackie Academic & Professional: Glasgow.

Friberg, S., K. Larsson, and J. Sjoblom (Eds.). 2003. *Food Emulsions (Food Science and Technology)*, 4th ed. CRC Press: Boca Raton, FL.

Shaw, D.J. 1980. *Introduction to Colloid and Surface Chemistry*, 3rd ed. Butterworth-Heinemann: Oxford.

Whitehurst, R.J. (Ed.). 2004. *Emulsifiers in Food Technology*. Blackwell: Oxford.

REFERENCES

Ambwani, D.S. and T. Fort Jr. 1979. Pendant drop techniques for measuring liquid boundary tensions. In: *Surface and Colloids Science*, vol. II, Eds. R.J. Good and R.R. Stronberg, Plenum: New York, 93–119.

Arai, S. and W. Watanabe. 1988. Emulsifying and foaming properties of enzymatically modified proteins. In: *Advances in Food Emulsion and Foams*, Eds. E. Dickinson and G. Stainsby, Elsevier Applied Science: New York, 189–220.

Bergenstahl, B.A. and P.M. Claesson. 1990. Surface forces in emulsions. In: *Food Emulsions*, Eds. K. Larsson and S.E. Friberg, Marcel Dekker: New York, 41–96.

Brooker, B.E. 1993. The stabilisation of air in cake batters—The role of fat. *Food Structure* 12: 285–296.

Carlin, G.T. 1944. A microscopic study of the behaviour of fats in cake batters. *Cereal Chemistry* 21: 189–199.

Clark, D.C., M. Coke, A.R. Mackie, A.C. Pinder, and D.R. Wilson. 1990. Molecular diffusion and thickness measurements of protein-stabilised thin liquid films. *Journal of Colloid Interface Science* 138: 207–218.

Clark, D.C., A.R. Mackie, P.J. Wilde, and D.R. Wilson. 1994. Differences in the structure and dynamics of the adsorbed layers in protein-stabilised model foams and emulsions. *Faraday Discussions* 98: 253–262.

Guy, R.C.E. and S.S. Sahi. 2006. Applications of a lipase in cake manufacture. *Journal of the Science of Food and Agriculture*. Special Issue: Enzymes in Grain Processing 86 (11): 1679–1687.

Hoerr, C.W. 1960. Morphology of fats, oils and shortenings. *Journal of American Oil Chemists Society* 37: 539–546.

Izydorczyk, M., C.G. Biliaderis, and W. Bushuk. 1991. Physical properties of water-soluble pentosans from different wheat varieties. *Cereal Chemistry* 68 (2): 145–150.

Knightly, W.H. 1988. Surfactants in baked foods: Current practice and future trends, *Cereal Foods World*, 33, 405–412.

Krog, N. and K. Larsson. 1968. Phase behaviour and rheological properties of aqueous systems of industrial distilled monoglycerides. *Chemistry Physics Lipids* 2: 129–143.

Montgomery, R. and F. Smith. 1956. Hemicelluloses in flour. *Journal of Agricultural and Food Chemistry* 47: 716–720.

Prins, A. 1988. Principles of foam stability. In: *Advances in Food Emulsions and Foams*, Eds. E. Dickinson and G. Stainsby, Elsevier Applied Science: London, 91–122.

Regenstein, J.M. and C.E. Regenstein. 1984. Sulphydral chemistry. In: *Food Protein Chemistry*, Academic Press: London, p83.

Sahi, S.S. 1994. Interfacial properties of the aqueous phases of wheat flour doughs. *Journal of Cereal Science* 20: 119–127.

Shaw, D.J. 1980. Liquid–gas and liquid–liquid interfaces. In: *Introduction to Colloid Chemistry*, 3rd ed. Butterworths: London, pp. 60–90.

Shepherd, I.S. and R.W. Yoell. 1976. Chapter 5: Cake Emulsions. In: *Food Emulsions*, Ed S. Friberg, Marcel Dekker: New York, pp/ 270–274. .

5 Cake Batter Rheology

Serpil Sahin

CONTENTS

5.1   INTRODUCTION

Understanding the rheological characteristics of food materials is necessary for plant and product design. It is important to determine the rheological properties of cake batter because the quality attributes of cakes such as volume and texture can be correlated with rheological properties of batter.

The basic ingredients in cake batters are flour, fat, egg, milk, sugar, and salt. Flour, egg white, milk solids, and salt are used to toughen the cake, while sugar, fat, and egg yolk are used to tenderize the cake. Cake batter can be considered as a complex oil-in-water emulsion with a continuous aqueus phase containing dissolved or suspended dry ingredients such as sugar, flour, salt, and baking powder. The oil phase remains dispersed in clumps throughout the continuous or liquid phase and does not become part of the liquid phase (Painter, 1981). The interaction of ingredients and structure development occurs during the mixing and baking stages.

The incorporation of air cells in the system during mixing gives rise to a foam. It is important to obtain a large number of small cells to provide higher volume (Handleman et al., 1961). During baking, an aerated emulsion of cake batter is converted to a semisolid porous, soft structure mainly due to starch gelatinization, protein coagulation, and carbon dioxide gas produced from chemicals dissolved in the batter, air occlusion during mixing, and the interaction among the ingredients.

Viscosity of cake batter is the controlling factor for the final cake volume. The rate of bubble rise due to buoyancy force is inversely proportional to the viscosity.

In the presence of less-viscous batter, carbon dioxide evolves, and the water vapor produced might not be trapped in the system during baking, thus resulting in cakes with low volume. Higher cake batter viscosities help to retain more air bubbles in the batter and retard the rise of bubbles to the surface. In addition, the velocity gradient in the batter during baking induces convection current at any given time depending on its viscosity, with lower batter viscosity resulting in more convection flow (Frye and Setser, 1991). It is also known that higher batter viscosity prevents the entrapped air from coalescing due to drainage of surrounding batter during baking and reduces shrinkage. In fact, there is an optimum viscosity of cake batter to achieve cakes with high volume. If the viscosity of the batter is too low, batter cannot hold the air bubbles inside and cakes collapse in the oven. Although a highly viscous batter can hold the air bubbles inside, the expansion of this batter is restricted because of its high viscosity (Sahi and Alava, 2003).

There are many studies in the literature in which rheological properties of cake batter are correlated with the quality of cakes (Vali and Choudhary, 1990; Sahi and Alava, 2003; Lakshminarayan et al., 2006). Batters with low specific gravity and high viscosity produced cakes with higher volumes (Vali and Choudhary, 1990). The increase in batter density (decrease in air content incorporated in batter) decreased the storage (elastic) and loss (viscous) moduli of cake batter and also the specific volume of cake (Sahi and Alava, 2003). Batter with low viscosity produced cake with low volume and firmer texture (Lakshminarayan et al., 2006).

In this chapter, rheological methods used in cake batter will be explained briefly, and then the parameters affecting the rheology of cake batter will be summarized.

5.2   RHEOLOGICAL METHODS

Food materials exhibit flow, deformation, or both under external force. Rheology is the science dealing with the description of the mechanical properties of food materials under well-defined deformation conditions. Basic rheology concepts can be classified into viscous flow, elastic deformation, and viscoelasticity.

Non-Newtonian shear thinning (pseudoplastic) behavior was observed in cake batters having different formulations, and the rheological behavior of cake batters was well described by the power law model (Baixauli et al., 2007; Gujral et al., 2003; Sakiyan et al., 2004; Shepherd and Yoell, 1976; Turabi et al., 2007). The power law model (Ostwald–de Waele equation) is expressed as

$$\tau_{yz} = k \left(\frac{dv_z}{dy} \right)^n = k(\dot{\gamma}_{yz})^n$$

(5.1)

where k is the consistency coefficient (Pa · s^n), n is the flow behavior index ($n < 1$ for shear thinning fluids), τ_{yz} is the shear stress (N/m^2), and $\dot{\gamma}_{yz}$ is the shear rate (1/s).

Baik et al. (2000) observed that the flow behavior of cake batters having different formulations was pseudoplastic with a yield stress, and the shear stress–shear rate relationship was fitted well with the Casson model:

$$(\tau_{yz})^{0.5} = (\tau_0)^{0.5} + k_c (\dot{\gamma}_{yz})^{0.5} \tag{5.2}$$

where τ_0 is the Casson yield stress (N/m²), K_c is the consistency coefficent (Pa·s)$^{0.5}$.

In rice cake batters, both power law model and Casson model were found to be suitable to explain their rheological behavior (Turabi et al., 2007).

Ideal elastic solids display Hookean behavior. When a force is applied to a solid material having this behavior, a straight-line relationship between stress and strain was observed. This relationship is known as Hooke's Law:

$$\tau = G\gamma \tag{5.3}$$

where G is the shear modulus (N/m²), τ is shear stress (N/m²), and γ is shear strain.

Cake batter, like many complex structured food materials, displays both viscous and elastic properties and is known as a viscoelastic material. Different methods, such as the dynamic (oscillatory) test, creep recovery test, stress relaxation, and so forth, can be used to study the viscoelastic behavior of food material (Sahin and Sumnu, 2006). In the study of Lee et al. (2004), the rheological properties of cake batters were studied using an oscillatory test.

In the stress relaxation test, stress is measured as a function of time as the material is subjected to a constant strain. Stress in viscoelastic solids will decay to an equilibrium stress σ_e, which is greater than zero, but the residual stress in viscoelastic liquids is zero.

If a constant load is applied to biological materials and if stresses are relatively large, the material will continue to deform with time. In the creep recovery test, an instantaneous constant stress is applied to the material, the resulting strain is measured as a function of time, called creep, and then the stress is removed while the strain is continued to be recorded as a function of time, called recovery. Ideal elastic materials show complete recovery, and ideal viscous materials show no recovery when the applied stress is removed. Viscoelastic materials are in between—that is, they show partial recovery after the removal of stress.

In the dynamic (oscillatory) test, material is subjected to deformation or stress that varies harmonically with time. Then, the transmitted shear stress or deformation in the sample, which also varies harmonically, is measured, respectively (Figure 5.1). Storage modulus (G′), which is high for elastic materials, and loss modulus (G″), which is high for viscous materials, are defined as follows:

$$G' = \frac{\tau_0 \cos\theta}{\gamma_0} \tag{5.4}$$

$$G'' = \frac{\tau_0 \sin\theta}{\gamma_0} \tag{5.5}$$

where τ_0 is shear stress (input or output), γ_0 is shear strain (output or input), and θ is time lag (phase shift).

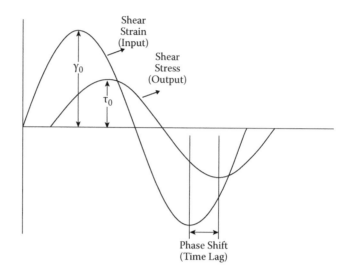

FIGURE 5.1   Harmonic shear stress versus strain for a viscoelastic material in dynamic test.

5.3   FACTORS AFFECTING RHEOLOGY OF CAKE BATTERS

Rheological properties of fluid foods are complex and depend on many factors such as composition, shear rate, duration of shearing, and previous thermal and shear histories. The most important parameters affecting the rheological properties of cake batters are type and concentrations of the ingredients, level of air incorporation, and temperature. Air incorporation is affected by time and speed of mixing, design of the mixer, and surface tension of the batter. Studies related to the effects of ingredients (flour, fat, fat replacer, emulsifiers, sugar, hydrocolloids, and egg), beating and mixing, dosing, and temperature on cake batter rheology are summarized in this section.

5.3.1   EFFECTS OF INGREDIENTS

5.3.1.1   Flour

Flour is one of the most important ingredients affecting the rheological properties of cake batter and, consequently, the quality of cakes. Freshly milled wheat flour has a pale yellow color due to its carotenoid content and yields sticky dough. During storage, as a consequence of oxidative reactions, flour gradually turns white, and the rheological properties of dough and, consequently, quality of the baked product are improved. The age of flour is known to affect the viscosity of cake batter. Shelke et al. (1992) observed that freshly milled flours had low water-binding capacity and produced batters with low viscosity at ambient temperature and also during heating. Batter viscosity at ambient temperature and during heating increased with flour age. Viscosity of batters at ambient temperatures increased as a function of postmilling time. However, the age of wheat at milling did not significantly affect batter viscosity.

In making cakes, doughnuts, cookies, crakers, wafers, pretzels, and similar products, soft wheat flour is used. Soft wheat flour has low gluten content, low water-absorption capacity, and low granulation size. It is commonly chlorinated for the production of cakes. Chlorination is usually done with chlorine gas and can be monitored by a drop in pH of flour.

Chlorine treatment of soft wheat flours improves cake volume and produces a stiffer, more resilient crumb (Donelson et al., 2000). Chlorine treatment of cake flour is functionally beneficial to the production of high-ratio cakes. White layer cakes baked using untreated flour are unsatisfactory in volume, contour, crumb grain, and texture. Chlorine-treated flour produces cake crumb with a drier, less sticky mouthfeel (Kissell and Yamazaki, 1979).

Starch, lipid, and proteins are all affected by chlorination. The oxidative depolymerization of starch which increases the water-binding capacity of starch is one of the important changes that occur during chlorination (Huang et al., 1982). At temperatures 90°C and above, swelling power and solubility of high starch fraction increased as a result of chlorination (Huang et al., 1982). Water activities of batters made with chlorinated or untreated flours were the same until the temperature reached 80°C. Batters containing chlorinated flour had higher water activity at 80°C. Storage modulus (G′) increased in the case of chlorine-treated batter, but it decreased in the case of untreated sample between 90 and 100°C (Ngo et al., 1985). The batter prepared with chlorine-treated flour had a much higher loss modulus (G″) than did the batter made from untreated flour. Alteration of starch accelerates the thickening of the viscosity of the batter which results in higher volume (Gaines and Donelson, 1982). Shelke et al. (1992) also observed that the chlorinated flour had higher viscosity than untreated flour. Freshly milled flours produced batters with low minimum viscosity, and minimum viscosity of flours increased during aging. Minimum viscosity of batter during heating is an important property because it reflects the ability of the batter to retain gas bubbles and to resist settling of starch.

Chlorination affects the hydrogen bonds of the proteins and causes greater solubility of the soft wheat proteins (Kissell, 1971). Hydrophobicity of proteins also increases with chlorination (Sinha et al., 1997). Chlorine treatment enhances the gel-forming properties (Frazier et al., 1974). Chlorination also changes the hexane-extractable flour lipids which results in higher batter expansion during baking (Kissell et al., 1979).

Gaines and Donelson (1982) studied the effects of bleaching of flour and batter liquid levels on cake batter viscosity and expansion during heating. Flours obtained from two different varieties of wheat were used in this study. Apparent viscosities of the cake batters prepared using different water levels were measured with a modified viscograph between 20 and 100°C continuously. Flour bleaching increased batter expansion in both flour types. There was an optimum water level between the extreme liquid levels giving maximum cake expansion for both flour types. Bleached flours at the optimum liquid levels achieved higher pasting viscosities more rapidly than unbleached flours. Faster pasting may contribute to the stability and reduced shrinkage of bleached-flour cakes on cooling (Kissell and Yamazaki, 1979). There was also an optimum viscosity for cake expansion. That is, cake expansion increased up to a certain level and then started to decrease as batter viscosity increased.

Rats were fed with cakes made with chlorine-treated flour at ingestion levels equivalent to the consumption of cake in the human diet in order to test the safety of consumption of chlorinated flour, and no adverse reaction was observed (Daniels et al., 1963). However, at higher ingestion levels, reduced growth rate, increased liver, kidney, and heart weights, and reduction in ovary weight (among female mice) were observed (Cunningham et al., 1977; Ginocchio et al., 1983). Therefore, production of baked products using untreated wheat flour is of great concern for safety aspects, and it has been removed from the permitted list of food ingredients in the European Union for cake manufacture. However, in order to produce cakes having similar properties to those produced using chlorine-treated flour, some additional ingredients may be added to untreated flour or heated flour may be used as an alternative to chlorine-treated flour. Improved cake volume and contour were observed when untreated flour–starch blend was used (Johnson and Hoseney, 1979). When a certain amount of (12 to 43% depending on the type of the untreated flour) starch was added to the untreated flour, it was possible to obtain pasting curve areas equivalent to those obtained in the case of chlorine-treated flours in Rapid Visco Analyzer (Donelson et al., 2000). Starch acts as a water sink during baking, contributing to the setting of the structure of the cake during baking.

Heat-treated flour has been introduced as an alternative to chlorinated flour. Heat treatment improves flour performance as in the case of chlorine treatment. Heat treatment of flour denatures wheat proteins and reduces their solubility in water. Therefore, batters made from heat-treated flour have higher viscosity compared with batters made with untreated flour, and higher viscosity helps to trap and subdivide air in the batter during mixing. The temperature of the first rise of viscosity decreases with heat-treated flour in Brabender Amylograph analysis (Seguchi, 1990). Heat treatment may split some of the linkages of starch amylose or amylopectin chains resulting in lowering the optimum viscosity of wheat flour. Thomasson et al. (1995) added xanthan gum, L-cysteine, or hydrogen peroxide plus peroxidase enzyme to heat-treated flour and obtained cakes having comparable volume with the ones produced using chlorine-treated flour.

5.3.1.2 Fat and Fat Replacer

Oils are refined and may be partially hydrogenated. Cake margarines and shortenings are formulated from blends of processed oils and fats. In cake margarines, the aqueous phase contains skim milk. Shortening normally contains a mixture of glycerides having widely different melting points. In a cake system, shortening serves three major functions: to entrap air during the creaming process, to physically interfere with the continuity of starch and protein particles, and to emulsify the liquid in formulation. Thus, shortening affects the tenderness and moisture content of the cake. The addition of lipid improves air incorporation and foam stability which affect baking quality. In addition, fats and emulsifiers are known to delay gelatinization by delaying the transport of water into the starch granule due to the formation of complexes between the lipid and amylose during baking (Larsson, 1980).

The functions of fat and oil on cake structure development are different. Increase in fat content increases shear modulus, and increase in oil content decreases shear modulus (Mizukoshi, 1985). The effect of different fats and oil on specific gravity and viscosity of cake batters and, consequently, the volume and tenderness of baked cake were studied by Vali and Choudhary (1990). Cake batter with margarine could hold a greater amount of air and showed low specific gravity, and that with oil and hydrogenated fat showed high specific gravity. Batters with low specific gravity showed high viscosity, and they produced cakes with greater volumes.

The effects of concentration of fat (0%, 12.5%, 25%, 37.5%, 50%) on the rheological properties of cake batter were studied by Sakiyan et al. (2004). It was found that cake batter with different fat concentrations exhibited shear thinning and time-independent behavior. The increase in fat content caused a decrease in the apparent viscosity. Figure 5.2 shows the decrease in apparent viscosity with applied shear rate for cake batters containing emulsifier blend, Lecigran®, and different amounts of fat. This trend was also observed for the samples containing no emulsifier. A decrease in viscosity with the increase in fat concentration was an expected result, because an increased amount of fat caused more air entrapment during the creaming process. The reduction of apparent viscosity with the addition of fat was also explained by the lubrication effect of uniformly dispersed fat particles.

Fat aids in the entrapment of air during mixing and, as a result, improves the leavening of product. Fat also imparts desirable flavor and softer texture to the cakes. Although fat plays an important role in improving the product quality, the trend is toward the reduction of fat for health concerns. Several types of fat replacers are available in the market for this purpose. Fat replacers can be categorized as carbohydrate

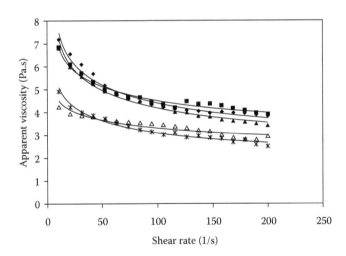

FIGURE 5.2 Effects of fat content on apparent viscosity of cake batter containing Lecigran®. (Line represents the power law model, ◆ 0% fat, ■ 12.5% fat, ▲ 25% fat, △ 37.5% fat, * 50% fat.) (Reprinted from Sakiyan, O., Sumnu, G., Sahin, S., and Bayram, G., *European Food Research and Technology*, 219, 635–638, 2004, Figure 1. With permission from Springer Science and Business Media.)

based (e.g., cellulose, dextrin, maltodextrin, fiber, gums, Oatrim, starch), protein based (e.g., microparticulated protein, modified whey protein, Simplesse®), and lipid based (e.g., caprenin, salatrim, olestra, emulsifiers, sucrose polyesters). Detailed information about the functions of fat replacers can be found in Chapter 12. Rheological studies on fat replacers in cake batter are limited to carbohydrate-based fat replacers in literature. Carbohydrate-based fat replacers, in the presence of water, form a smooth gel resulting in lubricant and flow properties similar to fats (Swanson et al., 1999). They increase viscosity and provide creamy, slippery mouthfeel similar to that of fat.

White layer cake batters containing starch-based fat replacers and no emulsifiers had higher specific gravities and lower viscosities both at ambient temperature and during heating (Bath et al., 1992). The low viscosity of batters made using fat replacers both at ambient temperature and during heating was responsible for their rapid rate of heating in electrical resitance ovens. Cakes without fat became flat and had low volume.

Khouryieh et al. (2005) studied the effect of incorporating xanthan gum, maltodextrin, and sucralose to make no-sugar-added and low-fat muffins and observed that removing the fat from the muffins was responsible for the increase in hardness and chewiness. Grigelmo-Miguel et al. (2001) used dietary fiber as an oil substitute in muffins and found that it increased the hardness and chewiness of muffins. Shearer and Davies (2005) observed an increase in batter viscosity and a decrease in firmness and elasticity of the muffin with the increase in flaxseed meal used as a source of fiber. Masoodi et al. (2002) studied the effects of using apple pomace at different concentrations and particle sizes as a source of dietary fiber on the quality of cakes. Batter viscosity increased with increasing concentration and decreasing particle size of pomace.

Lakshminarayan et al. (2006) investigated the effects of fat replacement by maltodextrin on cake batter viscosity and the quality of the baked product. The viscosity of batter was reduced significantly when fat was replaced with equal quantities of maltodextrin. Batter with low viscosity produced cake with low volume and firmer texture. When the amount of replacement was lower in the formulation, viscosity of the batter was relatively higher and quality of the resultant cakes was relatively better.

Kim et al. (2001) studied the effects of replacement of shortening with maltodextrin, amylodextrin, octenyl succinylated amylodextrin, or mixtures of them on yellow layer cake batter and baked product properties. The specific gravity and viscosity of cake batter and volume index of baked cake were significantly reduced by maltodextrin, but the cake with amylodextrin and octenyl succinylated amylodextrin showed higher volume index than the control cake containing shortening.

The soluble fiber in oat bran, β-glucan, is a well-recognized nutraceutical. Nutrim oat bran, which is a hydrocolloid obtained from oats, provides a viable source of β-glucan for use in the food industry. Flaxseed is a good source of omega-3 fatty acids, α-linoleic acid, dietary fiber, and lignans which have beneficial health effects. The effects of replacement of shortening with Nutrim oat bran and flaxseed powder on the physical and rheological properties of cakes were investigated (Lee et al., 2004). Cake batters showed shear thinning behavior. Slightly higher shear viscosity and oscillatory storage and loss moduli were observed in batters containing Nutrim oat bran than in the control, and the replacement of shortening with flaxseed powder reduced viscosity and oscillatory storage and loss moduli of the control cake batter. Cakes having similar quality parameters with the control cake could be obtained with up to 40% fat replacement.

In another study, the effects of, replacement of shortening with 20%, 40%, and 60% by weight of Oatrim (oat β-glucan amylodextrin) on the physical and rheological properties of cakes were investigated (Lee et al., 2005). The specific gravity of the cake batters increased as more shortening was replaced with the Oatrim, and no significant effect was observed in the case of 20% Oatrim content. Shear thinning behavior was observed in all types of batters. The control cake had the highest viscosity, and an increase in the levels of shortening replacement with Oatrim caused a reduction in viscosity of cake batters. The mean size of the air bubbles, 1.5×10^{-9} m^2, was not significantly affected, but the number of air bubbles incorporated in the sample was significantly affected from the replacement of shortening.

The cake batters containing more Oatrim displayed a higher gelatinization temperature due to amylodextrins in the Oatrim and they have lower volume. The rheological properties of the cakes were studied during heating to mimic the baking process. The oscillatory shear storage moduli decreased upon initial heating, then increased due to starch gelatinization, and finally reached a plateau value that varied based on the sample composition. Increased replacement of shortening with Oatrim resulted in higher observed oscillatory shear storage moduli. Correlations between oscillatory shear storage moduli and the differential scanning calorimetry (DSC) thermograms were investigated.

5.3.1.3 Emulsifiers

Emulsifiers are commonly used in the baking industry. They have the ability to provide the necessary aeration and gas bubble stability during the process. An emulsifier reduces the interfacial tension between oil and water and therefore facilitates the disruption of emulsion droplets during homogenization. The emulsifier adsorbs on the surfaces of emulsion droplets to form a protective coating that prevents the droplets from aggregating with each other. Emulsifiers aid in the incorporation of air and disperse the shortening in smaller particles to give the main number of available air cells (Painter, 1981). Air incorporation, volume, and dispersion of ingredients were affected by the amount and type of emulsifier (Cloke et al., 1984).

The addition of lipid-like emulsifiers to cake systems affects both the interfacial and bulk properties of the batter. The effects of two different types of emulsifiers (glyceryl monostearate [GMS] and polyglycerol ester [PGE]) on the structure of sponge batters were studied by Sahi and Alava (2003). The addition of an emulsifier improved water binding and decreased the fluidity of the batter. A dynamic oscillatory test was performed to determine the viscoelastic properties of the batters, and it was observed that the addition of an emulsifier resulted in an increase in elastic and viscous moduli (Table 5.1). The decrease in batter density (increase in air content incorporated in batter) increased the viscous and elastic moduli of cake batter and also the specific volume of the cake. Bubbles in batter samples immediately

TABLE 5.1

Results of Dynamic Oscillatory Measurements of Sponge Cake Batters at a Frequency of −1 Hz

Emulsifier Type	Concentration (% batter wt)	Elastic Modulus (Pa)	Viscous Modulus (Pa)	Phase Angle (deg)
Control	—	59 ± 8	83 ± 7	55 ± 2
GMS[a]	0.25	115 ± 16	142 ± 11	51 ± 2
GMS	0.75	487 ± 60	392 ± 57	39 ± 3
GMS	1.50	1223 ± 111	1037 ± 217	40 ± 4
PGE[b]	0.25	111 ± 11	158 ± 11	55 ± 1
PGE	0.75	306 ± 10	321 ± 12.9	46 ± 1
PGE	1.50	1047 ± 65	878 ± 70	40 ± 2

[a] Glyceryl monostearate.

[b] Polyglycerol ester.

Source: Reprinted from Sahi, S.S. and Alava, J.M., *Journal of the Science of Food and Agriculture*, 83, 1419–1429, 2003. Copyright 2003 Society of Chemical Industry. Reproduced with permission. Permission is granted by John Wiley & Sons Ltd. On behalf of the SCI.)

after mixing were imaged using an optical microscope and a charge-coupled device (CCD) video camera. It was observed that the addition of an emulsifier at low levels (0.25% glyceryl monostearate) resulted in an increase in bubble size and heterogeneity of bubbles. Increasing the concentration of the emulsifier (0.75% glyceryl monostearate) provided smaller bubbles with more uniform bubble size distribution.

When GMS or sodium steroyl lactate (SSL) was used as an emulsifier in cake formulations in which 25% fat was replaced with maltodextrin, the viscosity of cake batter increased (Lakshminarayan et al., 2006). However, a decrease in the cake batter viscosity was observed when fat replacement with maltodextrin was 50% or more in the case of GMS addition and 75% in the case of SSL addition (Figure 5.3). SSL is more hydrophilic in nature with a hydrophilic–lipophilic balance (HLB) value of 10 to 12, and it was more efficient in increasing batter viscosity in a system when the fat level was relatively low. GMS is a more lipophilic emulsifier with an HLB value of 3 to 4. Boyd et al. (1972) observed that oil-in-water emulsions were more stable when emulsifiers with higher HLB values were used.

The best result is obtained using a blend of emulsifiers with different hydrophilic–lipophilic properties in complex systems like cake. The effects of different types of emulsifier blends (Purawave® and Lecigran®) on the rheological properties of cake batter have been studied by Sakiyan et al. (2004). Purawave (Puratos, Belgium) was composed of lecithin, soy protein, mono- and diglycerides, and vegetable gums. Lecigran (Riceland Foods, Arizona) was composed of oil-free soybean lecithin, wheat flour, and hydrogenated vegetable oil. All types of cake batters prepared with and without emulsifiers exhibited shear thinning and time-independent behavior. Experimental data provided a good fit with the power law model. The addition of emulsifier caused a decrease in the apparent viscosity. Figure 5.4 shows the effects

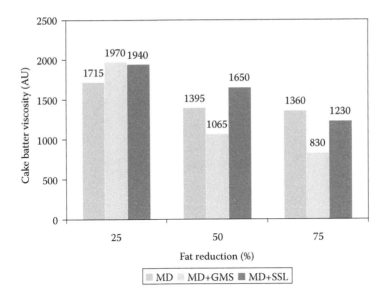

FIGURE 5.3 Combined effect of maltodextrin (MD) and emulsifiers (GMS and SSL) on the viscosity of cake batter. (Reprinted from Lakshminarayan, S.M., Rathinam, V., and Krishna-Rau, L., *Journal of the Science of Food and Agriculture*, 86, 706–712, 2006. Copyright 2006 Society of Chemical Industry. Reproduced with permission. Permission is granted by John Wiley & Sons Ltd. on behalf of the SCI.)

of emulsifiers on apparent viscosity of cake batter having 37.5% fat content. Effects of emulsifiers on apparent viscosity of cake batters having different fat concentrations were also studied, and the same trend was observed in other fat concentrations. A decrease in viscosity by emulsifier addition was explained by the increase in air entrapment in cake batter, because the emulsifier aids in the incorporation of air and disperses the shortening in smaller particles to give the main number of available air bubbles.

Guy and Sahi (2006) studied the effect of using lipase enzyme instead of emulsifiers to improve the performance of high-ratio layer cake batters made with heat-treated flours. It was reported that the commercial lipase, Lipopan F®, produces monoacyl lipids from lecithin and mono- and digalactosyl diglycerides from the triglycerides. As the lipase concentration increased, viscosity of the batter increased, which indicated the formation of surfactants. The surfactants produced by lipase may help to hydrate wheat proteins and increase their hydrodynamic volume and viscosity in the aqueous phase, as well as stabilize the air bubbles by forming new interfacial membranes (Guy and Sahi, 2006). The addition of enzyme reduced the mixing time, which was defined as the time taken for the batter density to fall to a target value of 0.85 kg/L while being mixed.

Gujral et al. (2003) studied the effect of the addition of sodium lauryl sulfate which is an anionic surfactant to egg albumen during the mixing stage on the rheology of sponge cake batter. The rheological behavior was well described by the power law equation. Sodium lauryl sulfate is used to improve the foaming properties of egg

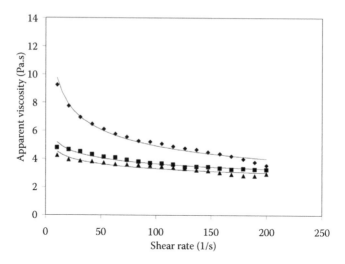

FIGURE 5.4   Effects of emulsifiers on apparent viscosity of cake batter at 37.5% fat content. (Line represents the power law model, ◆ no emulsifier, ■ Purawave®, ▲ Lecigran®. (Reprinted from Sakiyan, O., Sumnu, G., Sahin, S., and Bayram, G., *European Food Research and Technology*, 219, 635–638, 2004, Figure 4. With permission from Springer Science and Business Media.)

whites, marshmallows, and angel food cake mixes. Increasing the concentration of sodium lauryl sulfate lowered the specific gravity, consistency coefficient, and air bubble diameter of cake batter. The flow behavior index increased significantly with increasing sodium lauryl sulfate concentration.

5.3.1.4 Sugar

The addition of sugar reduces the available water for starch which affects rheological properties, delays starch gelatinization, retards structural development, and controls the heat-setting temperature of egg proteins during baking. Effects of cake ingredients on the shear modulus of cake were studied, and it was observed that sugar decreased the shear modulus of a degassed batter (Mizukoshi, 1985). Batter viscosity at ambient temperature increased with an increase in the sugar concentration (Shelke et al., 1990). An increase in sugar concentration also increased the onset temperature. Sucrose was more effective as compared to glucose and fructose when the onset temperatures of batters containing different types of sugars were compared at the same concentration.

5.3.1.5 Hydrocolloids

Hydrocolloids are high-molecular-weight water-soluble polysaccharides used for viscosity control in many food systems. They are added to cake batters in small amounts to improve product volume and texture, to increase moisture retention during baking, and to prevent staling.

The effect of xanthan gum on the rheology of white layer cake batter during heating was studied using an oscillatory probe viscometer in conjuction with electric resistance oven heating (Miller and Hoseney, 1993). During heating, batter viscosity decreased to a minimum and then increased sharply to a maximum. The addition of xanthan to the batter increased the minimum and maximum points. Batter viscosity at ambient temperature and the onset temperature of rapid increase in viscosity were not significantly changed with the addition of xanthan. The addition of xanthan gum improved the cake volume.

Rice is one of the most frequently used cereals as a wheat substitute in gluten-free baked products for patients who have celiac disease. However, some food additives such as starches, gums, hydrocolloids, or dairy products should be added to the gluten-free baked products to reduce the poor quality due to the lack of gluten in the formulation. Turabi et al. (2007) studied the effects of different gums (xanthan gum, guar gum, locust bean gum, κ-carrageenan, hydroxy-propyl-methylcellulose [HPMC], xanthan–guar gum blend, and xanthan–κ-carrageenan gum blend) and an emulsifier blend, Purawave® (Puratos, Belgium), which is composed of lecithin, soy protein, mono- and diglycerides, and vegetable gums, on rheological properties of rice cake batter and quality characteristics of rice cakes baked in an infrared–microwave combination oven. In this study, all the formulations containing different kinds of gums and emulsifier blend showed shear-thinning behavior, which means that apparent viscosity decreased as the shear rate increased. The flow behavior of the rice cake batters was described by the power law model. Table 5.2 shows the power law model constants for all the formulations containing different kinds of gums and emulsifier blend. The flow behavior index of batters ranged from 0.399 to 0.623. Batter containing HPMC gum had the lowest consistency index. The Casson model was also found to be a suitable model to explain the rheological behavior of rice cake batters. The coefficient of determination values and the model constants for the Casson model are given in Table 5.2. The lowest Casson yield stress was found for HPMC-containing batters. Using the values of flow behavior index and consistency index, apparent viscosities at a shear rate of 150 s^{-1} were calculated for different formulations for the Casson model and are given in Figure 5.5. The highest viscosities were obtained for batters containing xanthan and xanthan–guar blend. This result was explained by the xanthan's unique, rod-like conformation, which is more responsive to shear than a random-coil conformation (Urlacher and Noble, 1997). The higher viscosity values of xanthan-containing batters improved cake structure, and this resulted in higher volumes. In the study of Miller and Hoseney (1993), it was also observed that xanthan gum significantly improved cake volume. Synergistic interaction between xanthan and guar gum resulted in higher apparent viscosity as compared to other gums. This synergistic effect was not observed in the xanthan–κ-carrageenan blend which gave lower apparent viscosity values (Figure 5.5). When locust bean gum was used in the formulations, it gave lower apparent viscosity values as compared to the guar-gum-containing batters. This was explained by the higher molecular weight of guar gum. HPMC-containing batters had the lowest specific gravity values due to more air incorporation during mixing and the lowest apparent viscosity values (Figure 5.5). This resulted in the collapse of HPMC-containing cakes. The addition of an emulsifier blend to the xanthan–guar gum blend and the xanthan–carragenan

TABLE 5.2

Power Law and Casson Model Constants for Cake Batters with Different Formulations

Formulation	Power Law Model			Casson Model		
	n	$K(Pa \cdot s^n)$	r^2	Z_0 (Pa)	$K_c(Pa \cdot s)^{1/2}$	r^2
Control	0.421	38.356	0.977	63.03	0.817	0.999
Xanthan	0.563	61.870	0.997	107.848	1.798	0.997
Guar	0.399	100.524	0.987	170.459	1.171	0.997
Xanthan + guar	0.552	69.580	0.994	126.293	1.811	0.992
Carrageenan	0.418	52.568	0.965	84.339	0.957	0.997
Locust bean gum	0.496	35.730	0.994	63.915	1.047	0.994
HPMC	0.596	12.898	0.991	20.025	0.952	0.999
Xanthan + carrageenan	0.541	53.091	0.986	86.902	1.569	1.000
Xanthan + emulsifier	0.610	46.980	0.997	78.711	1.871	0.997
Guar + emulsifier	0.399	111.830	0.991	213.131	1.149	0.949
Xanthan + guar + emulsifier	0.545	59.377	0.997	114.833	1.595	0.981
Carrageenan + emulsifier	0.495	52.740	0.996	102.394	1.227	0.975
Locust bean gum + emulsifier	0.513	35.710	0.997	65.398	1.113	0.993
HPMC + emulsifier	0.611	12.780	0.997	20.994	0.983	0.999
Xanthan + carrageenan + emulsifier	0.623	29.520	0.997	47.967	1.557	0.999

Source: Reprinted from Turabi, E., Sumnu, G., and Sahin, S., *Food Hydrocolloids,* 22: 305–312, 2008. Copyright 2008, with permission from Elsevier.

gum blend decreased the apparent viscosity of cake batter significantly (Figure 5.5). When the firmness values of cakes were investigated, xanthan- and emulsifier-blend-containing cakes were softer than those prepared with other formulations, which were also correlated with apparent viscosity results.

Cake batters contain a large number of components that can interact with each other. Kim and Walker (1992) studied the interactions of starches, sugars, and emulsifiers in a high-ratio cake model system. Model cakes were prepared using different

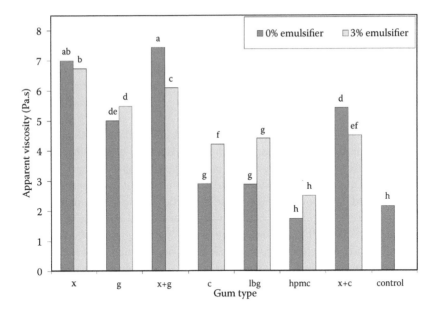

FIGURE 5.5 Apparent viscosities (Pa.s) of formulations at 150 s⁻¹ constant shear rate according to the Casson model. (x: xanthan, g: guar, x + g: xanthan + guar, c: carrageenan, lbg: locust bean gum, HPMC: hydroxyl-propyl-methylcellulose, x + c: xanthan + carrageenan, * bars with different letters are significantly different $p \leq 0.05$.) (Reprinted from Turabi, E., Sumnu, G., and Sahin, S., *Food Hydrocolloids*, 22:315–312, 2008. Copyright 2008, with permission from Elsevier.)

types of starch (wheat, corn, or potato starch), sugar (lactose or dextrose, or replacement of 50% of lactose or dextrose with sucrose), and emulsifiers (sucrose ester F-160 or polysorbate 60). The pototo starch, lactose, and polysorbate 60 combination produced a significant increase in batter viscosity. Increase in viscosity tended to aid in air incorporation.

5.3.1.6 Egg

Egg proteins are critical for the structure of cake batter. Egg white proteins may enhance foam stability and also act to coagulate during heat setting. Lipoproteins in egg yolk act as emulsifiers and assist in aeration and foaming (Shepherd and Yoell, 1976). Substitution of 10% of egg white in angel food cake by freeze-dried wheat water solubles, recovered from a by-product of a gluten-starch washing plant, decreased batter viscosity (Maziyadixon et al., 1994). Whipping time increased as percent of substitution increased.

5.3.2 EFFECTS OF MIXING AND DOSING

The rheological properties of batters depend not only on the types and concentrations of the ingredients but also on the beating and mixing process. Physical and structural changes during aerated batter processing may alter batter performance during baking or may alter the quality of the final product.

During mixing, the size of bubbles decreases. Providing necessary aeration by means of mixing and also the stability of the gas bubbles during the baking process until the structure is set are important factors in considering rheological properties of cake batters. Air incorporation during mixing reduces the specific gravity and apparent viscosity of batter.

Viscoelastic properties of manually dosed batters are different than those of batters passed through the automatic dosing unit. Baixauli et al. (2007) studied the effect of the use of an automatic dosing unit on the rheological properties of an aerated muffin batter. Flow and viscoelastic properties of batters dosed automatically were measured and compared with these properties of the batters dosed manually. In both cases, shear thinning behavior was observed and the data fit to the power law equation very well. The consistency index of the batters dosed automatically was significantly higher, and the flow index was not affected from the dosing type. Passing the batter through the automatic dosing unit produced an increase in both storage modulus (G′) and loss modulus (G″) at 25°C (Figure 5.6), but there was no significant difference between different dosing types at 85°C (Figure 5.7). It was also observed that using an automatic dosing unit affects the microstructure of batters. Batters showed greater compactness, smaller fat globules, and partial deformation in starch granules when passed through an automatic dosing unit.

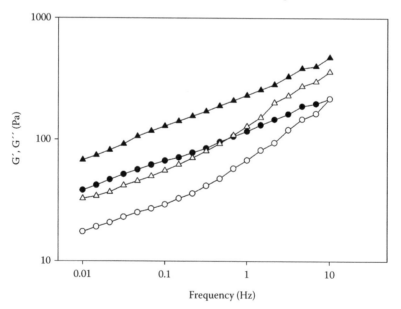

FIGURE 5.6 Mechanical spectra of manually dosed batter (circles) and batter that has passed through the automatic dosing unit (triangles) at 25°C. G′ values are represented by solid symbols; G″ are represented by open symbols. Shear stress wave amplitude: 0.1 Pa. (Reprinted from Baixauli, R., Sanz, T., Salvador, A., and Fiszman, S.M., *Food Hydrocolloids*, 21, 230–236, 2007. Copyright 2007, with permission from Elsevier.)

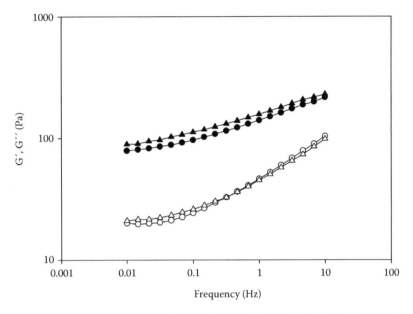

FIGURE 5.7 Mechanical spectra of manually dosed batter (circles) and batter that has passed through the automatic dosing unit (triangles) at 85°C. G' values are represented by solid symbols; G'' are represented by open symbols. Shear stress wave amplitude: 0.6 Pa. (Reprinted from Baixauli, R., Sanz, T., Salvador, A., and Fiszman, S.M., *Food Hydrocolloids*, 21, 230–236, 2007. Copyright 2007, with permission from Elsevier.)

5.3.3 EFFECT OF TEMPERATURE

Rheological behaviors of cake batters at different temperatures (27 to 50°C) have been studied by Shepherd and Yoell (1976). Non-Newtonian shear thinning behavior was observed. A drastic change in the flow behavior index was observed between 35°C and 40°C. This was explained by the melting of fat surrounding the air bubbles when the temperature was above 35°C which made the bubbles more mobile and able to migrate to the aqueous phase.

A change in batter rheology during heating is very important. Generally, the batter viscosity decreases at the beginning of heating and then starts to increase at the starch gelatinization temperature (Chang et al., 1990). Ngo and Taranto (1986) measured viscoelastic properties of cake batters using a dynamic (oscillatory) test. They observed that the storage modulus (G') and loss modulus (G'') increased when the cake batter temperature increased from 30 to 45°C and then gradually decreased, reaching a minimum value at a temperature around 85°C which varied with sugar content. When heating was continued, G' and G'' increased rapidly until the batter temperature reached 100°C. It was explained that the increases in G' and G'' when the temperature increased from 30 to 45°C might be caused by gluten, milk, or egg protein interactions, because this trend was not observed in the case of the starch paste system.

The baking process consists of three stages—initial, middle, and final stages—according to Mizukoshi (1986). The viscosity of the batter decreased with increasing temperature in the initial stages due to foam drainage and bubble coalescence. Then, in the middle stages of baking, viscosity started to increase due to starch gelatinization. The increased shear modulus of the continuous phase stabilized the bubble structure and reduced foam drainage (Mizukoshi, 1983). In the final baking stage, starch gelatinization and protein coagulation were accelerated. The loss modulus (G″) reached its maximum values as the foam structure of cake changed from discontinuous to continuous.

Yasukawa et al. (1986) measured the dynamic viscoeleastic properties of cake batter in a model cake system during baking. Both storage modulus (G′) and loss modulus (G″) started to increase at the gelatinization temperature. Initial structural development was due to starch gelatinization. G″ reached a maximum point at 88°C, which is very close to the gas release temperature. The temperature at which batter expansion ceased coincided closely with its gas release temperature and the temperature at which G″ reached its maximum value.

Major ingredients affect batter viscosity during heating in different ways. Effects of ingredients and additives (sugar type and concentration, shortening, egg white, hydrocolloids, emulsifier) on the dynamics of cake baking were studied (Shelke et al., 1990). Batter viscosity was determined using a continuous oscillatory rod viscometer in conjuction with electric resistance oven heating. Batter viscosity decreased as the temperature increased from ambient to 60°C during the early heating, but viscosity increased sharply with temperature as the heating continued above 60°C due to the gelatinization of starch. The onset temperature of the rapid viscosity increase that is, starch gelatinization was not affected significantly by shortening level. However, the increase in the rate of viscosity decreased with increased shortening levels, because shortening limited the starch swelling. Increasing shortening levels decreased cake volume. Batters containing surfactants had higher viscosity. No significant effect on the onset temperature was observed with the addition of surfactant, but the rate of viscosity increase decreased after the onset. Batter viscosity at ambient temperature increased with an increase in sugar concentration. However, viscosity of batter decreased with an increase in sugar concentration during heating. At ambient temperature, not all of the sugar was dissolved. A decrease in viscosity during heating was explained with dissolving sugar. The onset temperature increased as the sugar concentration increased. Sucrose was the most effective sugar when the onset temperatures of batters containing different types of sugars (glucose, fructose, and sucrose) were compared at the same concentration. The use of egg white in cake batter increased the batter viscosity at ambient temperature and minimum viscosity during heating. Viscosity of the batter at ambient temperature was higher when fresh egg white was used. Onset temperature was not affected by the addition of dried or fresh egg white. The addition of hydrocolloids (xanthan, guar, carboxy methyl cellulose [CMC]) increased the viscosity at ambient temperature. When the minimum viscosities of heated batters were compared, xanthan was able to maintain higher batter viscosities than guar and CMC at the same concentrations. Higher viscosity during heating would give the batters greater capacity to retain expanding air and resist settling of starch granules that improve cake volume and crumb grain.

5.4 CONCLUSION

Rheological data are required in product quality evaluation, engineering calculations, and process design. In rheological studies, cake batters mostly showed non-Newtonian shear thinning behavior, and a correlation was observed between rheological properties of cake batter and final cake quality. The rheological properties of cake batters are mainly affected by the type and concentrations of the ingredients, the level of air incorporation, and temperature. The viscosity of batters at ambient temperatures increases as a function of postmilling time. Soft wheat flour is commonly chlorinated for the production of cakes. Chlorinated flour had higher viscosity than untreated flour. Heat-treated flour can also be used as an alternative to chlorinated flour as they show similar rheological properties. The addition of fat improves air incorporation and foam stability and causes a decrease in apparent viscosities of cake batters. Using fat replacers, emulsifiers, sugars, and gums also affects the rheological behavior of cake batters. Air incorporation during mixing reduces the apparent viscosity of batter. Viscoelastic properties of manually dosed batters are different than those of batters passed through an automatic dosing unit. Generally, the batter viscosity decreases at the beginning of heating and then starts to increase at the starch gelatinization temperature.

REFERENCES

Baik, O.D., M. Marcotte, and F. Castaigne. 2000. Cake baking in tunnel type multi-zone industrial ovens. Part II. Evaluation of quality parameters. *Food Research International* 33:599–607.

Baixauli, R., T. Sanz, A. Salvador, and S.M. Fiszman. 2007. Influence of the dosing process on the rheological and microstructural properties of a bakery product. *Food Hydrocolloids* 21:230–236.

Bath, D.E., K. Shelke, and R.C. Hoseney. 1992. Fat replacers in high-ratio layer cakes. *Cereal Foods World* 37:495–500.

Boyd, J., P. Sherman, and C. Parkinson. 1972. Factors affecting emulsion stability and the HLB concept. *Journal of Colloid Interface Science* 41:359–370.

Chang, C.N., S. Dus, and J.L. Kokini. 1990. Measurement and interpretation of batter rheological properties. In *Batters and Breadings in Food Processing*, Ed. K. Kulp and R. Loewe, 199–226. American Association of Cereal Chemists, St. Paul, MN.

Cloke, J.D., E.A. Davis, and J. Gordon. 1984. Relationship of heat transfer and water-loss rates to crumb-structure development as influenced by monoglycerides. *Cereal Chemistry* 61:363–371.

Cunningham, H.M., G.A. Lawrence, and L. Tryphonas. 1977. Toxic effects of chlorinated cake flour in rats. *Journal of Toxicology and Environmental Health* 2:1161–1171.

Daniels, N.W.R., D.L. Frap, P.W. Russell Eggitt, and J.B.M. Coppock. 1963. Studies on the lipids of flour. II. Chemical and toxicological studies on the lipid of chlorine-treated cake flour. *Journal of the Science of Food and Agriculture* 14:883–893.

Donelson, J.R., C.S. Gaines, and P.L. Finney. 2000. Baking formula innovation to eliminate chlorine treatment of cake flour. *Cereal Chemistry* 77:53–57.

Frazier, P.J., F.A. Brimblecombe, and N.W.R. Daniels. 1974. Rheological testing of high-ratio cake flours. *Chemistry and Industry* 24:1008–1010.

Frye, A.M. and C.S. Setser. 1991. Optimizing texture of reduced-calorie yellow layer cakes. *Cereal Chemistry* 61:363–371.

Gaines, C.S., and J.R. Donelson. 1982. Cake batter viscosity and expansion upon heating. *Cereal Chemistry* 59:237–240.

Ginocchio, A.V., N. Fisher, J.B. Hutchinson, R. Berry, and J. Hardy. 1983. Long-term toxicity and carcinogenicity studies of cake made from chlorinated flour. *Food and Chemical Toxicology* 21:435–439.

Grigelmo-Miguel, N., E. Carreras-Boladeras, and O. Martin-Belloso. 2001. Influence of the addition of peach dietary fiber in composition, physical properties and acceptability of reduced-fat muffins. *Food Science and Technology International* 7:425–431.

Gujral, H.S., C.M. Rosell, S. Sharma, and S. Singh. 2003. Effect of sodium lauryl sulphate on the texture of sponge cake. *Food Science and Technology International* 9:89–93.

Guy, R.C.E. and S.S. Sahi. 2006. Application of a lipase in cake manufacture. *Journal of the Science of Food and Agriculture* 86:1679–1687.

Handleman, A.R., J.F. Conn, and J.W. Lyons. 1961. Bubble mechanics in thick foams and their effects on cake quality. *Cereal Chemistry* 59:500–506.

Huang, G., J.W. Finn, and E. Varriano-Marston. 1982. Flour chlorination. II. Effects of water binding. *Cereal Chemistry* 59:500–506.

Johnson, A.C. and R.C. Hoseney. 1979. Chlorine treatment of cake flour. II. Effect of certain ingredients in the cake formula. *Cereal Chemistry* 56:336–338.

Khouryieh, H.A., F.M. Aramouni, and T.J. Herald. 2005. Physical and sensory characteristics of no-sugar added/low fat muffin. *Journal of Food Quality* 28:439–451.

Kim, C.S. and C.E. Walker. 1992. Interactions between starches, sugars, and emulsifiers in high-ratio cake model systems. *Cereal Chemistry* 69:206–212.

Kim, H.Y.L., H.W. Yeom, H.S. Lim, and S.T. Lim. 2001. Replacement of shortening in yellow layer cakes by corn dextrins. *Cereal Chemistry* 78:267–271.

Kissell, L.T. 1971. Chlorination and water solubles content in flours of soft wheat varieties. *Cereal Chemistry* 48:102–108.

Kissell, L.T. and W.T. Yamazaki. 1979. Cake baking dynamics: Relation of flour chlorination rate to batter expansion and layer volume. *Cereal Chemistry* 56:324–327.

Kissell, L.T., J.R. Donelson, and R.L. Clements. 1979. Functionality in white layer cake of lipids from untreated and chlorinated patent flours. I. Effects of free lipids. *Cereal Chemistry* 56:11–14.

Lakshminarayan, S.M., V. Rathinam, and L. KrishnaRau. 2006. Effect of maltodextrin and emulsifiers on the viscosity of cake batter and on the quality of cakes. *Journal of the Science of Food and Agriculture* 86:706–712.

Larsson, K. 1980. Inhibition of starch gelatinization by amylose-lipid complex-formation. *Starch/Staerke* 32:125–126.

Lee, S., G.E. Inglett, and C.J. Carriere. 2004. Effect of nutrim oat bran and flaxseed on rheological properties of cakes. *Cereal Chemistry* 81:637–642.

Lee, S., S. Kim, and G.E. Inglett. 2005. Effect of shortening replacement with Oatrim on the physical and rheological properties of cakes. *Cereal Chemistry* 82:120–124.

Masoodi, F.A., B. Sharma, and G.S. Chauhan. 2002. Use of apple pomace as a source of dietary fiber in cakes. *Plant Foods for Human Nutrition* 57: 121–128.

Maziyadixon, B.B., C.F. Klopfenstein, and C.E. Walker. 1994. Freeze-dried wheat water solubles from a starch-gluten washing stream—Functionality in angel food cakes and nutritional properties compared with oat bran. *Cereal Chemistry* 71:287–291.

Miller, R.A. and R.C. Hoseney. 1993. The role of xanthan gum in white layer cakes. *Cereal Chemistry* 70:585–588.

Mizukoshi, M. 1983. Model studies of cake baking. IV. Foam drainage in cake batter. *Cereal Chemistry* 60:399–402.

Mizukoshi, M. 1985. Model studies of cake baking. VI. Effects of cake ingredients and cake formula on shear modulus of cake. *Cereal Chemistry* 62:247–251.

Mizukoshi, M. 1986. Rheological studies of cake baking. In *Fundamentals of Dough Rheology*, Eds. H. Faridi and J.M. Faubion, 73–76. American Association of Cereal Chemists, Inc., St. Paul, MN.

Ngo, W., R.C. Hoseney, and W.R. Moore. 1985. Dynamic rheological properties of cake batters made from chlorine-treated and untreated flours. *Journal of Food Science* 50:1338–1341.

Ngo, W.H. and M.V. Taranto. 1986. Effect of sucrose level on the rheological properties of cake batters. *Cereal Foods World* 31:317–322.

Painter, K.A. 1981. Functions and requirements of fats and emulsifiers in prepared cake mixes. *Journal of the American Oil Chemists' Society* 58:92–95.

Sahi, S.S. and J.M. Alava. 2003. Functionality of emulsifiers in sponge cake production. *Journal of the Science of Food and Agriculture* 83:1419–1429.

Sahin, S., and S.G. Sumnu. 2006. *Physical Properties of Foods, Food Science Text Series*, Springer: New York.

Sakiyan, O., G. Sumnu, S. Sahin, and G. Bayram. 2004. Influence of fat content and emulsifier type on the rheological properties of cake batter, *European Food Research and Technology* 219:635–638.

Seguchi, M. 1990. Effect of heat-treatment of wheat flour on pancake springiness. *Journal of Food Science* 55:784–785.

Shearer, A.E.H. and C.G.A. Davies. 2005. Physicochemical properties of freshly baked and stored whole-wheat muffins with and without flaxseed meal. *Journal of Food Quality* 28:137–153.

Shelke, K., J.M. Faubion, and R.C. Hoseney. 1990. The dynamics of cake baking as studied by a combination of viscometry and electrical resistance oven heating. *Cereal Chemistry* 67:575–580.

Shelke, K., R.C. Hoseney, J.M. Faubion, and S.P. Curran. 1992. Age related changes in the properties of batters made from flour milled from freshly harvested soft wheat. *Cereal Chemistry* 69:145–147.

Shepherd, I.S. and R.W. Yoell. 1976. Cake emulsions. In *Food Emulsions*, Ed. S. Friberg, 215–275. Marcel Dekker: New York.

Sinha, N.K., H. Yamamoto, and P.K.W. Ng. 1997. Effects of flour chlorination on soft wheat gliadins analyzed by reversed-phase high-performance liquid chromatography, differential scanning calorimetry and fluorescence spectroscopy. *Food Chemistry* 59:387–393.

Swanson, R.B., L.A. Carden, and S.S. Parks. 1999. Effect of a carbohydrate based fat substitute and emulsifying agents on reduced-fat peanut butter cookies. *Journal of Food Quality* 22:19–29.

Thomasson, C.A., R.A. Miller, and R.C. Hoseney. 1995. Replacement of chlorine treatment for cake flour. *Cereal Chemistry* 72:616–620.

Turabi E., G. Sumnu, and S. Sahin. 2008. Rheological properties and quality of rice cakes formulated with different gums and an emulsifier blend. *Food Hydrocolloids* 22:305–312.

Urlacher, B. and O. Noble. 1997. Thickening and gelling agents for food. In *Xanthan*, Ed. A. Imeson, 284–311, Chapman & Hall: London.

Vali, N.S.S.A. and P.N. Choudhary. 1990. Quality characteristics of cakes prepared from different fats and oil. *Journal of Food Science and Technology–Mysore* 27:400–401.

Yasukawa, T., M. Mizukoshi, and K. Aigami. 1986. Dynamic viscoelastic properties of cake batter during expansion and heat setting. In *Fundamentals of Dough Rheology*, Eds. H. Faridi and J.M. Faubion, 63–72. American Association of Cereal Chemists, St. Paul, MN.

6 Cookie Dough Rheology

Meryem Esra Yener

CONTENTS

6.1   INTRODUCTION

Cookies and biscuits are products made from soft flours. Low content of protein (8 to 10% in the grain), low water absorption, and low resistance to deformation are the characteristics that describe the suitability of wheat for biscuit production (Pedersen et al., 2004). Cookies are characterized by a formula high in sugar and fat and low in water. Cookie dough is cohesive but to a large degree lacks the extensibility and elasticity characteristics of bread dough. Relatively high quantities of fat and sugar in dough provide dough plasticity and cohesiveness without the formation of the gluten network, and they produce less elastic dough (Faridi, 1990). A highly elastic dough is not desirable in biscuit making because it shrinks after lamination. In addition, and again depending on the formulation, cookie dough tends to spread (become larger and wider) as it bakes rather than to shrink as does cracker dough. Spread is an important quality parameter for cookies.

Determining rheological properties of dough yields valuable information concerning the quality of the raw materials, the machining properties of the dough, and possibly the textural characteristics of the finished product (Faridi, 1990). Mixing time or mixing protocol has an important effect on rheological properties of biscuit dough (Maache-Rezzoug et al., 1998a; Manohar and Rao, 1999a). Generally, a long mixing time results in the softening of dough and reductions in both viscosity and relaxation time. However, the effect of mixing time cannot be differentiated from the composition of cookie dough and will be discussed as integrated to the effects of ingredients. The main ingredients that affect the rheology of cookie dough are water, sugar, fat, and the protein content of the flour. The effects of these ingredients on rheological properties of cookie dough and spread of cookies are discussed in this chapter. Brief descriptions of the rheological methods commonly used for this purpose are also given.

6.2   RHEOLOGICAL METHODS

6.2.1   EMPIRICAL MEASUREMENT METHODS

Empirical instruments are commonly used to determine the flow behavior of food products. Because they do not measure fundamental rheological properties, they are mostly indexers. However, the results are used both in quality control and in correlation to sensory data (Steffe, 1996).

6.2.1.1 Dough Testing Equipment

Cereal chemists are familiar with dough testing equipment and use it widely to investigate dough behavior. This equipment includes the farinograph, mixograph, extensograph, and alveograph.

The farinograph and mixograph are torque-measuring devices that provide empirical information about mixing properties of flour by recording the resistance of dough to mixing. In the farinograph, there is a kneading type of mixing; the mixograph involves a planetary rotation of vertical pins. The information derived from farinogram (consistency versus time curve) and mixogram (torque versus time curve) is given by Sahin and Sumnu (2006) in detail. In cookie dough, the height of the farinogram band is used as a measure of dough consistency, and bandwidth is used as a measure of dough cohesiveness (Jacob and Leelavathi, 2007; Olewnik and Kulp, 1984). Deduced parameters of the mixograms such as instantaneous specific energy and total specific energy were used to analyze rheological behavior of biscuit dough as well (Maache-Rezzoug et al., 1998a, 1998b). Instantenous specific energy is defined as the energy (J/kg.s) transmitted to the biscuit dough during the mixing cycle. Total specific energy is the integral of the instantaneous specific energy during the entire mixing operation. Soft and sticky dough is reported to have higher total specific energy than firm and inhomogeneous dough, because soft dough sticks and wraps itself round the mixer blades, increasing the transmitted energy.

The extensograph and alveograph measure rheological properties of dough after mixing. The extensograph measures the dough resistance to stretching and extensibility; the alveograph measures the pressure required to blow a bubble in a sheeted piece of dough. The typical extensogram and alveogram are given by Sahin and Sumnu (2006). Extensibility of cookie dough (in mm) was obtained by extensograph by Pedersen et al. (2004). Alveograph P (maximum overpressure needed to blow a dough bubble) is an index of resistance to deformation, L (the average abscissa at bubble rupture) is an index of dough extensibility, W (the deformation energy) is an index of dough strength, and the P/L (curve configuration ratio) is an index of the elastic to viscous component of dough (i.e., gluten behavior) (Agyare et al., 2004).

6.2.1.2 Texture Profile Analysis

In texture profile analysis (TPA), a bite size of food (usually a 1-cm cube) is compressed two times between two plates, usually to 80% of its original height. Because this test is intended to reflect the human perception of texture, the first and second compressions are referred to as the first and second bites. A texture profile curve, which is force versus time, is analyzed to give textural properties (Sahin and Sumnu, 2006; Steffe, 1996). Two-bite TPA was performed to measure consistency, hardness, cohesiveness, adhesiveness, and stickiness of biscuit dough using either the Instron Universal Testing Machine (Manohar and Rao, 1997a, 1997b, 1999a, 1999b, 1999c) or a texture analyzer (Zoulias et al., 2000).

6.2.1.3 Compression

A compression test measures the distance that a food is compressed under a standard compression force or the force required to compress a food a standard distance (Sumnu and Sahin, 2006). The Instron Universal Testing Machine was used to measure the force required to compress cookie dough 50% (Gaines, 1990) or 80% (Jacob and Leelavathi, 2007), and this force was used as a measure of consistency and hardness of the cookie dough, respectively.

6.2.1.4 Penetration

Penetrometers are designed to measure the distance that a cone or a needle sinks into a food under the force of gravity for a standard time (Sumnu and Sahin, 2006). A texture analyzer was used to measure the force that a 6-mm-diameter cylinder probe penetrates to a distance of 20 mm in cookie dough (Zoulias et al., 2000).

6.2.2 Fundamental Measurement Methods

Fundamental measurement methods include measurement of strain (ε) when a stress (σ) is applied to a material, or vice versa. Stress is defined as the force applied per unit area; strain is defined as the amount of deformation relative to the initial dimensions (height, length, or volume). Generally, the relation between the two is expressed as modulus (σ/ε) or as compliance (ε/σ). For viscoelastic materials like dough, the fundamental tests are performed under unsteady-state conditions, like transient and dynamic tests.

6.2.2.1 Modified Penetrometer

Generally, the penetrometer is used to determine the consistency of fats on the basis of distance moved by the cone or needle through the material in a particular time (Steffe, 1996). The penetrometer was modified by Manohar and Rao (1992) to perform a uniaxial compression between two plates. The optimum compression weight was found to be 410 g for biscuit dough (2.2 cm diameter × 1 cm height). The initial height (h_1) and the height of the dough after compression of 10 sec (h_2) were recorded. The compression plate was lifted up, the dough was allowed to recover for 1 min, and the recovered height (h_3) was measured. The compliance and elastic recovery were calculated (Manohar and Rao, 1992) as

$$\text{Compliance} (\%) = \frac{h_1 - h_2}{h_1} \times 100 \qquad (6.1)$$

$$\text{Elastic recovery} (\text{mm}) = \left(h_3 - h_2\right) \times 10 \qquad (6.2)$$

The reader should realize that Equation 6.1 is the definition of strain but not compliance. However, defining compliance as in Equation 6.1 does not change the interpretation of the physical meaning.

6.2.2.2 Transient Tests

In transient tests, the response of a material as a function of time is measured after subjecting the material to an instantanenous change in either strain or stress. Creep recovery and stress relaxation are the two common transient tests. In creep recovery, a constant stress is applied to a material, and its strain is recorded as a function of time (creep). Then the stress is removed, and the strain is recorded as a function of time, called recovery. In cookie dough, the maximum strain is a measure of its extensibility, and percent recovery is the measure of its elasticity. In stress relaxation, a constant strain is applied to a material, and the stress required to keep this strain constant is recorded as a function of time. An elastic solid never relaxes, and an ideal liquid immediately relaxes to zero. The larger the relaxation time, the more elastic is the material. A mechanical analog of an elastic solid is spring, and that of an ideal liquid is dashpot. The behavior of viscoelastic materials is expressed by mechanical models such as Maxwell, Kelvin-Voight, and Burger models. The reader can obtain more detailed information about these models in Steffe (1996).

Maache-Rezzoug et al. (1998a, 1998b) performed a stress relaxation test under lubricated uniaxial compression and used the Maxwell model to determine viscosity (η) and relaxation time (λ_{rel}). Pedersen et al. (2004, 2005) performed creep recovery measurements with a creep and recovery time of 300 s and a stress value of 10 Pa to determine maximum strain and percent recovery of cookie dough.

6.2.2.3 Dynamic Tests

In dynamic tests, an oscillatory strain is applied to a material and the resulting stress is measured. When viscoelastic materials are deformed, part of the energy is stored (as in an elastic solid) and part of it is dissipated as heat (as in a liquid). Therefore, when a linear viscoelastic material is subjected to a periodically varying (frequency, ω) stress, the strain will also vary periodically but out of phase with stress. The result is expressed with a stress equation (Steffe, 1996) as

$$\sigma = G'\gamma + \left(G''/\omega\right)\dot{\gamma} \tag{6.3}$$

where γ is shear strain, $\dot{\gamma}$ is shear rate, G' is the storage modulus which is the measure of elastic behavior of a material, and G'' is the loss modulus which is the measure of liquid behavior of a material. Storage modulus (G') and loss modulus (G'') are functions of the phase lag (δ) between stress input (σ_0) and strain output (γ_0), as follows:

$$G' = \left(\frac{\sigma_0}{\gamma_0}\right)\cos\left(\delta\right) \tag{6.4}$$

$$G'' = \left(\frac{\sigma_0}{\gamma_0}\right)\sin\left(\delta\right) \tag{6.5}$$

δ is 0° for an ideal elastic solid and 90° for a Newtonian fluid. Tangent of the phase lag is (tanδ) also a popular material function to describe viscoelastic behavior and is defined as

$$\tan\delta = \frac{G''}{G'} \tag{6.6}$$

The dynamic tests should be performed within the linear viscoelastic region of a material and determined by a stress sweep. Frequency sweep within the linear viscoelastic region is used to determine the effects of different ingredients in a cookie dough formulation. Temperature sweep explains the dough behavior during baking.

Rheological properties of short doughs (standard, firm-fat, lowfat, liquid oil, sugar-free, and starch doughs) were determined by using a controlled stress rheometer at small deformation (Baltsavias et al., 1997). The rheometer was operated in a stress sweep at a fixed angular frequency of 6.28 rad/s (1 Hz), in time sweep at a fixed strain amplitude of 2×10^{-4} and an angular frequency of 6.28 rad/s, in frequency sweep at a fixed strain amplitude of 2×10^{-4}. The linear viscoelastic region of short doughs were found to be very limited—nonlinearity started at a strain about 2×10^{-4} except with liquid-oil dough where this value was 4×10^{-4}.

Agyare et al. (2004) determined the effect of substituting canola oil/caprylic-acid-structured lipid for partially hydrogenated shortening on rheology of soft wheat flour dough by performing frequency sweep from 0.01 to 20 Hz at 25°C using a strain of 0.02. Lee and Inglett (2006) studied the effect of replacing shortening in cookies by 20% jet-cooked oat bran, which is a carbohydrate-based fat replacer. Viscoelastic properties of dough were determined by using a controlled strain rheometer, by performing oscillatory testing over a frequency range of 0.01 to 10 Hz at a strain of 5×10^{-3}, which was in the linear viscoelastic region for all samples.

Pedersen et al. (2004) studied rheological properties of semisweet biscuit dough from different cultivars and the effect of chemical and enzymatic modification on dough rheology (Pedersen et al., 2005) by using a controlled stress rheometer. The linear viscoelastic region was determined to be up to strain of 1.5×10^{-3} by performing a strain sweep at a frequency of 1 Hz. A target strain of 1×10^{-3} corresponding to a stress value of 10 Pa was used in all the experiments. A frequency sweep was performed in the range from 0.1 to 60 Hz; scattering of data was observed above 30 Hz.

6.2.2.4 Extensional Viscosity

Extensional flow is an important aspect of dough processing during sheeting, and it does not involve shearing. Although there are three basic types of extensional flow—uniaxial, planar, and biaxial (Steffe, 1996)—biaxial extensional viscosity is commonly measured in cookie dough (Baltsavias et al., 1999a, 1999b; Lee and Inglett, 2006; Manohar and Rao, 1997a, 1997b, 1999a, 1999b, 1999c). Biaxial extensional flow is achieved in a lubricated squeezing flow between parallel plates (uniaxial compression). The sample between the plates is compressed with a constant velocity (constant deformation rate) of the top plate. This is called crosshead speed, and the typical values used for cookie dough are 6, 18, and 60 mm/min (Lee and Inglett, 2006); 50 mm/min (Manohar and Rao, 1997a, 1997b, 1999a, 1999b, 1999c); and 1, 10, and 100 mm/min (Baltsavias et al., 1999a, 1999b). When compressing the sample with a constant velocity, the force exerted on the sample is recorded as a function of time $\left(F\left(t \right) \right)$. The stress exerted is defined as

$$\sigma = \frac{F(t)}{\pi R^2} \tag{6.7}$$

where R is the radius of the sample in the case of constant area (Steffe, 1996) as used by Baltsavias et al. (1999a, 1999b), or

$$\sigma = \frac{F(t)}{A(t)} \tag{6.8}$$

in the case of constant volume (Steffe, 1996) as used by Manohar and Rao (1997a, 1997b, 1999a, 1999b, 1999c) and Lee and Inglett (2006). The biaxial strain rate, $\dot{\varepsilon}_B$ (1/s), is given (Steffe, 1996) by

$$\dot{\varepsilon}_B = \frac{v}{2h(t)} = \frac{v}{2(h_0 - vt)} \tag{6.9}$$

where v is the crosshead speed (m/s) and h_0 is the initial height of the sample. The apparent biaxial extensional viscosity (Pa.s) is given by

$$\eta_B = \frac{\sigma}{\dot{\varepsilon}_B} \tag{6.10}$$

The apparent biaxial extensional viscosity of cookie dough is expressed by the power law model (Baltsavias et al., 1999a, 1999b; Lee and Inglett, 2006) at large deformation, as

$$\eta_B = K\dot{\varepsilon}_B^{n-1} \tag{6.11}$$

where K is the consistency index, and n is the flow behavior index.

6.3   EFFECTS OF INGREDIENTS

6.3.1   WATER

Water is an essential ingredient in dough formation, because it is necessary for solubilizing other ingredients for hydrating proteins and carbohydrates and for the development of gluten network. The dough, having 12.5% fat containing 13.3% to 15.5% water, was not consistent because it lacked hydration. The dough with high water content (>21%) was extremely soft and sticky, making it impossible to work. An increase in water content led to a significant decrease in dough viscosity (η) and a slight reduction of the relaxation time (λ_{rel}), indicating reduction of elasticity (Table 6.1). The biscuits expanded lengthwise, with a smaller thickness (Maache-Rezzoug, 1998b). Similarly, increasing water content of biscuit dough by 3% increased compliance and decreased extrusion time, apparent biaxial extensional viscosity (η_B), and consistency, indicating a decrease in dough viscosity. Dough became more cohesive but soft, adhesive, and sticky. On the contrary, an increase in elastic recovery indicated increased elastic properties of dough (Manohar and Rao, 1999b).

The effect of water content on dough properties was reported to vary with the quantities and relative distribution of fat (Olewnik and Kulp, 1984). In deposit dough,

---

TABLE 6.1

Effect of Water Content on Viscosity (η) and Relaxation Time (λ_{rel}) of Biscuit Dough

Water Content (%)	$\eta \times 10^{-4}$ (Pa.s)	λ_{rel} (s)
17.0	8.8	2.05
18.0	4.7	1.85
19.0	3.8	1.63
20.0	3.2	1.45
21.0	2.8	1.30
22.5	1.7	1.03

Source: Adapted from Maache-Rezzoug, Z., Bovier, J.M., Allaf, K., and Patras, C., *Journal of Food Engineering,* 35, 23, 1998b. With permission.

which has 63% fat, water (9 to 13%) did not affect the consistency and cohesiveness very much, due to the dominating effect of fat. In wire-cut dough that has 45% fat, water (14 to 20%) increased consistency, because the addition of water reduced the percentage of fat, allowing the hydration of flour and development of a gluten network. In rotary-molded dough that has 27% fat, the smearing of fat over the particles of sugar and flour was retarded during mixing because the dough was low fat. The incomplete formation of fat film increased the accessibility of sugar crystals to water, resulting in faster, more extensive formation of sugar syrup. Dough consistency was reduced as the quantity of water increased from 13 to 16%. More dough liquid decreased dough consistency and increased cookie spread of sugar-snap cookies, as well (Gaines, 1990).

6.3.2 SUGAR AND SUGAR REPLACERS

Sugar is an important ingredient of short dough biscuits. It contributes to texture, flavor, sweetness, and color of biscuits (Manohar and Rao, 1997). The effect of sugar on dough behavior is an important factor in biscuit making. Sugar causes softening of the dough, due in part to competition between the added sugar and the availability of water in the system. Sugar restricts the development of gluten network by competing for water that otherwise would have been absorbed by gluten. The limited amount of water used in biscuit formulation, and also its nonavailability to protein and starch, partially contributes to the crispness of biscuits.

In the farinograph, consistency of wire-cut dough remained fairly constant at 30%, 45%, and 60% added sugar, but cohesiveness increased at the 60% level. In deposit cookie dough, added high levels of sugar up to 55% increased consistency and cohesiveness sharply. However, in rotary-molded dough, high levels of sucrose up to 50% caused a sharp reduction in dough consistency and cohesiveness. This is because of extra water (10% weight of dough) added into the farinograph because low fat and water levels in this cookie formula resulted in crumbly doughs that were unsuitable for direct farinography (Olewnik and Kulp, 1984).

Mixograms of biscuit dough showed that the dough changed from a solid and consistent texture to an extremely soft texture as sugar content increased from 20 to 50% (Maache-Rezzoug et al., 1998b). The addition of sugar to the formula decreased dough viscosity (η). At sugar contents less than or equal to 30%, relaxation time (λ_{rel}) of dough was constant; at sugar contents between 30 and 50%, relaxation time was reduced (Table 6.2).

The effects of mixing time, sugar level, and type on rheological properties of biscuit dough prepared from weak flour (8.8% gluten) were determined by Manohar and Rao (1997a, 1997b). Prolonged mixing of dough having 300 g sugar/kg of flour resulted in increased elastic properties. The optimum mixing time was selected as 180 s among 90, 180, and 300 s (Manohar and Rao, 1997a). Increasing sugar content decreased extrusion time, elastic recovery, apparent biaxial extensional viscosity, (η_B), consistency and hardness, and increased compliance, cohesiveness, adhesiveness, and stickiness of biscuit dough (Table 6.3). These results indicated that increasing the sugar content resulted in soft (with less viscosity and elasticity) but cohesive dough that was adhesive and sticky. Incorporation of 20 g of reducing sugars per

TABLE 6.2

Effect of Sugar Content on Viscosity (η) and Relaxation Time (λ_{rel}) of Biscuit Dough

Sugar Content (%)	$\eta \times 10^{-4}$ (Pa.s)	λ_{rel} (s)
20	5.1	1.52
30	4.8	1.51
35	3.9	1.38
40	3.5	1.14
50	3.0	1.03

Source: Adapted from Maache-Rezzoug, Z., Bovier, J.M., Allaf, K., and Patras, C., *Journal of Food Engineering,* 35, 23, 1998b. With permission.

kilogram of flour, instead of 300 g sugar per kilogram of flour, like dextrose, liquid glucose (LG), invert syrup (IS), and high-fructose corn syrup (HFCS), affected the rheological characteristics in the same way as increasing the sugar content in the formulation (Table 6.4). However, the syrups had a greater influence on the rheological characteristics (Manohar and Rao, 1997b). Spread as well as thickness of the biscuits increased, but density decreased significantly by the addition of sugar.

Zoulias et al. (2000) studied the effect of sucrose replacement by fructose or polyols on dough rheology in low-fat cookies. Polydextrose was used to replace 35% of fat in low-fat cookies. Mannitol dough was very firm and difficult to sheet; however, it presented moderate values of hardness and consistency and the lowest adhesiveness. Maltitol and fructose resulted in dough with high values of hardness and consistency and low adhesiveness and cohesiveness, while lactitol, sorbitol, and xylitol had the opposite effects. The xylitol dough presented some problems in sheeting for the formation of cookies due to its high adhesiveness. The rheological properties of doughs prepared by lactitol and sorbitol were similar to those prepared by sucrose (Table 6.5). Cookies with fructose or polyols were less sweet than sucrose-containing ones, but supplementation with acesulfame-K, which did not interfere with the dough rheology, increased sweetness and improved perceived flavor and general acceptance. The properties of cookies prepared with maltitol, lactitol, and sorbitol were acceptable. Mannitol restricted spread of cookies.

When rheological behavior of standard and sugar-free short doughs was compared at small deformation, sugar-free dough had higher storage modulus (G') and lower phase angle (tanδ) than the standard dough (Table 6.6), indicating that sugar-free dough was more solid and elastic (Baltsavias et al., 1997). The frequency dependency of storage modulus (G') and phase angle (tanδ) of the sugar-free dough were similar to those of the standard dough. They decreased with increasing frequency up to 3 rad/s and then remained constant. At large deformation, short doughs behaved like strain rate thinning liquids (Baltsavias et al., 1999a).

When the apparent biaxial extensional viscosity of the doughs (η_B) expressed with the power law (Equation 6.11), the consistency index (K) of the sugar-free dough

TABLE 6.3
Effect of Sugar Content on Rheological Characteristics[1] of Biscuit Dough

Sugar (g/kg)	Extrusion Time (s)	Compliance (%)	Elastic Recovery ×10 (mm)	$\eta_{IB}[2] \times 10^{-5}$ (Pa.s)	Consistency (N.s)	Hardness (N)	Cohesiveness	Adhesiveness (N.s)	Stickiness (°)
250	70[a]	35.9[a]	5.51[a]	2.54[a]	866[a]	814[a]	0.138[a]	35.8[c]	44.8[c]
300	45[b]	40.6[b]	4.60[b]	1.91[b]	786[b]	662[b]	0.194[b]	54.3[b]	50.3[b]
350	23[c]	49.8[a]	3.80[c]	1.41[c]	616[c]	512[c]	0.223[a]	57.6[a]	69.5[a]

[1]Values for a particular column followed by different letters differ significantly ($p < 0.05$). [2]η_{IB}: Apparent biaxial extensional viscosity, at 50% compression.
Source: Adapted from Manohar, R.S., and Rao, P.H., *Journal of the Science of Food and Agriculture,* 75, 383, 1997b. With permission.

TABLE 6.4
Effect of Different Types of Reducing Sugars on Rheological Characteristics[1] of Biscuit Dough

Sugar Type[2]	Extrusion Time (s)	Compliance (%)	Elastic Recovery ×10 (mm)	$\eta_{IB}[3] \times 10^{-5}$ (Pa.s)	Consistency (N.s)	Hardness (N)	Cohesiveness	Adhesiveness (N.s)	Stickiness (°)
Control	45[a]	40.6[c]	4.60[a]	1.91[a]	786[a]	662[a]	0.194[d]	54.3[c]	50.3[c]
Dextrose	36[b]	44.2[b]	4.10[b]	1.64[b]	654[b]	586[b]	0.202[cd]	55.8[bc]	54.8[b]
LG	26[c]	48.8[a]	3.80[d]	1.42[c]	524[c]	492[c]	0.220[ab]	57.8[b]	70.5[a]
IS	24[c]	48.6[a]	3.95[c]	1.41[c]	538[c]	476[c]	0.210[bc]	62.5[a]	69.8[a]
HFCS	22[c]	49.8[a]	3.80[d]	1.36[d]	530[c]	474[c]	0.230[a]	64.3[a]	71.5[a]

[1]Values for a particular column followed by different letters differ significantly ($p < 0.05$).
[2]Control is 300 g sugar/kg of flour; reducing sugars are at 20 g/kg of flour level; LG, liquid glucose; IS, invert sugar; HFCS, high-fructose corn syrup.
[3]η_{IB}: Apparent biaxial extensional viscosity, at 50% compression.
Source: Adapted from Manohar, R.S., and Rao, P.H., *Journal of the Science of Food and Agriculture,* 75, 383, 1997b. With permission.

TABLE 6.5

Effect of Sweeteners on Rheological Properties[1] of Low-Fat Cookie Dough

Sweetener	$F_{penetration}$ (N)	F_{TPA} (N)	Consistency (N·s)	Adhesiveness (N·s)	Cohesiveness
Sucrose	2.4[a]	66.8[ab]	69.6[ab]	6.79[c]	0.656[b]
Fructose	3.0[a]	69.9[b]	78.3[b]	4.42[b]	0.548[a]
Maltitol	3.2[a]	69.3[b]	74.5[b]	4.83[b]	0.534[a]
Lactitol	2.9[a]	58.2[ab]	60.2[ab]	6.40[c]	0.693[b]
Sorbitol	2.9[a]	51.3[ab]	51.0[ab]	6.06[c]	0.665[b]
Xylitol	2.4[a]	46.2[a]	47.5[a]	7.38[c]	0.638[b]
Mannitol	9.7[b]	59.0[ab]	55.9[ab]	1.55[a]	0.643[b]

[1]Values for a particular column followed by different letters differ significantly ($p < 0.05$).

Source: Zoulias, E.I., Piknis, S., and Oreopoulou, V., *Journal of the Science of Food and Agriculture,* 80, 2049, 2000. With permission.

TABLE 6.6

Dynamic Rheological Properties[1] of Short Doughs at Small Deformation

	$G' \times 10^{-5}$ (N/m²)	$\tan\delta$
Standard dough	6.69	0.255
Firm-fat dough	8.73	0.277
Low-fat dough	4.34	0.390
Liquid-oil dough	0.22	0.525
Sugar-free dough	7.50	0.238
Starch dough	5.36	0.202

[1]Results were taken 1 h after the end of mixing, T = 20°C, ω = 6.28 rad/s, γ_{max} = 2 × 10⁻⁴.

Source: Baltsavias, A., Jurgens, A., and van Vliet, T., *Journal of Cereal Chemistry,* 26, 289, 1997. With permission.

was higher, but its flow behavior index (n) was lower than those of standard and sucrose syrup doughs (Table 6.7). The effects of sucrose content and mixing time on the rheological properties of short doughs were further studied by Baltsavias et al. (1999b) at large deformation. Doughs with different sugar levels were prepared by dissolving sucrose in water before adding to the dough. The doughs denoted as sucrose syrup 1, 2, and 3 corresponded to 64.6, 50, and 40% sucrose solutions, respectively. The doughs called sugar-free 1, 2, and 3 contained no sucrose, but the compositions of the other ingredients were the same in the formulations (Table 6.8 and Table 6.9). Regardless of the dough type, mixing time decreased

TABLE 6.7

Consistency Index (K) and Flow Behavior Index (n) for Apparent Biaxial Extensional Viscosity (η_B)[1] of Short Doughs

	K × 10^{-4} (Pas^n)	n
Standard dough	1.1	0.10
Firm-fat dough	2.9	0.15
Low-fat dough	12.1	0.31
Sucrose syrup dough	1.2	0.12
Sugar-free dough	1.4	0.03
Starch dough	0.8	0.09

[1] Values were calculated at $\varepsilon_B = 0.2$

Source: Baltsavias, A., Jurgens, A., and van Vliet, T., *Journal of Cereal Science,* 29, 33, 1999a. With permission.

TABLE 6.8

Effect of Mixing on Consistency Index (K) and Flow Behavior Index (n) for Apparent Biaxial Extensional Viscosity (η_B)[1] of sucrose syrup and sugar-free doughs

	Mixing Time (min)	K × 10^{-4} (Pas^n)	n
Sucrose syrup 1 (64.6%)	2	1.8	0.12
	8	1.3	0.14
	20	0.7	0.09
Sugar-free 1	2	4.7	0.08
	8	1.6	0.04
	20	1.0	0.02

[1] Values were calculated at $\varepsilon_B = 0.2$

Source: Baltsavias, A., Jurgens, A., and van Vliet, T., *Journal of Cereal Science,* 29, 43, 1999b. With permission.

the dough consistency significantly (Table 6.8). In addition, it drastically changed the shape of the stress–strain curve for sugar-free dough (Figure 6.1). The stress–strain curves for the sugar-free doughs indicated a stronger elastic contribution to the deformation than did those for sucrose syrup doughs. Sucrose syrup doughs exhibited prominent yielding and flow behavior. Their apparent biaxial extensional viscosity(η_B) decreased with increasing sucrose content (Table 6.9).

6.3.3 FAT AND FAT REPLACERS

Fat forms one of the basic components of a cookie formulation and is present at relatively high levels. Fat acts as a lubricant and contributes to the plasticity of cookie dough. It prevents excessive development of the gluten network during

TABLE 6.9

Consistency Index (K) and Flow Behavior Index (n) for Apparent Biaxial Extensional Viscosity $(\eta_B)^1$ of Short Doughs

	K × 10⁻⁴ (Pasⁿ)	n
Sucrose syrup 1 (64.6%)	1.3	0.14
Sucrose syrup 2 (50%)	1.2	0.08
Sucrose syrup 3 (40%)	1.4	0.08
Sugar-free 1	1.6	0.04
Sugar-free 2	1.6	0.07
Sugar-free 3	2.2	0.09

[1] Mixing time = 8 min; Values were calculated at $\varepsilon_B = 0.2$

Source: Baltsavias, A., Jurgens, A., and van Vliet, T., *Journal of Cereal Science,* 29, 43, 1999b. With permission.

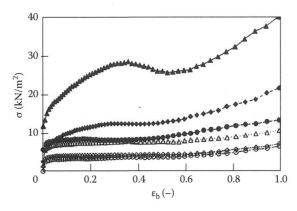

FIGURE 6.1 Stress–strain curves for various cookie dough formulations: sucrose-syrup 1 (open symbols), sugar-free 1 (closed symbols); mixing time = 2 min (triangles); mixing time = 8 min (diamonds); mixing time = 20 min (circles). Initial $\dot{\varepsilon}_b$ = 2.8×10⁻⁴ 1/s. (From Baltsavias, A., A. Jurgens, and T. van Vliet, *Journal of Cereal Science,* 29, 43, 1999b. With permission.)

mixing. The presence of fat contributes to the reduction of the elastic nature of dough and therefore the shrinking of the dough during molding. Although in technological respects (consistency of dough), fat has similar effects to sugar, their physicochemical roles are different. The globules of fat surround the proteins and starch, isolate them, and prevent the formation of polymers, thereby reducing the density of the network (Maache-Rezzoug et al., 1998b).

Mixograph results showed that an increase in the fat content of biscuit dough from 5 to 25% decreased the mixing energy. Dough was more homogenous and soft when the fat content increased from 15 to 20%. The addition of fat softened the dough and decreased viscosity (η) and relaxation time (λ_{rel}) (Table 6.10). Fat contributed to an increase in length and to a reduction in thickness and weight of the biscuits

TABLE 6.10

Effect of Fat Content on Viscosity (η) and Relaxation Time (λ_{rel}) of Biscuit Dough

Fat Content (%)	$\eta \times 10^{-4}$ (Pa.s)	λ_{rel} (s)
5	8.7	1.58
10	5.7	1.60
12.5	5.3	1.50
15	4.0	1.20
20	2.5	1.10
25	2.1	1.07

Source: Adapted from Maache-Rezzoug, Z., Bovier, J.M., Allaf, K., and Patras, C. *Journal of Food Engineering*, 35, 23, 1998b. With permission.

(Maache-Rezzoug et al., 1998b). Increasing the level of fat from 150 g/kg flour to 250 g/kg flour increased compliance, cohesiveness, and adhesiveness and decreased extrusion time, elastic recovery, apparent biaxial extensional viscosity (η_B), consistency, hardness, and stickiness (Manohar and Rao, 1999c). Therefore, increasing fat content produced soft, less viscous, and less elastic dough.

As the amount of fat increased in wire-cut doughs from 40 to 50%, consistency of the dough varied within a fairly narrow range; however, cohesiveness was considerably reduced. Deposit doughs had a greater reduction in consistency with increased fat content from 58 to 68%, as well as cohesiveness. Rotary-molded doughs having 27 to 29% fat had the highest consistency and cohesiveness than the other two but followed the same pattern as the fat content increased (Olewnik and Kulp, 1984).

The effect of fat type on cookie dough rheology was studied by preparing sugar-snap cookies with four different fat types: emulsified bakery fat, emulsified margarine, nonemulsified hydrogenated fat, and sunflower oil (Jacob and Leelavathi, 2007). The moisture, protein, and gluten contents of the flour were 11.9, 9.7, and 7.3%, respectively. At the beginning of mixing, the most consistent dough was that containing bakery fat; the least consistent dough was the one containing sunflower oil. However, consistency of the dough containing sunflower oil increased while the consistency of the doughs containing margarine and bakery fat decreased during mixing. The consistency of the dough containing hydrogenated fat remained almost constant. At the end of mixing, the dough containing sunflower oil had the highest consistency. The consistency of the dough containing hydrogenated fat remained almost constant (Table 6.11). The increase in consistency of the dough containing sunflower oil was explained by its lacking ability to smear all the flour particles leading for development of gluten network during mixing. The decrease in consistency of the dough containing both margarine and bakery fat was due to the well aeration of the dough during mixing. Although the dough containing sunflower oil was the least

TABLE 6.11

Effect of Fat Type on Rheological Properties of Sugar-Snap Cookie Dough

Fat Type	Farinograph Dough Consistency (BU)		Farinograph Dough Cohesiveness (BU)		Compression Force (hardness) (kg)
	0 min	10 min	0 min	10 min	
Bakery fat	440	360	60	80	3.0
Margarine	380	270	60	60	2.1
Hyodrogenated fat	310	300	80	80	4.5
Sunflower oil	200	400	20	120	2.9

Source: Adapted from Jacob, J. and Leelavathi, K., *Journal of Food Engineering,* 79, 299, 2007. With permission.

cohesive at the beginning, its cohesiveness increased and remained constant during mixing. The least cohesive dough was that containing margarine after mixing. When the hardness was considered, the dough containing hydrogenated fat was the highest (Table 6.11). Manohar and Rao (1999c) reported that cookie dough prepared by hydrogenated fat was the hardest among the doughs prepared by other fat types, as well. Cookies made by emulsified bakery fat and emulsified margarine that showed consistency decrease during mixing had the smaller spread ratio than the others. The addition of emulsifiers with a level of 5 g/kg flour to biscuit dough was reported to decrease elastic recovery, indicating their contribution to the shortening effect on gluten (Manohar and Rao, 1999c). Furthermore, emulsifiers decreased extrusion time, consistency, and apparent biaxial extensional viscosity (η_B), indicating decrease in dough viscosity. The doughs became more cohesive but soft with lower adhesiveness and stickiness.

The rheological properties of short doughs prepared with different fat types, firm-fat, low-fat, and liquid oil, were determined both at small deformation (Baltsavias et al., 1997) and at large deformation (Baltsavias et al., 1999a). At small deformation, loss modulus (G'') tended to be linear over a wider range of strain compared with storage modulus (G'). Regardless of composition, phase angle (tanδ) increased with increasing strain, indicating a more liquid-like behavior. The storage modulus (G') and loss modulus (G'') curves of the standard dough crossed over at a strain rate of 10^{-1}, showing that the material changed from solid-like to more liquid-like. The changes induced by shear were reported to be less severe in the case of the low-fat and the liquid-oil dough. But both had a significantly lower storage modulus (G') and higher phase angle (tanδ) than the standard dough, indicating that they were more liquid-like and more easily deformed (Table 6.6). The firm-fat dough was relatively more liquid-like than the standard dough, with the phase angle (tanδ) being slightly higher. But on the contrary, higher storage modulus (G') was explained by higher air volume fraction in the standard dough. Reducing the fat content or replacing solid fat with liquid oil brought about substantial changes in the resulting dough. The elastic component became more frequency dependent, and the phase angle (tanδ) increased with increasing angular frequency above 3 rad/s. This behavior was explained by the formation of more fat-dispersed dough by lowering the fat content or replacing solid fat with liquid oil. At large deformation, all doughs behaved more like strain-rate thinning liquids (Table 6.7). The doughs showed large differences in apparent biaxial extensional viscosity (η_B) depending on fat type. Low-fat dough had the highest consistency index (K) and flow behavior index (n).

Agyare et al. (2004, 2005) studied the effect of substituting canola oil/caprylic acid structured lipid (SL) for partially hydrogenated shortening (at 0, 25, 50, 75, and 100%) on the rheology of soft wheat flour dough (28.4% total lipid on flour basis, 43% moisture content). Figure 6.2 shows that the addition of partially hydrogenated shortening to untreated dough resulted in a significant decrease in resistance to deformation (Alveograph P), dough extensibility (Alveograph L), and dough strength (Alveograph W). This was attributed to the lubrication action of the added shortening. SL substitution for shortening did not significantly affect dough deformation

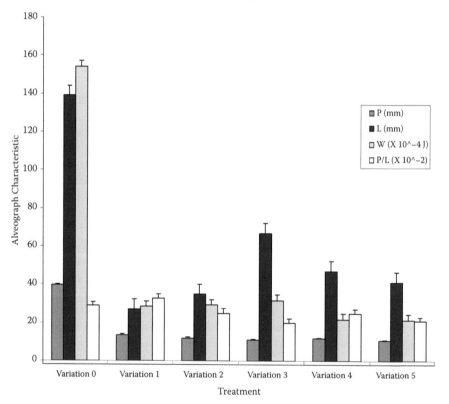

FIGURE 6.2 Effect of substituting varying levels of structured lipid (SL) for partially hydrogenated shortening in soft wheat flour dough on Alveograph characteristics: Variation 0: formulation without shortening; Variation 1: formulation with 100% shortening, 0% SL; Variation 2: formulation with 75% shortening, 25% SL; Variation 3: formulation with 50% shortening, 50% SL; Variation 4: formulation with 25% shortening, 75% SL; Variation 5: formulation with 0% shortening, 100% SL. (From Agyare, K.K., Addo, K., Xiong, Y.L., and Akoh, C.C. *Journal of Cereal Science,* 42, 309, 2005. With permission.)

(Alveograph P) and dough strength (Alveograph W). However, dough extensibility (Alveograph L) was greater for formulations with 50% and 75% SL. In addition, the ratio of elastic to viscous component (Alveograph P/L) was lower in formulation with 50% SL compared to the standard shortening. This indicated that dough with formulations 50% and 75% SL had greater extensibility and low elasticity than the others. This was also proved with large diameters (high spread) with the cookies baked from formulations with 50% and 75% SL (Agyare et al., 2005). In dynamic testing, frequency sweep showed that all the dough formulations, regardless of the content of SL, gave higher values of storage modulus (G') and loss modulus (G'') at higher frequencies, indicating that the recovery of stressed dough was a slow process—that is, the network was not completely elastic. The addition of shortening to dough lowered storage modulus (G') and loss modulus (G'') over the entire frequency range, and increased SL substitution further decreased storage modulus (G') and loss modulus (G''), indicating that the dough was being less viscous and less elastic. This was the result of shortening being solid fat and SL being liquid at room temperature.

Lee and Inglett (2006) studied the effect of replacing shortening (10, 20, and 30%) in cookies with 20% jet-cooked oat bran, also called Nutrim OB (NU), which is a carbohydrate-based fat replacer. For all doughs, storage modulus (G') and loss modulus (G'') increased with frequency (Figure 6.3). All doughs had higher values of storage modulus (G') than those of loss modulus (G''), suggesting that they had more elastic properties than viscous properties. The control had the highest storage modulus (G'). Increasing replacement of shortening with NU from 10 to 30% caused a decrease in both storage modulus (G') and in loss modulus (G''). This decrease in the dynamic viscoelastic properties of cookie doughs containing NU was attributed to their increased moisture content. The same behavior was observed in extensional

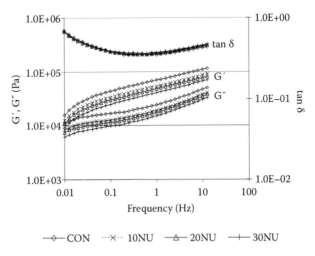

FIGURE 6.3 Effect of Nutrim OB (NU) on the dynamic viscoelastic properties of cookie dough. (From Lee, S. and Inglett, G.E., *International Journal of Food Science and Technology,* 41, 553, 2006. With permission.)

viscosity. At all strain rates tested, the biaxial extensional viscosity (η_B) decreased as shortening was replaced with more NU. At large deformation, the doughs behaved as shear thinning liquids. Consistency index (K) decreased, but the flow behavior index (n) was not significantly decreased with shortening replacement with NU (Table 6.12). However, during baking, after starch gelatinization around 80 to 90°C, storage modulus (G′) increased with temperature. At 120°C, while the control had the lowest storage modulus (G′), increasing NU replacement from 10 to 30% gave rise to the increase in storage modulus (G′), indicating that the dough containing NU was more elastic than the control, which resulted in a decrease in diameter of the cookies made by NU compared to the control (Figure 6.4). This reversed elastic

TABLE 6.12

Consistency Index (K) and Flow Behavior Index (n) for Apparent Biaxial Extensional Viscosity (η_B) of Cookie Doughs[1] Containing Shortening (CON) and Different Levels of Nutrim OB (NU)

	K (Pas^n)	n
CON	3.77[a]	0.19[a]
10NU[2]	3.67[ab]	0.20[a]
20NU[3]	3.34[bc]	0.20[a]
30NU[4]	3.11[c]	0.21[a]

[1]Values with the same superscript in the same column are not significantly different at the 5% level;
[2]10% Nutrim OB; [3]20% Nutrim OB; [4]30% Nutrim OB

Source: Lee, S. and Inglett, G.E., *International Journal of Food Science and Technology*, 41, 553, 2006. With permission.

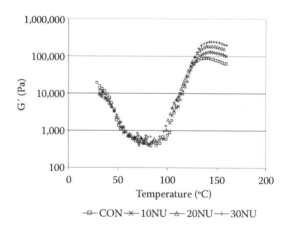

−□−CON −✳−10NU −△−20NU −+−30NU

FIGURE 6.4 Changes in the storage modulus of cookies with shortening (CON) and different levels of Nutrim OB (NU). (From Lee, S. and Inglett, G.E., *International Journal of Food Science and Technology,* 41, 553, 2006. With permission.)

behavior during baking was not seen in the cookies made by replacing shortening with SL (Agyare et al., 2004).

6.3.4 PROTEIN CONTENT IN FLOUR AND PROTEIN MODIFIERS

Generally, gluten is responsible for the rheological properties of dough. The effect of protein content was studied by using flour containing 11.2% protein (0% added gluten) to 21.9% protein (15% added gluten) (Maache-Rezzoug et al., 1998a). Changing the protein content of the flour from 14 to 20% increased the dough viscosity (η) because it favored the structuring of the gluten network during mixing. Up to 14% and beyond 20%, the protein content appeared to have no effect on relaxation time (λ_{rel}), whereas it increased linearly between 14 and 20%, indicating that within this range protein content increased dough elasticity (Table 6.13). Consequently, protein content produced an overall reduction in the spread of biscuits.

Pedersen et al. (2004) studied the rheological properties of semisweet biscuit doughs from different cultivars. Among the used cultivars (Table 6.14), Galatea has no high-molecular-weight (HMW) glutenins, and hence it produces dough with almost no gluten structure. Although Ritmo is a hard endosperm cultivar classified as bread wheat, it is frequently used as biscuit wheat. For all cultivars, a decrease in maximum strain and increase in percent recovery were observed with increasing aging time (Figure 6.5). Galatea had the highest extensibility due to its low HMW-glutenin content which results in a higher proportion of gliadins. It is well established that gliadins contribute to the extensibility and viscous properties of wheat flour dough. Extensibility of Ritmo, the hard endosperm cultivar, was significantly less than the extensibility of Galatea. However, it showed the highest percent recovery, being more elastic due to the elasticity in its gluten network. The cultivars Galatea and Encore,

TABLE 6.13

Effect of Protein Content on Viscosity (η) and Relaxation Time (λ_{rel}) of Biscuit Dough

Protein Content (%)	Added Gluten (%)	$\eta \times 10^{-4}$ (Pa.s)	λ_{rel} (s)
11.2	0	3.3	1.10
12.8	2	3.5	1.15
14.4	4	3.8	1.11
15.8	6	7.1	1.22
17.3	8	7.6	1.40
18.7	10	9.0	1.60
20.0	12	9.4	1.70
21.2	14	9.4	1.72
21.9	15	9.4	1.70

Source: Adapted from Maache-Rezzoug, Z., Bovier, J.M., Allaf, K., and Patras, C. *Journal of Food Engineering,* 35, 23, 1998b. With permission.

TABLE 6.14

Creep and Recovery Characteristics of Semisweet Biscuit Dough from Different Cultivars

Cultivar	Protein in Flour (%dry matter)	Gluten (% of flour)	SED[1] (ml)	WA[2] (% of flour)	E[3] (mm)	R[4] (BU)	Maximum Strain (%)[5]	Recovery (%)[5]
Encore	9.0 ± 0.7	17.7 ± 0.3	21 ± 5	54.8 ± 2.0	117 ± 11	187 ± 59	1.19[b]	35.7[c]
Ritmo	9.1 ± 0.9	21.2 ± 2.7	32 ± 6	56.8 ± 3.0	128 ± 16	232 ± 108	0.87[c]	48.5[a]
Galatea	9.2 ± 0.5	20.5 ± 1.5	<10	53.7 ± 3.5	—	—	1.48[a]	35.7[c]
Claire	9.2 ± 0.4	21.2 ± 2.7	19 ± 2	51.5 ± 2.7	135 ± 10	262 ± 68	0.77[c]	40.3[bc]
Banker	9.4 ± 0.6	21.6 ± 4.3	21 ± 3	53.7 ± 3.3	128 ± 4	177 ± 49	1.26[b]	38.3[bc]

[1]SED, sedimentation value.

[2]WA, water absorption (measured by Farinograph).

[3]E, extensibility after 45 min dough resting (measured by Extensograph).

[4]R, resistance after 45 min dough resting (measured by Farinograph).

[5]Values for particular column followed by different letters differ significantly ($p < 0.001$); all years 1998, 1999, 2000.

Source: Adapted from Pedersen, L., Kaack, K., Bergsøe, M.N., and Adler-Nissen, J., *Journal of Cereal Science*, 39, 37, 2004. With permission.

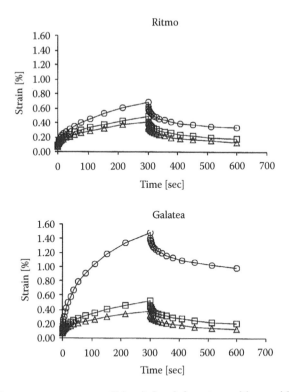

FIGURE 6.5 Creep recovery curves of biscuit dough from two cultivars with different resting times: (○) 10 min, (□) 25 min, (△) 35 min (From Pedersen, L., Kaack, K., Bergsøe, M.N., and Adler-Nissen, J. *Journal of Cereal Science, 39*, 37, 2004. With permission.)

with very weak gluten structure, recovered less than the others did. Claire was less extensible than the other soft endosperm cultivars and recovered to a high degree. Results of frequency sweep tests for two distinctive cultivars, Ritmo and Galatea, are shown in Figure 6.6. Increasing frequency increased both storage modulus (G') and loss modulus (G''). Correlation between frequency and phase angle (δ) depended on the frequency range. The phase angle (δ) decreased at frequencies less than 0.5 Hz, was nearly constant between 0.5 and 1 Hz, and increased at frequencies higher than 1 Hz. These differences imply that the dough acted more like a solid when imposed to slow changes in stresses, but very fast changes would make the dough act as a liquid. Measurements of storage modulus (G') were unable to distinguish between doughs, which have different strength of gluten network. However, phase angle (δ) distinguished between the cultivars that were different in the gliadin–glutenin ratio.

When flour was replaced by starch in a short dough formulation, the storage modulus (G') of the dough decreased (Table 6.6) because of the absence of gluten. (Baltsavias et al., 1997). However, the frequency dependency of storage modulus (G') was similar to that of the standard dough.

Pedersen et al. (2005) examined the effect of adding sodium metabisulfite (SMS), which is a disulfite cleaving agent, and protease to semisweet cookie

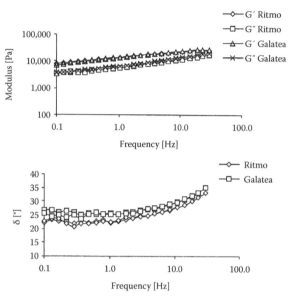

FIGURE 6.6 Frequency dependence of dynamic viscoelastic properties of biscuit doughs from two different cultivars. (From Pedersen, L., Kaack, K., Bergsøe, M.N., and Adler-Nissen, J. *Journal of Cereal Science,* 39, 37, 2004. With permission.)

dough prepared with the same cultivars in long-time behavior (creep recovery), as well as in the short-time behavior (dynamic testing). The main results of creep recovery and oscillation from cultivars 1998 and 1999 are shown in Table 6.15. There was a significant ($p < 0.001$) difference between the two years for all the rheological characteristics with exception of loss modulus (G'') and percent recovery due to the higher protein content (10.2 % average protein, 25.1% average gluten) of wheat from 1999 compared with the one from 1998 (8.9% average protein, 10.2% average gluten). The addition of SMS or protease had a significant effect on the rheological characteristics, with the exception of percent recovery in 1998. Maximum strain (i.e., extensibility of the doughs) highly increased because SMS and protease reduced the disulfite bonds. Consequently, elasticity decreased as indicated with the decrease in percent recovery of 1999 cultivars. Similarly, storage modulus (G') and loss modulus (G'') decreased whereas phase angle (tanδ) increased, indicating less elastic and liquid-like behavior. There was a significant interaction between the cultivars and addition of SMS and protease. The largest effect of the addition of modifiers was observed for the hard endosperm Ritmo. Cultivars with a high ratio of glutenins are supposed to be affected by SMS or protease addition more than cultivars with a lower ratio. This agreed well with the low content of gliadins measured for Ritmo. Maximum reduction in the length of the biscuits was observed for Ritmo, whereas Galatea was slightly reduced in length.

Gaines (1990) studied the effect of modifying agents on the consistency of sugar-snap cookies made from different cultivars. Potassium iodate (the sulf-

TABLE 6.15
Rheological Properties of Semisweet Biscuit Dough with Addition of Sodium Metabisulfite (SMS) and Protease, Means of Cultivars 1998 and 1999[1]

		No Addition	Addition of SMS (360 mg/kg)	Addition of Protease (300 mg/kg)
Maximum strain	1998	1.23[a]	2.46[b]	2.00[b]
(%)	1999	1.19[a]	4.36[c]	3.16[b]
Recovery (%)	1998	36.80[a]	33.88[a]	35.66[a]
	1999	41.07[a]	33.21[b]	35.82[b]
G' (kPa)	1998	17.20[a]	10.68[c]	13.36[b]
	1999	16.79[a]	8.48[b]	10.04[b]
G'' (kPa)	1998	7.28[a]	5.02[c]	6.03[b]
	1999	7.79[a]	4.54[b]	5.07[b]
tanδ	1998	0.42[a]	0.46[b]	0.45[b]
	1999	0.46[a]	0.52[b]	0.50[b]
log G''/log frequency	1998	0.19[a]	0.22[c]	0.21[b]
	1999	0.22[a]	0.28[c]	0.26[b]

[1]Means within a row followed by different letters are significantly different (p≤ 0.05).
Source: Pedersen, L., Kaack, K., Bergsøe, M.N., and Adler-Nissen, J. *Journal of Food Science*, 70, 152, 2005. With permission.

hydryl oxidizing agent that increases gluten elasticity), L-cystein (the disulfate cleaving agent that decreases gluten elasticity), N-ethylmaleimide (which blocks sulfhydryl groups—initially increasing gluten elasticity), and dithioerythritol (the disulfide cleaving agent that bonds with the resulting disulfide bonds, removing them from further thiol interchange) were studied. L-cystein did not significantly change the consistency of the doughs made from any cultivars. There was no specific pattern of response of dough consistency or cookie diameter to the modifiers. Dithioerythritol had the most effect on the dough consistency. It reduced the consistency and spread of cookies made from most of the cultivars.

The effects of protein-modifying agents on the rheological characteristics of biscuit doughs were studied by Monahar and Rao (1997a). Incorporating oxidizing agents potassium bromate (PB) and ascorbic acid (AA) or the sulfhydryl blocking agent N-ethylmaleimide (NEMI), produced significant change in any of the rheological characteristics studied (Table 6.16). This suggests that the conditions in the biscuit dough were not favorable to oxidizing agents, and there was little involvement of interchange reactions of sulfhydryl and disulfite groups in this type of biscuit dough. However, L-cystein hydrochloride (LCS) and dithioerythritol (DTE), two disulfite cleaving agents, reduced the extrusion time, apparent biaxial extensional viscosity, consistency, and hardness and increased the compliance, adhesiveness, and cohesiveness. The reduction of elastic recovery in the presence of the disulfide cleaving agents indicated a weakening of the gluten network.

TABLE 6.16

Effect of Protein Modifying Agents on the Rheological Characteristics[1] of Biscuit Dough

Modifying Agent[2]	Extrusion time (s)	Compliance (%)	Elastic Recovery ×10 (mm)	$\eta_B{}^3 \times 10^{-5}$ (Pa.s)	Consistency (N.s)	Hardness (N)	Cohesiveness	Adhesiveness (N.s)	Stickiness (°)
Control	45[a]	40.6[b]	4.60[a]	1.91[a]	786[a]	662[a]	0.194[b]	54.3[b]	50.3[b]
PB	44[a]	40.2[b]	4.55[a]	1.92[a]	792[a]	656[a]	0.198[b]	52.2[bc]	52.3[b]
AA	45[a]	40.1[b]	4.65[a]	1.89[a]	764[b]	652[a]	0.192[b]	51.6[c]	51.6[b]
LCS	35[b]	43.2[a]	3.60[b]	1.62[b]	718[c]	598[b]	0.232[a]	58.6[a]	54.6[c]
NMI	45[a]	40.4[b]	4.50[a]	1.92[a]	772[ab]	660[a]	0.196[b]	54.6[b]	50.4[b]
DTE	38[b]	43.3[a]	3.60[b]	1.54[c]	708[c]	602[b]	0.240[a]	60.4[a]	56.4[a]

[1] Values for a particular column followed by different letters differ significantly ($p < 0.05$).

[2] Control (300 g/kg sugar), mixing time 180 min; PB, potassium bromate; AA, ascorbic acid; LCS, L-cystein hydrochloride; NEMI, N-ethylmaleimide; DTE, dithioerythritol.

[3] η_B: Apparent biaxial extensional viscosity, at 50% compression.

Source: Adapted from Manohar, R.S. and P.H. Rao. *Journal of Cereal Science*, 25, 197, 1997a. With permission.

6.4　INTERRELATIONSHIP BETWEEN RHEOLOGICAL PROPERTIES OF DOUGH AND QUALITY OF COOKIES

Rheological properties of biscuit dough as influenced by ingredients, processing conditions such as mixing, and additives were related to the quality of biscuits by Manohar and Rao (2002). Extrusion time ($r = -0.54$, $p < 0.01$), elastic recovery ($r = -0.84$, $p < 0.01$), apparent biaxial extensional viscosity ($r = -0.62$, $p < 0.01$), consistency ($r = -0.73$, $p < 0.01$), and hardness ($r = -0.68$, $p < 0.01$) of the dough were significantly correlated to the spread of the biscuits. Elastic recovery ($r = -0.64$, $p < 0.01$) and cohesiveness ($r = 0.67$, $p < 0.01$) of dough mainly influenced the thickness of biscuits. Extrusion time ($r = 0.53$, $p < 0.01$), elastic recovery ($r = 0.78$, $p < 0.01$), apparent biaxial extensional viscosity ($r = 0.59$, $p < 0.01$), consistency ($r = 0.62$, $p < 0.01$), and hardness ($r = 0.60$, $p < 0.01$) were positively correlated to the density of biscuits. Among the various rheological characteristics studied, elastic recovery was reported to be the best index in predicting the quality of biscuits. Similarly, Pederson et al. (2005) concluded that shrinkage ($r = 0.70$) and spread ($r = 0.43$) of biscuits were correlated to percent recovery of the dough, and protein and gluten content.

6.5　CONCLUSION

Dough ingredients had a major effect on rheological properties and quality of cookies and biscuits. Increases in water, sugar, and fat content of cookie dough decreased consistency and elasticity and therefore increased spread of cookie dough. Reducing sugars had a similar effect as increasing the sugar content. Sucrose could be replaced by maltitol, lactitol, or sorbitol in low-fat cookies with supplementation of acesulfate-K for the improvement of taste. Sugar-free dough was more elastic than the sucrose syrup dough. Emulsified fat and margarine provided better rheological properties than nonemulsified hydrogenated fat and sunflower oil. Incorporation of hydrogenated fat to the formulation produced hard cookie dough. The addition of emulsifiers reduced the elasticity of cookie dough. Low-fat and liquid-oil doughs were more liquid like, and firm-fat dough was more elastic than the standard dough. Substituting shortening with 50% to 75% canola oil/caprylic acid structured lipid (SL) or with a carbohydrate-based fat replacer, Nutrim OB (NU) lowered the elasticity of the dough. An increase in protein and gluten content increased elasticity and decreased spread of cookies because of gluten network formation. Disulfite cleaving agents and protease increased extensibility and decreased elasticity of cookie doughs by weakening the gluten network.

REFERENCES

Agyare, K.K., Y.L. Xiong, K. Addo, and C.C. Akoh. 2004. Dynamic rheological and thermal properties of soft wheat flour dough containing structured lipid. *Journal of Food Science* 69: 297–302.

Agyare, K.K., K. Addo, Y.L. Xiong, and C.C. Akoh. 2005. Effect of lipid on alveograph characteristics, baking and textural qualities of soft wheat flour. *Journal of Cereal Science* 42: 309–316.

Baltsavias, A., A. Jurgens, and T. van Vliet. 1997. Rheological properties of short doughs at small deformation. *Journal of Cereal Science* 26: 289–300.

Baltsavias, A., A. Jurgens, and T. van Vliet. 1999a. Rheological properties of short doughs at large deformation. *Journal of Cereal Science* 29: 33–42.

Baltsavias, A., A. Jurgens, and T. van Vliet. 1999b. Large deformation properties of short doughs: Effect of sucrose in relation to mixing time. *Journal of Cereal Science* 29: 43–48.

Faridi, H. 1990. Application of rheology in the cookie and cracker industry. In *Dough Rheology and Baked Product Texture*, Eds. H. Faridi and J.M. Faubion, 363–384. New York: AVI.

Gaines, C.S. 1990. Influence of chemical and physical modification of soft wheat protein on sugar-snap cookie dough consistency, cookie size, and hardness. *Cereal Chemistry* 67: 73–77.

Jacob, J. and K. Leelavathi. 2007. Effect of fat type on cookie dough and cookie rheology. *Journal of Food Engineering* 79: 299–305.

Lee, S. and G.E. Inglett. 2006. Rheological and physical evaluation of jet-cooked oat bran in low calorie cookies. *International Journal of Food Science and Technology* 41: 553–559.

Maache-Rezzoug, Z., J.M. Bovier, K. Allaf, and C. Patras. 1998a. Study of mixing with rheological properties of biscuit dough and dimensional characteristics of biscuits. *Journal of Food Engineering* 35: 43–56.

Maache-Rezzoug, Z., J.M. Bovier, K. Allaf, and C. Patras. 1998b. Effect of principle ingredients on rheological behaviour of biscuit dough and quality of biscuits. *Journal of Food Engineering* 35: 23–42.

Manohar, R.S. and P.H. Rao. 1992. Use of a penetrometer for measuring rheological characteristics of biscuit dough. *Cereal Chemistry* 69: 619–623.

Manohar, R.S. and P.H. Rao. 1997a. Effect of mixing period and additives on the rheological characteristics of dough and quality of biscuits. *Journal of Cereal Science* 25: 197–206.

Manohar, R.S. and P.H. Rao. 1997b. Effect of sugars on rheological characteristics of biscuit dough and quality of biscuits. *Journal of the Science of Food and Agriculture* 75: 383–390.

Manohar, R.S. and P.H. Rao. 1999a. Effect of mixing method on the rheological characteristics of biscuit dough and the quality of biscuits. *European Food Research and Technology* 210: 43–48.

Manohar, R.S. and P.H. Rao. 1999b. Effects of water on the rheological characteristics of biscuit dough and the quality of biscuits. *European Food Research and Technology* 209: 281–285.

Manohar, R.S. and P.H. Rao. 1999c. Effects of emulsifiers, fat level and type on the rheological characteristics of biscuit dough and the quality of biscuits. *Journal of the Science of Food and Agriculture* 79: 1223–1231.

Manohar, R.S. and P.H. Rao. 2002. Interrelationship between rheological characteristics of dough quality of biscuits; use of elastic recovery of dough to predict biscuit quality. *Food Research International* 35: 807–813.

Olewnik, M.C. and K. Kulp. 1984. The effect of mixing time and ingredient variation on farinograms of cookie doughs. *Cereal Chemistry* 61: 532–537.

Pedersen, L., K. Kaack, M.N. Bergsøe, and J. Adler-Nissen. 2004. Rheological properties of biscuit dough from different cultivars, and relationship to baking characteristics. *Journal of Cereal Science* 39: 37–46.

Pedersen, L., K. Kaack, M.N. Bergsøe, and J. Adler-Nissen. 2005. Effect of chemical and enzymatic modification on dough rheology and biscuit characteristics. *Journal of Food Science* 70: 152–158.

Sahin, S., and S.G. Sumnu. 2006. *Physical Properties of Foods.* New York: Springer.
Steffe, J.F. 1996. *Rheological Methods in Food Process Engineering.* East Lansing, MI: Freeman Press.
Zoulias, E.I., S. Piknis, and V. Oreopoulou. 2000. Effect of sugar replacement by polyols and acesulfame-K on properties of low fat cookies. *Journal of the Science of Food and Agriculture* 80: 2049–2056.

7 Technology of Cake Production

Suzan Tireki

CONTENTS

7.1 INTRODUCTION

It is difficult to define cakes in a precise manner due to their wide variety and the broad range of their formulations. Cake products contain relatively high amounts of enriching ingredients like sugar, shortening, eggs, milk, and flavors in addition to soft wheat flour, and they have sweet taste, a tender and a short texture, and pleasing flavors and aromas. Cakes may be grouped into two broad categories: shortening-based cakes and foam-type cakes. In shortening-based cakes, crumb structure is derived from the fat–liquid emulsion created during the processing of batter. In foam-type cakes, structure and volume are primarily dependent on the foaming and aerating properties of eggs. Sweet dough products can also be classified as a third

149

group of cakes, which are yeast leavened and involve various jams, fruit, nut filling, and toppings.

Cake quality is affected by the ingredients used, an appropriate and properly balanced formulation, and mixing and baking procedures. In addition to these, correct preparation of hoops and other containers in order to provide adequate protection during baking, careful preparation of all of the ingredients before mixing (especially with regard to temperature and to fruit if being used), and careful batter handling during scaling and depositing are the other factors contributing to cake quality (Desrosier 1977; Pyler 1988). A general flow diagram of cake production is shown in Figure 7.1. Cake production includes the steps of mixing, depositing, baking, cooling, and packaging.

This chapter focuses on cake production technology and the equipment used in cake production.

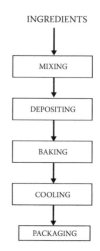

FIGURE 7.1 Cake production flow diagram.

7.2   MIXING

7.2.1   MIXING METHODS

Mixing method depends on whether shortening or foam-type cake is mixed.

7.2.1.1   Cake Mixing

Bringing about a complete and uniform dispersion and homogeneous mutual emulsification of the various ingredients, generally accompanied with the entrapment and size reduction of air cells and a minimum gluten development in flour, are the main purposes of cake mixing. The mixing process, taking place in mixers with various mixer accessories (Figure 7.2 and Figure 7.3), differs in the order of ingredient incorporation, duration, and rate of mixing action during different stages of multistage methods, temperature of the ingredients, and other factors according to the nature of the cake being produced.

The creaming (sugar batter) method combines shortening with the granulated sugar and usually with some of the dry ingredients at slow or medium mixing speed until the components are thoroughly blended and the resulting mixture becomes aerated. This step is followed by the incorporation of eggs as the creaming action is continued. With adding milk and flour, cake mixing is completed. Large volumes of air incorporation in the minuscule cells in the fat phase of the batter, coating of the flour and sugar by fat, delaying hydration and solubilization, and near absence of gluten development in flour are the main advantages of the creaming mixing method. The creaming method takes 15 to 20 min with an 8 to 10 min initial creaming stage, a 5 to 6 min second stage of egg incorporation, and 5 to 6 min final stage of milk and flour addition.

In the blending (flour batter) mixing method, flour and shortening are creamed to a fluffy mass in one bowl, while, at the same time, the eggs and sugar are whipped

FIGURE 7.2   Cake batter beater.
(See color insert after p. 158.)

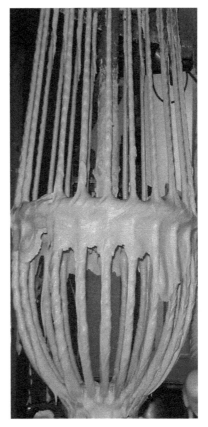

FIGURE 7.3   Cake batter whip.
(See color insert after p. 158.)

at medium speed to a foam that is semifirm in a second bowl. The separate mixing in the two bowls takes about 10 min. The sugar–egg foam is next combined with the creamed mixture of flour and shortening, and then milk is added in small increments. The blending mixing method has the advantage of achieving a very thorough shortening dispersion throughout the batter and producing an extremely fine grain and uniform texture in the cake, and this permits the use of higher sugar and liquid levels than is possible with the creaming method. However, in addition to the usage of two mixing bowls, there is a somewhat less amount of air incorporation with a following loss in product volume and a more pronounced gluten development causing perceptible toughness in the cake in this type of mixing.

In the single-stage cake mixing method, all of the major ingredients are put into the mixing bowl at one time and mixed into a homogeneous mass. The single-stage mixing method is usually composed of 1 to 3 min of blending the ingredients into a homogeneous mixture with a flat beater at low speed, followed by 3 to 5 min mixing at medium speed and 2 min final mixing at low speed—the total mixing time is 8 to 10 min. Incorporating the baking powder is generally done during the final mixing stage.

In addition to the above mixing methods, several other mixing procedures are used. According to a batter mixing method, all of the sugar and about one half its weight of water are put in the bowl and mixed at medium speed for about 30 sec. Then, emulsifier, shortening, flour, nonfat dry milk solids, baking powder, and salt are added, and mixing is continued for 5 min at medium speed. The remaining water, eggs, and flavoring are added finally and mixed for an additional 1 min at the low speed. Either leavening amount should be decreased by 25% or the water should be increased by 15% because this method promotes good aeration. Cakes produced with this mixing method are stated to develop a better crust color, a more tender crust with less indication of undissolved sugar, and greater volume. The cake quality improvement might be due to the initial solution of the sugar.

The sugar and shortening are creamed together for 2 to 3 min into a smooth mass in the emulsion mixing method. This mixing method is especially suited for large-volume cake mixers. Then, the milk is added in several portions with continuous beating for about 5 min at medium speed, and following this, the flour is added over 2 min and the eggs are added and mixed for an additional 4 to 5 min. The total mixing time of the emulsion method takes 12 to 15 min.

Fluid cake shortenings improve cake mixing efficiency. This might be due to the greater ease of dispersion of fluid fat in both the dry and liquid batter ingredients, and partly due to its more effective lubrication of the mixing bowl walls, reducing the need for their frequent scraping down. In addition, inclusion of newer emulsifier types in the shortening provides faster air incorporation in the batter and permits the ready emulsification of higher levels of liquids. All the mentioned factors contribute to a decrease in the mixing time with respect to the requirements in the case of plastic shortening usage.

7.2.1.2 Foam Cake Mixing

The structure and volume of foam cakes are dependent on the ability of eggs in the formulation to occlude air and to form stable foams. As egg white is beaten, air is incorporated, and the air cells become smaller in an increasing manner as beating is continued. The egg foam stability is dependent on the beating time, with more stable foams being formed as the time is extended. The beating time is affected by the beating conditions, such as egg temperature, beater type (whether of the wire whip or the blade beater type), beating speed, and sugar addition time.

The mixing of sponge cake batters may be conducted in several ways. Sometimes, people prefer to separate the whites and yolks of eggs and whip them separately with a portion of the sugar to the desired density prior to the recombination of them. The purpose of this approach is to obtain maximum batter volume. The beating of eggs (tempered to a temperature of about 26.7°C) with a wire whip or blade beater at medium speed is the most common procedure. Sugar may be added at the outset of mixing or in a slow stream during beating in order to counteract the tendency of whites to overwhip. After the egg foam attains the proper density, the flour and the liquid are folded in as lightly as possible to prevent foam structure breakdown. Fat must be added at the final mixing stage in order to minimize the loss in volume in short sponge cakes.

7.2.2 CAKE MIXING MACHINES

There are many types of equipment for the purpose of cake mixing. These are horizontal mixers, vertical mixers, Morton pressure whisk, Tonelli mixing system, Oakes continuous mixer, Mondomix continuous aerating equipment, Oakes and Strahmann mixers for pastes, high-speed mixers, Tweedy mixer and Mono, Cresta, Stephan, and Gilbert (Bennion and Bamford 1997).

7.2.2.1 Horizontal Mixers

Horizontal mixers are of very strong construction, and the troughs of this type of mixer are so arranged that the ingredients can be readily put in and discharged. The beaters are designed for mixtures containing butter or shortening, mixing into the mass of any ingredients or fruit without damage. An automatic safety lid is fitted to the machine as part of the standard design in order to avoid the machine being opened while in motion. Two speeds are available in horizontal mixers, high speed for mixing the light batter and slow speed for mixing the flour, fruits, and so forth.

7.2.2.2 Vertical Mixers

Vertical mixers have been developed for all mixing types in the bakery (dough, batters, and sponges and foams). Three or four speeds are available, changing gear is easy, and automatic timing devices are also fitted. The bowls of vertical mixers are safe during the time the machine is used, and they can be detached and removed from the machine easily either by placing on an appropriate trolley or by manhandling. The equipment is fitted with whisks for sponge work, with beaters for cake batters and dough hooks for bread or bun dough or for pastry. The beaters and whisks revolve in a planetary motion in order to scrape the sides and the bottom of the bowl. An automatic scraping apparatus is provided in some models in order to keep the sides of the bowl clean during the mixing process. Variable-speed motors are fitted in some models to eliminate the gearbox, whereas other models have three-speed, constant-mesh, preselective gearboxes with complete automatic lubrication.

7.2.2.3 Morton Pressure Whisk

This mixer type was used almost universally for large-scale production of sponge goods until the introduction of continuous mixing. It has been replaced now in modern production plants. However, it is still used as a premixer linked to the continuous mixer. This mixer type is still used in semiautomatic plants for sponge work. It has a Morton whisk in the center container and two hoppers, one at each side, where the eggs and sugar are fed into the mixing compartment. Two hoppers are present in order to prevent the wetting of sugar and hence prevent a stoppage in the hopper. The larger-scale equipment is motor driven, and there is an air compressor. It is not required to remove the lid until the end of the day with this mixer type. The compressor has an automatic regulator being set and locked in order to ensure that the pressure will not increase above that needed to work the equipment. A safety valve on the air receiver and a special type of safety cock on the machine are also fitted. For observing the working pressure, pressure gauges are fitted on all machines. There is a piston-type ejector at the bottom of the container so that the air pressure in the container ejects the batter into a bowl or mixing machine as needed when the valve is opened. By using a Morton pressure whisk, a sponge batter can be produced in 3 min instead of from 15 to 30 min, and the sponge obtained has a very fine, even texture of great uniformity (Bennion and Bamford 1997).

7.2.2.4 Tonelli Mixing System

The Tonelli mixing system, which is not a continuous system, is stated to be the first batch system to integrate completely with bulk systems and programmable control. It is a unique mixing system using a twin-tool variable-speed planetary mixing action said to be able to cut mixing times up to 75%. A constant bowl scraper can be involved in this mixing system, and it can mix creams, cake batters, and dough with suitable change of mixing tools. The sealed mixing bowl permits mixing under pressure, and a jacketed bowl is available for cooling or chilling during mixing. Many mixing tools are available from various whisk types for creams and foams, light and heavy batters, in addition to the tools for pie, short pastry, and cookies.

7.2.2.5 Oakes Continuous Mixer

Oakes continuous mixers are used for the production of all types of confectionery. The mixing head is the operational part of the machine, and the ingredients are fed in the form of liquid batter continuously with an air stream into the back stator of the mixing head where they then follow in radial direction outward to the periphery, and then flow in radial direction inward along the front stator prior to being discharged through the outlet. The rotor speed and the distribution of teeth are arranged in a way that gives the optimum intensity for a particular product. A pressure-regulating valve is located at the outlet enabling the pressure in the mixing head to be regulated to ensure that the air bubbles are incorporated completely within the liquid batter before the mixture is released through the delivery pipe. The mixing head is equipped with cooling jackets; however, the temperature increase rarely exceeds -16.7 to $-16.1°C$ due to the fact that the material is only in the mixing head for seconds. The mixing degree attained produces a completely homogeneous product where the air is dispersed uniformly; hence, the finished product has uniform cell structure and texture and good keeping qualities. The operation method generally used in mixing cake batter is composed of dumping all the ingredients (wet and dry) into the bowl of a batch mixer, mixing for 1 to 3 min to disperse the materials within the liquid batter uniformly, and transferring it to a holding tank that is adjacent to the mixer. By gravity, the liquid batter flows from the holding tank to the product pump suction on the mixer, and this delivers the material through a pipeline to the mixing head under pressure and then to the depositor. It is possible to control the speed of the rotor, speed of the pump, back pressure, the amount of air, and the throughput of the mixing (Bennion and Bamford 1997).

7.2.2.6 Mondomix Continuous Aerator

A dosed amount of air is injected on the input side of the mixing head of Mondomix. The rotor and stator pins then cut the product and the air under constant pressure until a homogeneous mixture is obtained. A mass airflow meter combined with an automatic air dosing system makes the foam density remain constant. Mondomix supply complete production units including manifolds, rotating and static extruders, depositors, and all other equipment.

7.2.2.7 Other Types of Mixers

Oakes and Strahmann continuous bread mixing equipment is used for the continuous production of short paste for pies and tarts (Bennion and Bamford 1997).

High-speed mixers were originally introduced for bread dough mixing purposes, but were then adapted for use in mixing cake batters and pastes (Bayliss 1967).

The Tweedy mixer has been adapted for all cake-making types. All of the ingredients are put in the mixing bowl at once, and mixing is completed in various times depending on the type of the product (e.g., mixing is completed in 1 min for slab cakes, whereas in 3.5 min for angel cake). In addition, when the Tweedy mixer is used, no vacuum is needed in cake processing.

The Mono multipurpose machine, Cresta machine, and Stephan (for smaller mixing) machine are similar equipment to the Tweedy machine. The Gilbert machine, which is an ultra-high-speed mixer, differs inasmuch as the bowls are removable and transferable. All types of goods can be mixed without the need to change beaters.

7.3   DEPOSITORS

Cake batter should be deposited into cake pans and conveyed to the oven with a minimum time loss. This is due to the fact that the leavening agents, having entered into solution during mixing, begin to interact and evolve carbon dioxide gas. In more fluid batters, carbon dioxide gas tends to rise upward with the small bubbles coalescing in the process into larger cells having greater buoyancy. There is an inevitable carbon dioxide gas escape from the batter while resting in the open hopper of the depositor, as well as a coarsening of the cell structure with the passage of time. This aeration loss and associated detrimental effects are avoided with depositors.

There is a wide range of equipment for the purpose of depositing accurately predetermined quantities of mixes (Bennion and Bamford 1997).

7.3.1   Copeland Depositor

The Copeland depositor handles any mix exiting freely from the hopper under the effect of suction from the depositor plunger. It is possible to deposit a wide range of mixes from light sponge to heavy fruitcake. This depositor type involves a hopper and depositing head arranged to give a lift motion to break the deposit. The head can be stationary if there is a wire cut-off alternatively. Sheets or tins are transported to a position under the depositing head by an intermittent chain conveyor, and there is fast handwheel changeover on the conveyor when changing from one pan size to another (Bayliss 1967).

7.3.2 Mono Electronic Depositor

Mono electronic depositors are very flexible and popular, and they are ideally appropriate for many single and mixed applications. The recipe, mixing method, size of template, and speed of deposit are several factors affecting the minimum deposit. All of the electronics of this depositor type are microprocessor controlled, and an instant program change can be done with a simple two-digit code for each product. A variety of templates are available in various configurations from two to ten nozzles. There are also cluster and novelty heads, and a heated hopper is used in order to maintain optimum temperatures of fondant and icings.

7.4 CAKE BAKING

Baking is probably the most important factor governing the quality of the final cake product. Incorrect baking can offset the effect of all the other factors like correct formulations, good raw materials, and correct processing methods. All of the cakes should be baked at a suitable temperature, consistent with the nature of the ingredients and the shape and size of the cakes. A lean mixing having few enriching agents should be baked at a much higher temperature than a mix very rich in fats, sugars, fruits, and syrups. An oven that is too hot causes high crust color, small volume, peaked tops, close or irregular crumb, and probably all the faults because of underbaking. On the other hand, an oven that is too cold causes poor crust color, large volume, and often weak crumb. The optimum cake-baking conditions are determined by several factors such as level of sweetener in the formulation, milk amount in the batter, batter fluidity, pan size, and so forth. Batters with high sugar content need lower baking temperatures than leaner formulations. Large cakes require lower baking temperatures and, hence, longer baking times. The baking time is inversely related to the baking temperature. Baking time should not be extended beyond the limit needed to ensure a thorough bake, because otherwise, the evaporative losses will exceed accepted norms, and the shelf life of the cake will be impaired (Pyler 1988).

7.4.1 Small-Scale Cake Production

Small-scale cake production works on the basis of the baking of cakes requiring the highest temperature first, working through to those requiring lower temperatures and longer time at the end of the day's run. This approach comes from the days when large coke- or coal-fired brick ovens were commonly used. These ovens were not so flexible; however, they were very reliable and many bakery products were baked by residual heat on a falling temperature. This is still ideal for many cake types (Bennion and Bamford 1997).

7.4.2 LARGE-SCALE CAKE PRODUCTION

Arranging output in such a way as to allow long runs of a single product is the tendency of large-scale production of today, which makes for higher efficiency and better control of the oven. Many big manufacturers still require a multipurpose plant; hence, ovens contending with a mixed range of products have variable baking requirements (Bennion and Bamford 1997).

Large- or small-scale cake production needs maximum oven usage, and batches and production runs should be of such a size to achieve this if it is to be economical. A half-empty oven or an oven not in use for long time periods is very uneconomical, and furthermore, the baking of cakes is better when the oven is full.

7.5 COOLING

Cooling is crucial in cake production in terms of the final texture and appearance of the products.

Automatic coolers may be employed in some cake production. Automatic coolers are traveling coolers and can be built to have either one or two swings suspended from chains, and they are driven by a small motor incorporating a variable-speed device for the regulation of cooling times for different product types. Output depends on the cooler size. Natural or open air cooling is generally used. On the other hand, conditioned air can be used if the cooler has a totally enclosed main structure. Coolers are valuable for cakes when baked on trays, because coolers enable the products to cool off prior to finishing and packaging, and hence they eliminate congestion in the bakery. Moreover, coolers reduce the wear of the baking floor. In addition, coolers do not need to have the feed and the delivery points at the same level. Actually, the cakes can be placed in the cooler on any floor where the ovens may be located and delivered at the ground floor or suitable level of the dispatch department.

7.6 PACKAGING AND WRAPPING EQUIPMENT

Mechanical wrapping methods have been improved with the advances taking place in the sale of prepacked cakes. Forgrove BW6 and BW-6P universal overwrapping machines and the Forgrove 84-H are the most popular wrapping machines used in cake production. The BW6 and BW-6P are simple, adjustable equipment and are suitable for a wide range of cakes and cartons of cakes. The principle property is the self-measuring paper feed, giving substantial savings in wrapping materials. In the 84-H model, the wrapping material is formed into a tube around the article inside a folding box. Then, rotary crimpers form the cross seals, and the packages are separated by integral knives. This wrapping equipment is versatile and uses a wide range of wrapping materials with special attachments for sealing, printing, and type coding, and it can wrap from 40 to 100 packages per minute depending on the cake type (Bennion and Bamford 1997).

REFERENCES

Bayliss, E.A. 1967. Automatic cake production. ASBE Conference Proceedings, Nov. 17–25.

Bennion, E.B. and G.S.T. Bamford. 1997. *The Technology of Cake Making*. London: Blackie Academic and Professional.

Desrosier, N.W. 1977. *Elements of Food Technology*. Westport, CT: AVI.

Pyler, E.J. 1988. *Baking Science and Technology Volume II*. Meriam, Kansas City: Sosland Publishing Company.

| 10 min | 15 min | 20 min | 25 min | 30 min |

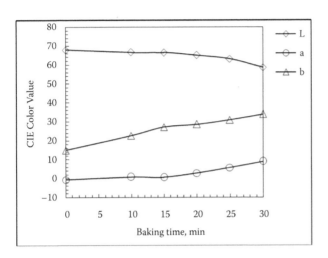

FIGURE 3.3   Change of color in cookies during baking at 160°C.

FIGURE 7.2 Cake batter beater.

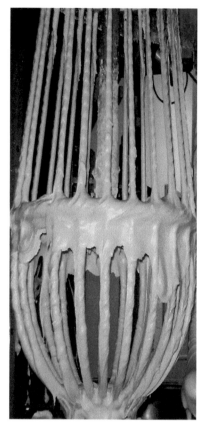

FIGURE 7.3 Cake batter whip.

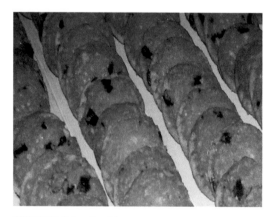

FIGURE 8.1 Cookies.

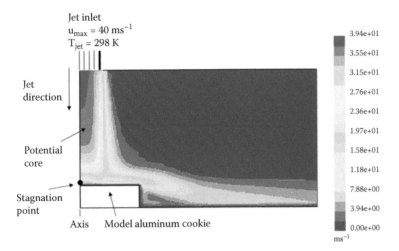

FIGURE 11.3 Contour plot of total velocity in a turbulent impinging jet on a model cookie at z/d = 3. (From Nitin, N. and Karwe, M.V., *Journal of Food Science* 69(2), 59–65, 2004. With permission from Blackwell Publishing.)

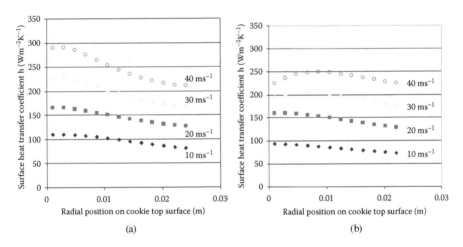

FIGURE 11.4 Variation of local surface heat transfer coefficient on the top surface of the model cookie as a function of position and jet velocity at jet temperature of 450 K for (a) z/d = 2, (b) z/d = 5. (From Nitin, N. and Karwe, M.V., *Journal of Food Science* 69(2), 59–65, 2004. With permission from Blackwell Publishing.)

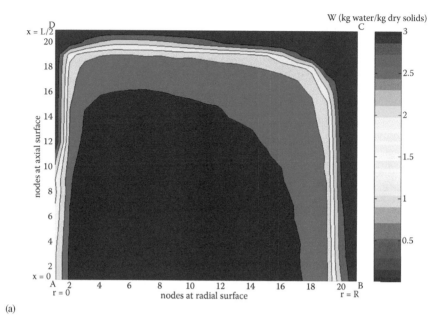

(a)

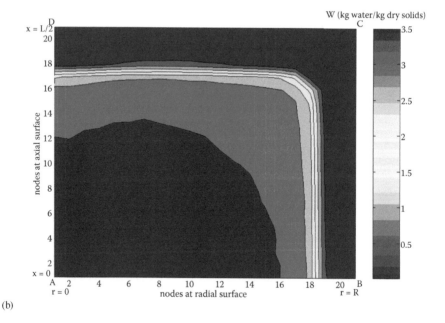

(b)

FIGURE 11.9 Moisture contours in potato for (a) case 1, (b) case 2. (From Kocer, D., PhD dissertation. Rutgers, The State University of New Jersey, New Brunswick, 2005.)

8 Technology of Cookie Production

Suzan Tireki

CONTENTS

8.1 INTRODUCTION

Cookies (Figure 8.1) may be defined as small cake-like products from a dough or batter made from raw materials such as flour, fat, sugar, milk, eggs, salt, starch, cocoa, leavening agents, emulsifier, and essences, which is viscous enough to allow the pieces of dough to be baked on a flat surface. They come in an infinite variety of sizes, shapes, texture, composition, tenderness, tastes, and colors (Pyler 1988). A general flow diagram of cookie production is shown in Figure 8.2.

Explained in this chapter will be different cookie-dough-making methods, mixer types, dough processing, baking and ovens, cooling of cookies, and the product packaging process.

8.2 PRODUCTION PROCESSES

8.2.1 DOUGH-MAKING PROCESS AND MIXERS

Cookies are generally categorized according to the equipment used in their production as cutting-machine cookies, rotary-molded cookies, wire-cut cookies, bar-machine cookies, and deposited cookies (Desrosier 1977).

The equipment type used limits the rheological qualities of dough and hence the composition.

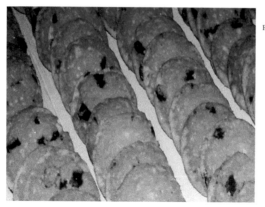

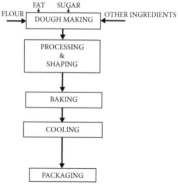

FIGURE 8.1   Cookies. (See color insert after p. 158.)

FIGURE 8.2   Cookie production flow diagram.

8.2.1.1 Dough-Making Process

The dough-making process depends on the type of the dough. The fat and sugar contents of cutting-machine dough are low, but their water content is high. As the formation of the gluten network is desired in cutting-machine dough, the mixing duration is long. The mixing period can be reduced by using sodium metabisulfite or a commercial protease. During mixing, the protein in flour contacts with water and in the following steps it takes time for the protein to absorb water and swell. If mixing is continued, the protein containing water forms a three-dimensional gluten network. For cutting-machine dough, all the ingredients are put into the mixer at the same time, and the duration of mixing is long as stated before because formation of a gluten network is desired. The sugar should be dissolved by the water in the dough. Otherwise, the sugar crystals caramelize during baking, causing brown spots. The temperature of the dough is important in terms of the fat used. At high temperatures, fat melts and dough becomes fatty. Dough should be plastic, and the shape given should be maintained.

In rotary-molded doughs, the amount of sugar and fat is high, and the amount of water is low (sugar: 20 to 45%, fat: 10 to 40%, water: 5 to 15 for 100% flour [flour weight basis]). This kind of dough can be crumbled easily, the maturity of gluten is undesired, and hence mixing duration is less. There are two steps in rotary-molded dough preparation:

1. Creaming (premix) step
2. Addition of flour

In the creaming step, all ingredients except flour are mixed and converted to cream. The creaming step should be prolonged as much as possible, which plays a crucial role in the density of dough. The second step of dough preparation, the addition of flour, takes a short time as formation of a gluten network should be prevented. If the duration of mixing is longer, gluten network formation starts and the dough will gain elasticity. As a consequence, a reduction in volume will be seen in dough during leaving of the molds. In the creaming step, sugar is dissolved and fat is softened, so they surround the protein molecules in flour and this prevents the interaction with water, and therefore, the formation of a gluten network is more difficult and slow. Fat is the most important input because it binds the ingredients in rotary-type doughs. Unless the dough is mixed enough, it cannot be shaped and exit from the mold will be difficult.

The composition of wire-cut cookies can be varied over a wider range in terms of formulation and final shape than any other type. In these doughs, it is required to have the material cohesive enough to hold together as it is extruded through an orifice, but it must not be sticky and it must be sufficiently short so that it separates cleanly as it is cut by wire (Pyler 1988). Wire-cut cookie dough is fatty and soft. Its water content is high and crystal sugar is generally used for this type of cookie dough. The development of gluten is avoided strictly, and the dough is mixed for about 2.5 to 6 min. Wafer doughs can be included in the wire-cut varieties. Wafer doughs have very low fat and sugar content. Wafer dough is a fluid dough having emulsion properties with high water content. It contains about 35 to 40% dry materials and 90% of the dry materials consist of flour. The main function of the mixing process is to obtain a homogeneous mixture. Gluten development and formation of a thread-like structure are not permitted in wafer dough. The mixing period is about 5 min. If thread-like structures are seen, the dough is not pumped, as it causes choking, and the dough does not spread on the wafer mold well. The temperature is as important as the amount of water added. If the temperature of water is high, the moisture problem comes out.

A bar-machine cookie is a relatively rich, highly flavored soft cookie-type product that is processed in a similar manner to that of wire-cut cookies except that bar-type cookies are of rather soft dough.

The batters of deposited cookies are very soft and lacking cohesiveness. Creaming up fat (or other fats) with the sugar, eggs, milk, and water and adding the flour later with quite a shorter mixing time (to achieve homogeneity) is usually best. The temperature of the dough is crucial in order to maintain required or higher consistency and correct dispersion of fat. Cooling the flour may be needed, and any water or milk used should be very cold. A temperature range of 10 to 16°C should be aimed at for the dough. In addition to the minimum amount of water, minimum mixing with flour is also required because of the fact that the tough dough must be avoided (Manley 2000).

Flour strength, product type, machinability of dough, dough temperature, mixer speed, and batch size are the most important factors affecting the mixing time for all of the cookie types. The ingredients that are intended to be present in the finished product as visible pieces like nuts, raisins, and chocolate bits should be added at the end of the mixing cycle and blended at slow speed for the minimum time consistent with enough distribution throughout the dough. Although doughs that are too cold can cause machining difficulties, high temperature developed during mixing is the usual problem. In any case, uniform temperatures are crucial for making uniform cookies as dough temperature affects spread, texture, and surface appearance of the cookie. The temperature should be below the upper limit of the plastic range of shortening for the best results.

8.2.1.2 Mixers

Mixing or blending can be performed by many different types of equipment, but all of them rely on one or more of the following;

1. Pushing parts of the mixture through other parts by blades, paddles, helical metal ribbons, etc.
2. Elevating and dropping all or a part of a batch so that random rebounding of individual particles results in redistribution of the particles
3. Gas or liquid injecting currents into a nonuniform material body in order to create turbulent movement of particles.

Cookie dough is prepared with the help of big mixers. Mixers can be classified as horizontal-fixed bowl and tilting bowl (high speed, low speed), vertical, reciprocating agitator, and continuous (agitator-in-tube, rotor and stator heads).

Horizontal mixers are useful for a wide variety of doughs from sugar wafer batters to extremely tough or dry dough. When gluten development is required, this mixer type is essential, because vertical mixers are too inefficient and slow for this purpose and spindle mixers lack the correct kind of action.

For discharging the dough, two methods are employed. The bowl can be tilted and hence the top is brought to a forward-facing position in some models. In other models, there is a tightly fitting door at the front of the bowl that can be raised and lowered independent of the immobile section of the bowl.

Mixer bowls are generally jacketed and a refrigerant can be circulated in the jackets. Some jackets are equipped in order to use direct expansion of cooling refrigerant like Freon-12 or ammonia, others use cooling liquid like propylene glycol, brine, and water.

There are various forms of agitators. One set of two, three, and four cylindrical bars, parallel to the front of the mixer, may be mounted on spiders connected to the axles at the point where they enter the jacket in the high-speed mixers that are designed for developing gluten. These arms may be attached by bearings in order to make them rotate, or they may be affixed in a rigid manner. The dough mass is stretched repeatedly and kneaded in a single direction so that gluten fibers tend to be oriented and the dough develops due to the limited clearance of the jacket wall. Slow-speed horizontal mixers have various forms of agitators and may be equipped with one or two sets of axles. The mixer arm configuration and the speed affect the action. For stiffer doughs, double-armed mixers are used; however, they may require heavier motors and drives.

The use of movable bowls or troughs is the unifying feature of vertical mixers. The other properties may be quite diverse. There may be one or more beater shafts that may be stationary or move in a planetary design. In addition, agitator design can be varied.

The agitator movement is described as planetary because it revolves around its own vertical axis at relatively high speed and, at the same time, the axis also moves in circles as it is rotated around the bowl. These combined movements guarantee that the mixer bowl at the bottom center is raised.

Spindle mixers are most often used for cracker dough, although they can also be used for cookie dough. They have few advantages for cookie dough. The main advantage for saltine dough is due to its adaptation to mixing in the special trough that is mobile for fermenting sponges. Due to this purpose, dough and sponges do not have to be transferred in and out of the mixer and trough between the various stages.

In reciprocating agitator mixers, a pair of agitator arms moves through intersecting elliptical-shaped paths in a shallow and slowly revolving bowl. Temperatures can be held near room temperature without the use of bowl jackets due to the relatively slow rate of energy input. If an intensive blending action is not needed, these types of mixers are useful in mixing temperature-sensitive dough. For equivalent times, adjunct breakdown is less in these types of mixers than in any other types.

The ingredients must be metered out of a bulk supply and fed at a set rate into the mixing zone in a continuous manner for a (timed) continuous mixing system. Air or another gas may be one of the components. Batch premixes are used by many setups. Single or multiple units may be used. For cake batters, the mixers composed of a high-speed disc-shaped rotor intermeshing with disc-shaped stators are used widely, and this equipment has been adapted for use on cookies. Continuous mixing would be possible for stiff mixtures, yet a different kind would be required.

The Oakes continuous automatic mixer can be used for mixing wafer dough, batters, and marshmallow. The mixing chamber is composed of a rear stator, the rotor, and the front stator and the two stators are bolted together and supported on the frame. They are furnished with blades projecting into the chamber, and the rotor fits between the stators and is mounted on a revolving shaft. Blades on the rotor mesh with the blades on the stators, and the speed can be varied over a wide range.

In continuous mixers, the products to be mixed are forced into the head by a positive displacement pump through an orifice at the center of the rear stator. The products then flow between the blades of the rear stator and the rotor to the outer circumference of the stator cavity. The mixture flows between the blades of the front stator and the rotor to the discharge vent located at the center of the front stator. Controls, motor and gear train, power supply, and the necessary pumps and piping are the remaining parts of the mixer. For temperature control, the mixer head can be jacketed. The revolving element imparts intense shear to the components of the mixture, creating a turbulence that quickly mixes them. Gas can be dispersed in liquids readily.

8.2.2 PROCESSING AND SHAPING

Rotary dough should be given to the machine regularly in order to prevent accumulation at the ends causing drying of the dough. Rotary dough is processed by the use of molds indented to the inside. Metal cylinder, mold cylinder, and rubber cylinder should be parallel to each other in all axes. The procedure applied by the mold cylinder to the rubber cylinder should be sufficient for the cookies to stick on the infinite cloth. Otherwise, wearing occurs on the mold and rubber cylinder and hence shape disorders can be seen on the cookies.

The most important step in processing cutting-machine dough is the thinning of dough. There should be more than one thinner cylinder. If dough is not passed through two or three thinning cylinders, hardening in the dough consistency and production of defective cookies are observed.

After leaving the squeezing cylinder, the dough is passed through two or three thinning cylinders in order to obtain a thin layer of the dough. When this process is finished, the dough is converted into two to six folds and passed through two or three cylinders again. This process is called lamination.

The aims of laminating are to provide a way to repair a poor dough sheet tending to be prepared with a simple pair of rolls; to turn the folded dough through with an angle of 90° in order to make the stresses more uniform in two directions; to work on gluten, making it more suitable for baking with rolling, folding, and following more rolling; and to make a flaky structure after baking with the addition of another material such as fat between layers of dough.

A vertical laminator with a continuous lapper and a one-sheeter, vertical laminator with a continuous lapper and two sheeters, horizontal laminators, and cut-sheet laminators are the types of automatic laminators (Manley 2000).

Molds that are indented outwards are used for cutting-machine doughs. Doughs get the desired shape between the mold and the rubber cylinder. Crumbed dough is returned back to the squeezing cylinder by mechanical ways.

Infinite cloth is the first cloth after the mold. Cookies stick to it easily, and it is impossible to take samples from here. The sticking is provided by applying vapor to the cookies.

Dough for processing on deposit, wire-cut, and bar machines varies in the degrees of softness, which are

1. Very soft: A batter needs to be deposited directly onto the steel oven band invariably. A wire-mesh oven band is not suitable for baking deposit goods (deposit machine).
2. Soft and usually sticky: Needs to be extruded and cut by a taut wire prior to being dropped onto a canvas belt carrying the dough pieces to the oven band. Direct cut onto the oven band is used in some cases (wire-cut machine).
3. Fairly soft but stiffer than the second degree of softness: Extruded in endless strips cut into suitable lengths with a reciprocating guillotine, on a canvas belt conveying to the oven band (bar machine or rout press machine).

In a deposit machine, which extrudes batter intermittently from a hopper through shaped nozzles, it is necessary to have a cut-off mechanism that is incorporated in the feed hopper to ensure a controlled flow and deposit, because the very soft dough has a fairly viscous movement with its own gravitational flow due to its relatively high liquid content. The deposit nozzles are connected to the feed hopper outlets with flexible tubes, hence allowing a patterned deposit to be made in some types of depositors.

Wire-cut machines are more complex when compared with deposit and bar machines, as they involve a device cutting off extruded dough pieces emerging from the die orifice, and a cut-off device is composed of a blade or a wire drawn through the dough fast by a harp moving back and forth below the orifice. Dough is fed to the hopper manually or by gravity from a holding trough. Vertical separator plates can be inserted in the hopper in order to make the feeding of two or more colors or flavors possible at the same time. For a uniform pressure and constant extrusion rate, it is crucial to keep the feed rate steady and to maintain hopper contents at about the same height at all times as initial steps. Hoppers are jacketed and warm air or water is circulated for the improvement of extrusion rate uniformity. Hoppers' ends are curved in order to decrease the dough tendency of stagnating in this area. Maintaining uniform weight and size in the finished cookies depends on the properties of the rollers pressing dough through the die cups. The delay between mixing and forming influences the cookie dough response to forming. It is important to process batches quickly, and to form a uniform schedule and follow it in order to avoid noticeable differences in size, weight, and appearance of the finished product.

As there is no need to separate the dough into cookie-size parts at the extruder, bar machines have a simple forming method. Continuous strings or dough strips onto the oven band are directly extruded by bar machines. These bands can be separated into individual bars prior to or after baking with usual cutting devices. When the die plate is inclined in the extrusion direction, the ribbon is supported for a longer time. This decreases breaking or thinning of the dough strand because of the gravitational pull. Die orifices of bar machines are generally with straight lower-edge slots for giving a flat bottom to the cookie and with grooved top-edge slots for giving a ribbed upper surface to the cookie.

8.2.3 BAKING

8.2.3.1 Baking Principles

Cooking can be described as the art of preparing foods by heating until they are changed in flavor, appearance, tenderness, and chemical composition. Baking is a form of cooking performed in an oven (Desrosier 1977).

Every baking process depends on heat transfer from one body to another in the direction from hot to cold. There are three modes of heat transfer: conduction (in solids or liquids at rest), convection (in liquids or gases in a state of motion), and radiation (which does not involve a material carrier). In addition to the oven design, the conformation of the dough pieces, size, shape, and container construction materials, pan, or band, and dough pieces distribution on the hearth affect the relative effectiveness of these heat transfer modes.

Conduction entails a direct close contact between the heat source and the material being heated. If the dough is baked in a band oven, heat conduction to the dough occurs only through the band. The band receives its energy store from heat conducted through the supports where it rides and from convection and radiation. Due to the localized nature of conductive transfer, steep temperature gradients can be set up within the dough piece, the hottest regions being the ones contacting with the band or pan.

Convection involves heat transfer by either fluids or gases in a state of motion. A hot body gives up some of its heat and hence increases the temperature of the gaseous media around it. If this heated gaseous media flows around a cold object, heat will be absorbed by the cold object. The top and side surfaces of an oven become hot and thus heat up the oven atmosphere, which in turn gives up some of its heat to a dough piece. Molecules of air gases, water vapor, or combustion gases circulate throughout the oven, constantly mixing with other gases and transferring heat by conduction when they contact with solid surfaces in the oven chamber. Convection occurs due to the movement of water vapor and other gases within the dough piece. In addition, translocation of liquid water, melted shortening, and other liquids causes heat to be transferred from one region of the dough to another. The convection mode of heat transfer can be generalized as a smoothing or evening effect on heat distribution within the dough piece.

Heat transfer by infrared radiation is a significant factor in most ovens. These radiations are converted into heat through absorption by and interaction with absorbing materials; they are not in themselves heat. The radiation mode of heat transfer has two properties making its action different from other heat transfer modes:

- It is subject to shadowing or blocking by intervening substances that are opaque to radiation.
- It is very responsive to changes in absorptive capacity of the dough.

Radiation energy comes from the burner flames and all hot metal parts in the oven, and it is not required that the oven part be red hot or otherwise visibly heated for the radiation of infrared rays. This radiant energy travels in a straight line, and much of it never reaches the dough piece because some substances that are not transparent to the radiation intercept the rays. Shadowing or blocking out of radiation by some intervening material can occur from parts of the oven, from pan walls, or from parts of the dough pieces.

During baking, it can be said that conduction and radiation tend to cause localized temperature differentials in a way that conduction acts to raise the temperature of the bottoms and radiation acts to raise the temperature of exposed surfaces, while convection tends to even out temperature gradients.

8.2.3.2 Changes in Dough during Baking

Dough pieces undergo physical and chemical changes within the oven. Crust formation, melting of shortening in the dough, conversion of water to steam, gas expansion, and escape of carbon dioxide, other gases, and steam are the physical changes occurring by heat treatment.

The outer surface of the dough soon becomes coated with a film or crust on entering the oven. Crust thickness develops as moisture is evaporated from the outside skin. Crust formation starts at 26.7°C and proceeds quickly at around 37.8°C. It is necessary that the crust achieve sufficient thickness in order to allow it to become elastic. The moisture content of the product and humidity of the oven atmosphere affect the degree of elasticity, particularly in the first two zones of a four-zone oven. The crust film becomes too thick if the heat of the top oven is too high because the water vapor and the gases are subsequently formed within the dough. This top crust having lost its elasticity will burst open. Due to this, in some cookie types, collapse of the internal structure is evident. Crust film elasticity is directly related to the oven atmosphere humidity. If the film crust is formed too rapidly, the flowability of the dough is limited; there is a suppression of the leavening action and hence texture formation.

Shortenings do not have sharp melting points as they are mixtures of compounds. The aggregates of shortening particles melt as soon as their immediate area in the dough reaches the melting, fusion, or slip-point temperature of shortening structure. Although the lower melting point fractions seep into the enveloping structure, shortening pockets remain more or less within their original position in the dough structure and thus contribute to cookie texture.

Water used for dough preparation is converted to steam during heating. Steam formation makes the dough pieces expand. Expansion due to steam formation is much greater than the expansion due to carbon dioxide or ammonia, despite carbon dioxide being evolved much earlier in the oven than the steam.

Carbon dioxide formation due to the chemical reactions within the dough piece under the effect of increasing temperatures increases the volume and stretchability of the dough. The mass will be opened up by the expanding gases to help to create the crumb texture depending on the strength of the structure of the gluten/starch/sugar/fat matrix.

The overall dough volume is reduced as carbon dioxide and other gases and steam are removed. If the loss is too great, this will cause the structure to collapse, resulting in hollowed tops and cracked surfaces. Not all of the internal gases must be allowed to escape until the structure becomes more or less set, otherwise the texture through the final baking stages and the cooling will not be maintained.

All of the physical changes (especially the internal ones) must be encouraged to take place in an order, environmental conditions, temperature, and a time optimum for the particular dough makeup and the desired attributes of the cookie to be produced.

Gas formation, protein changes, starch gelatinization, caramelization of sugar, and dextrinization are the chemical changes taking place in the oven.

Chemical leavening systems involve a gas source, almost always sodium bicarbonate, and one or more acid reacting substances. The function of a leavening acid is to promote a controlled and nearly complete evolution of gas from a dough where carbon dioxide exits in its dissolved form. The reaction of acid and carbonate (or carbonate alone) can be controlled by the solubility of the particular acid or carbonate in the dough moisture, temperature, and decomposition range of the carbonate.

Gluten and other proteins derived from milk and eggs begin to coagulate at temperatures from 62.8°C. This protein coagulation imparts strength to the cookie structure. At around a temperature of 73.9°C, the proteins undergo an irreversible denaturation—they become less soluble and the protein fibers become less extensible. Hence, the vesicle walls of the dough structure achieve a more or less fixed position, where the expansion of dough practically stops. The coagulated protein is the drier region in the baked cookie, the starch holds most of the present moisture, the shortening gives tenderness, and all combined give shortness. The moisture in the cookie migrates from the starch to the protein gradually, even without loss of moisture from the biscuit to the atmosphere during the shelf life of the cookie. This is not so evident in cookies like bread, where the moisture is greater with respect to cookies.

Starch gelatinizes to form viscous solutions or rigid gels when heated in the presence of sufficient water. If the gelatinized starch is allowed to cool, it becomes more viscous. When starch is mixed with cold water, it absorbs 30 to 35% of water with slight swelling of starch granules. Water removal leaves the starch in its original state. However, when the starch–water mixture is heated at a temperature above 54.4°C, water absorption is greater, and the starch granules swell to many times their original size. The starch cannot be recovered in its original state because this reaction is irreversible. Hence, starch gelatinization probably plays a crucial role in producing cookie structure during baking.

Sugar caramelization takes place around 148.9°C accompanied by melanoidins, and this is the reason for the degree of brown crust development. Caramelization is the consequence of sugar molecules such as maltose, fructose, and dextrose to produce the colored substances classified as caramels. Associated with this reaction, the Maillard reaction due to the interaction of reducing sugars with proteins and other nitrogenous substances gives rise to attractive colors, flavors, and aromas. At about 176.7°C, the brown color seems and tastes like caramel, and around 246.1 to 260°C, the melanoidins become black, bitter, and insoluble.

Starch is started to be converted to dextrin at temperatures slightly higher than 148.9°C. If a slight degree of dextrin can be formed on the dough surface during baking undue caramelization, then a surface brightness is developed, which is desired.

8.2.3.3 Ovens

Direct-fired, indirect-fired, and hybrid ovens are the main types of cookie oven-heating systems.

In direct-gas-fired ovens, many ribbon or strip burners are located above and below the baking band. Each burner is supplied with carbureted gas and air, and the pressure of this mixture determines delivered power. In order to provide even heating across the band, there are various arrangements for adjusting the flame size across the width of the oven. Direct-gas-fired ovens may additionally have a turbulence system improving the heat transfer rate. The top of the baking chamber is usually low, and the burners are as near to the band as is practicable, meaning that there is a high radiant heat component in the heat transfer profile reaching the product.

Electric-fired ovens are similar to direct-gas-fired ovens, but each burner is supplied with electricity.

Each zone of the forced convection direct-fired ovens has one large burner, and the combustion products are blown to plenum chambers above and below the band. It is possible to control blowing velocity and the ratio of hot air circulated above and below the band. The baking chamber roof of a forced convection direct-fired oven is higher than that of a direct-fired oven in order to maintain even air flows. This means that forced convection ovens contribute a lower proportion of radiated heat to the heat transfer profile yet allow more uniform temperature and heat transfer conditions across the baking chamber width.

In convectoradiant ovens, hot gases from the burner in a zone pass through tubes above and below the band, and then they are released from further tubes to blow over the first tubes in the band direction. The first tubes radiate heat to the cookies, and convection currents of air are given by the released air. In order to maximize the radiant heat effect, the radiant tubes are located as near as is practicable to the baking band.

Indirectly fired forced convection ovens are similar to the direct-fired forced convection ovens, but a heat exchanger near the zone burner heats the air passing through plenum chambers in the baking chamber. Hot gases pass through tubes above and below the baking band and circulate back to the burner in indirectly fired Cyclotherm. No combustion products pass into the baking chamber, and there is a separate air circulation system moving air in the baking chamber and over the hot tubes.

Hybrid ovens are the combination of two of the direct-fired and indirectly fired ovens. A very common hybrid oven is composed of a first zone of direct gas fired followed by two or more zones of forced convection type. Maximum power and much radiant heat are available early in the bake, and then much convected heat is provided for the drying part of the oven.

Indirectly fired ovens generally have a few large burners with the oven divided into large zones along the length. Direct-fired ovens have a large number of small burners grouped in similar large zones for control purposes. In direct-fired ovens, it is possible to turn off individual burners either above or below the band. Dampers are present to control and divert the hot gases' passage to various parts of the oven chamber or up to the flues to the atmosphere. In addition to the direct and indirect oven types, there are designs promoting convection or radiant heat transfer. Successive zones might have a different effect on heat transfer type, and a hybrid oven type may have different fuels in different zones.

The length of an oven and the baking time required to bake a cookie to the desired structure, color, and moisture content define the production rate of an oven. For most of the products, the time needed to dry the product satisfactorily determines the baking speed. It is possible to calculate the oven efficiency by measuring the amount of fuel of known calorific value burned in a time compared with the weight loss representing the quantity of the water evaporated and temperature rise of the cookie ingredients.

There is a terminal drum at each end of the oven. The drum is driven at the oven exit, and there is a tension device holding the band taut at the feed end. The drums have sufficient diameter in a way that the bands and their joints are not strained in flexing, and their axles can be inclined to facilitate tracking. It is sometimes required to coat the drum with fibrous and fire-resistant material to eliminate slip. The bands are supported on metal or graphite rollers spaced closely enough in order to prevent appreciable sagging of the band between them through the oven.

The distance that the oven band extends beyond the oven chamber at each end is related to the way a product is placed on it and also the amount of cooling needed before the baked product can be removed. Wire-cut and deposited cookies need lead-in space in order to allow room to locate the forming equipment over the band. Sugar-rich cookies need long run-out lengths, possibly with fan-assisted air or water spray cooling under the band, to allow them to set hard prior to stripping from the band.

A stripping knife, which may be a thin blade of steel or a hard synthetic substance or a comb of wire fingers, is used to remove the baked cookies from the band.

The knife should be designed to lift the cookie clear and transfer it with minimum disruption to the relative positioning of the cookies. This provides good feeding to the wrapping equipment or for further processing. Different types of knives are used for lifting different types of cookies. It is crucial that the knife does not bear so firmly on the band that it damages the surface of the band.

8.2.4   COOLING

The cookies are cooled on a cloth band after leaving the oven, and they are very hot, very soft, and generally very moist as they emerge from the oven. Hence, even though cooling is a must for packaging, it may be the least crucial aspect, for so many other things are taking place as the cookie cools.

Cookies having a crust temperature of 115.6°C and a crumb temperature of about 99°C coming from the oven are still in a somewhat plastic state. Some wire-cut cookie types are so soft and molten that they cannot be picked off the steel band near the oven mouth. In addition to high temperature and moisture content, there are other factors to be considered in cookie cooling. However, two factors are effective in how the other ingredients react to the cooling cycle. It seems reasonable to suggest that the flour starch is still in some gelatinous paste form, and dextrins are still in partial solution due to the relatively high moisture content. Sugars are in at least partial solution as well, and the shortening will be present as oil rather than as fat. The protein is probably in a firmer state with respect to other ingredients. Thus, every ingredient is in an unset state, meaning that each is hot, moist, and soft.

As cooling continues, with consequent moisture loss due in part to the using up of some of the internal heat of the cookie, the change is from a gel to a paste and to a dry complex structure. This may seem to depend on protein, which may attract water from those ingredients so ready to give it up, for example, dextrins setting to a brittle condition, sugars crystallizing out, and at least in wetted films of starch drying out. Thus, the cookie becomes rigid (i.e., set). The shortening does not crystallize out until the cookie reaches a temperature within the setting range of its makeup glycerides, and it might even solidify in fractions.

Moisture loss, temperature decrease, and the changes in the state of the main ingredients affect cookie dimensions, giving rise to shrinkage and maybe causing stresses to be set up within the cookie in reaching the set, nonmolten state. The mentioned stresses may cause cracking of the biscuits to a greater or a lesser degree under adverse conditions. Sudden cooling can be a reason for cracking, as it might firm up the crust and retard the moisture migration rate from the center crumb to the edges. This happens due to the excessive moisture gradient between these areas.

8.2.5   PACKAGING PROCESS AND EQUIPMENT

Packaging and storage are the last stages of cookie production. This stage is important in terms of its protection purpose. The time period from when the cookies are packaged to consumption is influenced by packaging and storage methods, and the flavor, taste, and appearance of the cookies should be protected during this time period.

The packaging material should protect the cookie from harmful environmental effects. The product must be protected from undue moisture change during its normal storage life as a primary requirement. When the packaging film protects against moisture transfer in an adequate manner, it likely excludes dirt, dust, mold spores, and other foreign particles, and in addition, it gives some protection against the absorption of off-odors.

As most cookies are very susceptible to crushing, mechanical strength should be present in the container if the cookie is to survive storage and transportation. The package should contribute to the dimensional stability of the cookie.

The packaging material of the cookies should be appropriate for being formed into the finished package easily and fast by mechanical ways. A fundamental need for packaging films is that the structure heat-seal readily. Moreover, the packaging material should not tear, crack, or stretch during the rapid transfers and foldings in the wrapping equipment.

The package should also help sell the cookie. Transparent films are used when the visibility of the product is important. A glossy surface enhances consumer acceptance, and printability is needed in most cases.

In addition to the above factors, the packaging film should be relatively low in price, and the supplier's plant or warehouse should be located so as to make transportation costs acceptable.

Packaging materials have different degrees of resistance to water vapor transfer and differ in their barrier properties to oxygen and other gases, hydrocarbons, and light. Cellulosic materials, plastic resins, metal films, and laminates are used generally for packaging materials for cookies. The cost of the packaging material changes according to the content of base stock or resin, film thickness, and coating application. The plastic film thickness is more or less directly related to the rate at which water vapor and oxygen can diffuse through the film.

To improve the sealing characteristics, to establish a better substrate for printing inks, and to improve barrier properties, coatings can be applied.

With the modification of base material, thickness, and coatings, the barrier properties of packaging materials can be changed to the desired directions. Lamination of two or more films may also be employed in order to modify barrier, sealing, and visual properties.

There are two general types of packaging schemes employed in the production of cookies: dump packaging and registered packaging. In dump packaging, the small pieces are allowed to fall into the pack in no particular order. In registered packaging, the pieces are kept in some predetermined relationship to each other throughout the packaging process and in the container.

For dump packaging of small- to medium-sized cookies, vertical form-fill seal equipment is used widely. Vertical form-fill seal packaging equipment takes a flexible film strip and wraps it around a metal tube open at both ends. Two vertical edges of the plastic strips are overlapped and heat sealed, and this makes the web into a vertical cylinder. Across the mentioned cylinder and just below the forming tube, a heat seal is made as soon as a weighed amount of cookies is dropped down the forming tube into the closed-off area. The clamp draws the web downward and pulls more of the film along the forming tube. The sealing jaw returns to its original position for another sealing cycle when it has drawn down a predetermined length. Next, it makes the top seal of the bottom bag and cuts it off as it is making the bottom seal of the next bag.

The form-fill sealer can be replaced with manual filling of premade bags when the cookie production is in small scale.

For registered packaging of cookies (where the cookies are kept in relationship to each other during packaging and in the container), the kinds of machinery available are more varied. The process is old but has been refined to the point where breakage is minimal with high speed.

Cartoning machines used in the cookie production can be grouped according to mode of operation (semiautomatic or fully automatic), the direction of loading (vertical or horizontal), and motion type (continuous or intermittent). Semiautomatic equipment, which requires that the operator put the cookie in the carton manually, is said to be more suitable if many different sizes are loaded and frequent changeovers are needed. The cookies are loaded into cartons automatically in the fully automatic mode.

REFERENCES

Desrosier, N.W. 1977. *Elements of Food Technology.* Westport, CT: AVI.

Manley, D.J.R. 2000. *Technology of Biscuits, Crackers and Cookies*, 3rd ed. Boca Raton, FL: CRC Press.

Pyler, E.J. 1988. *Baking Science and Technology*, Vol. 2. Kansas City, MO: Sosland.

9 Heat and Mass Transfer during Baking of Sweet Goods

Weibiao Zhou, Nantawan Therdthai

CONTENTS

9.1 INTRODUCTION

For sweet bakery goods, baking is a key process to develop desired product characteristics including structure, texture, flavor, and color. Each product has its own recipe that is supposed to yield the distinct characteristics of the product. However, those characteristics are often the direct consequence of heat and mass transfers during baking in an oven.

Typical sweet goods including cakes, biscuits, crackers, pies, and some breads have different baking profiles. If the baking temperature is too high for a product, crust may be formed too early, resulting in a much smaller volume. In addition, the crust might become too dark while the interior of the product is still underbaked. In contrast, if a product is baked under a low temperature, the baking time may need to be extended to develop a desired brown crust color. However, a longer baking time could develop a thicker crust. Some European bread varieties have thick crust characteristics. In general, bread can be baked at a temperature in the range of 200 to 240°C. For cakes, a relatively lower baking temperature in the range of 175 to 215°C is generally required. If the baking temperature is too low, both moisture loss and volume can be increased, which results in weak crumb and dry mouthfeel. On the other hand, if the baking temperature is too high, the quality of the cake may be poor. Not only does this result in underbaked crumb and small volume, but it also causes peaked top or irregular crumb (Conforti, 2006). In the case of fruit cakes, when the batter viscosity is too low, fruit pieces have a greater tendency to sink to the bottom of the cake during baking. Therefore, a slightly higher temperature should be used to shorten the time period of low viscosity (Cauvain and Young, 2001). Crackers require a higher baking temperature in the range of 220 to 260°C. Slow baking rates can produce a coarse texture in the case of Graham cracker. On the other hand, an oven temperature that is too high may cause blisters in cream crackers. With a very high temperature on the top, a baked cream cracker can be domed. In addition, with a very high temperature at the bottom, dishing of a baked cream cracker is possible. Therefore, a balance between the top and bottom heating plays an important role in producing the right cracker characteristics (Yoneya and Nip, 2006). In the case of pies with fillings, a long baking time might cause a boil-out of the pie fillings. To avoid boil-out, it is better to bake them at a high temperature for a short time. Otherwise, the total soluble solid in the pie fillings has to be increased to increase its boiling point as well as decrease the equilibrium relative humidity of the fillings (Cauvain and Young, 2001).

It is clear that all sweet goods require a well-designed and often unique baking condition that could produce the correct product characteristics. However, from an engineering viewpoint, during baking, high temperature profiles are created in dough and batter through convective, radiative, and conductive heat transfers. At the same time, mass transfers including water diffusion, evaporation, and condensation occur. The properties of sweet good products including density, specific heat, and thermal conductivity have been reviewed by Baik et al. (2001). This chapter focuses on the heat and mass transfer mechanisms during the baking process of sweet goods. Their impact on physicochemical changes of the products will also be discussed.

9.2   HEAT TRANSFER MECHANISMS DURING BAKING

Baking is a thermal process that carries out under high temperature. Generally, heat is supplied to the product mainly from oven walls through radiative heat transfer. In addition, convective heat is transferred to the product from hot air in the oven. Within the product, conductive heat transfer is often the main mechanism.

9.2.1 CONDUCTIVE HEAT TRANSFER

It is well known that conductive heat is transferred within dough and batter by direct contact between molecules without a macro movement of the materials. Temperature gradient is the driving force for heat transfer. As illustrated in Figure 9.1, the distance between two sides of area A (m²) in the dough and batter is denoted by dx (m), and the

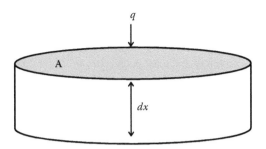

FIGURE 9.1 Conductive heat transfer within sweet good products.

corresponding temperature change across the distance is denoted by dT (°C). Heat flow rate can be described using Fourier's law as follows:

$$q = -kA \cdot \frac{dT}{dx} \tag{9.1}$$

where q is heat flow rate (W), dT/dx is temperature gradient (K·m⁻¹), and k is thermal conductivity (W·m⁻¹·K⁻¹).

Thermal conductivity is a physical property of dough and batter. It can vary considerably, depending on product composition. The thermal conductivity of dough and batter can be estimated from the quantity of constituents including water, fat, protein, carbohydrate, and ash. Singh and Heldman (2001) presented an empirical model developed by Sweat (1986) for predicting the thermal conductivity of solid and liquid food materials as follows:

$$k = 0.25 X_c + 0.155 X_p + 0.16 X_f + 0.135 X_a + 0.58 X_w \tag{9.2}$$

where X with subscripts of c, p, f, a, and w are mass fractions of carbohydrate, protein, fat, ash, and water, respectively.

During baking, the moisture content of dough and batter changes gradually and continuously; as a result, its thermal conductivity may be described as a function of moisture content (Zanoni et al., 1994):

$$k = k_d \cdot \frac{1}{1+W} + k_w \cdot \frac{W}{1+W} \tag{9.3}$$

where k_d is thermal conductivity of dry matter, whose value can be taken as 0.40 W·m⁻¹·K⁻¹; k_w is thermal conductivity of water, whose value can be taken as 0.60 W·m⁻¹·K⁻¹; and W is moisture content on dry basis (kg water/kg dry matter).

Rigorously speaking, the thermal conductivity of various constituents of dough and batter and therefore the thermal conductivity of dough and batter are also a function of temperature. However, in practice, constant values are normally used in system evaluations.

Table 9.1 lists values of the thermal conductivity of some sweet goods and bakery products that have been reported in the literature.

9.2.2 CONVECTIVE HEAT TRANSFER

Convective heat transfer involves the movement of fluids including air in the oven and water in the dough and batter. Similar to heat conduction, difference in temperature plays a key role. In a baking oven with a convective fan, heat is transferred by a forced convection mechanism which is very efficient. The heat transfer rate at the surface of dough and batter as well as at the surface of various oven parts (duct, wall, and ceiling) depends on airflow velocity.

Without a convective fan, heat is transferred from air to a solid surface by a natural convection mechanism, where the temperature difference creates a density gradient and thereby the movement of air. As such, the natural convection rate is dependent on the coefficient of thermal expansion of the fluid.

Convective heat transfer takes place when oven air at temperature T is moving past a surface at temperature T_s. At the surface, air velocity is practically zero. The area adjacent to the surface is a thin layer where the air velocity can be very low and is classified as streamline. Away from the surface, velocity increases, thereby heat transfer rate increases. From Newton's law of heating, heat flow per unit area is proportional to the temperature difference between the surface and oven air, as follows:

$$q = hA\left(T_s - T\right)$$

$$(9.4)$$

where h is the heat transfer coefficient ($W \cdot m^{-2} \cdot K^{-1}$).

According to a study on the apparent heat transfer in an oven (Sato et al., 1987), heat transfer coefficients were found to vary between 9 and 20 $W \cdot m^{-2} \cdot K^{-1}$ when the air velocity was in the range of 0.4 to 1.5 $m \cdot s^{-1}$. Watson and Harper (1988) stated that heat transfer coefficients of air under natural convection and forced convection were approximately 2.8 to 28 $W \cdot m^{-2} \cdot K^{-1}$ and 11 to 110 $W \cdot m^{-2} \cdot K^{-1}$, respectively. For viscous fluids (e.g., dough and batter) under forced convection, heat transfer coefficients are approximately 56 to 560 $W \cdot m^{-2} \cdot K^{-1}$.

Heat transfer coefficients may be estimated using dimensional analysis (Singh and Heldman, 2001). For an industrial continuous baking oven, Therdthai et al. (2003) estimated the heat transfer coefficient inside heating ducts to be at 100 $W \cdot m^{-2} \cdot K^{-1}$, and the overall heat transfer coefficient for heat loss through the oven walls and ceiling to be approximately 0.3 $W \cdot m^{-2} \cdot K^{-1}$.

9.2.3 RADIATIVE HEAT TRANSFER

Radiant heat is transferred to dough and batter as a combination of absorption, reflection, and transmission. The following equation holds:

$$\alpha + \rho_{re} + \tau = 1 \qquad (9.5)$$

where α is absorptivity, ρ_{re} is reflectivity, and τ is transmissivity.

TABLE 9.1

Thermal Conductivity of Sweet Goods and Bakery Products

Product	Temperature (°C)	Moisture Content (% wet basis)	Thermal Conductivity (W·m⁻¹·K⁻¹)	Ref.
Bread dough	−22.0	43.5	0.880	Lind (1988)
	23.0	43.5	0.460	Lind (1988); Sumnu et al. (2007)
	25.0	40.0	0.290	Sumnu et al. (2007)
Bread	25.0	40.0	0.070	Thorvaldsson and Janestad (1999); Sumnu et al. (2007)
	90.0	33.0	0.110	
French bread	Room temperature	42.0	0.0989	Sweat (1985)
White bread	25	—	0.158 ± 0.012	Tou et al. (1995)
	60–70	—	0.304 ± 0.040	Tou et al. (1995)
	80–90	—	0.353 ± 0.075	Tou et al. (1995)
Biscuit dough	29.8	4.1	0.405 ± 0.22	Kulachi and Kennedy (1978)
Biscuit	—	—	0.07–0.16	Standing (1974)
Yellow cake batter	—	41.5	0.223	Sweat (1973)
Yellow cake	—	35.5	0.121	Sweat (1973)
Cupcake batter	20 ± 1.2	34.6 ± 1.93	0.206 ± 0.007	Baik et al. (1999)
Cupcake (19.5 min baking time)	20 ± 1.2	27.6 ± 2.06	0.0683 ± 0.004	Baik et al. (1999)
Muffin	—	17	0.70	Tou and Tadano (1991)
Tortilla dough	55–75	50–60	0.0366–0.1079	Griffith (1985)
Tortilla	—	—	0.25 ± 0.020	Alvaro-Gil et al. (1995)
Tandoori roti	—	—	0.128	Saxena et al. (1995)
Chapati	58.5	43.0	0.330	Gupta (2001)

Source: Except those data from Thorvaldsson, K. and Janestad. H., *Journal of Food Engineering*, 40, 167–172, 1999; Gupta, T.R., *Journal of Food Engineering*, 47, 313–319, 2001; and Sumnu, G., Datta, A.K., Sahin, S., Keskin, S.O., and Rakesh, V., *Journal of Food Engineering*, 78, 1382–1387, 2007, all other data are adapted from Baik, O.D., Marcotte, M., Sablani, S.S., and Castaigne, F., *Critical Reviews in Food Science and Nutrition*, 41, 321–352, 2001.

For blackbodies, the absorptivity is maximized (i.e., $\alpha = 1$). However, dough or batter is not a blackbody. Emissivity of the surface of dough or batter, which is defined as the ratio of the emissive power of the surface to the emissive power of the blackbody, is assumed to be around 0.90 (De Vries et al., 1995; Thorvaldsson and Janestad, 1999) or 0.95 (Zanoni et al., 1994). The radiant heat to dough and batter can be estimated as follows:

$$q = \varepsilon \, \sigma \, A \left(T_1^4 - T_2^4 \right)$$
(9.6)

where ε is emissivity and σ is the Stefan–Boltzmann constant (5.6697×10^{-8} $W \cdot m^{-2} \cdot K^{-4}$).

9.3   MASS TRANSFER MECHANISMS DURING BAKING

During baking, water in dough and batter is transferred through pores within the dough and batter to the surface and further to the air outside the dough and batter. This causes a structure transformation from dough and batter to crumb and crust, as well as a moisture loss. The moisture loss during baking is the highest compared to those in other processing steps of sweet goods. The mass transfer mechanisms involved in the structure transformation and moisture loss include mass diffusion, evaporation, condensation, and mass convection. These mechanisms also interact with the heat transfer modes.

9.3.1   Mass Diffusion

By Fick's law, mass diffusion is driven by the concentration difference as follows:

$$\frac{m}{A} = -D \cdot \left(\frac{dC}{dx} \right)$$
(9.7)

where m is diffusion rate (kg.s^{-1}), A is area (m^2), C is concentration (kg.m^{-3}), and D is diffusivity (m^2.s^{-1}). Diffusivity is a physical property depending on temperature, pressure, and system composition.

Thorvaldsson and Janestad (1999) proposed the following model to estimate the diffusivity of water vapor (D_v) in dough as a function of temperature (T):

$$D_v = 9.0 \times 10^{-12} \cdot T^2$$
(9.8)

Typically, the diffusivities of liquid water and water vapor are approximately 1.35×10^{-10} m^2.s^{-1} and 8×10^{-7} m^2.s^{-1} at 25°C, respectively.

To count for the effect of moisture content, Zanoni et al. (1994) proposed the following model to estimate variations in the diffusivity of liquid water. When the moisture content is below 0.43,

$$D_w = W \cdot \frac{D_0}{0.43}$$
(9.9)

where W is the mass fraction of unbound water (kg water/kg dry matter), D_w is diffusivity, and D_0 is the diffusivity when the mass fraction of unbound water is more than 0.43 and its value is approximately 1.0×10^{-9} m².s⁻¹.

9.3.2 EVAPORATION ON THE SURFACE AND MASS CONVECTION

During the early stage of baking, water is vaporized from the surface of dough and batter to air using latent heat of evaporation. When the mass fraction of unbound water is more than 0.43, latent heat of evaporation (ΔH_0) is approximately 2.3339 MJ.kg⁻¹. However, during baking, unbound water is gradually decreased. When the mass fraction of unbound water (W) is reduced to less than 0.43, the latent heat (ΔH) can be estimated from the following equation:

$$\Delta H = \frac{L}{100W} + \Delta H_0 \qquad (9.10)$$

where L is 2.4948 MJ.kg⁻¹ (Zanoni et al., 1994).

During baking, there exists a concentration gradient of water vapor between the product surface and air. Therefore, water vapor is transferred through mass convection. It can be described by

$$\frac{m}{A} = k_m \cdot \left(C_s - C_b \right) \qquad (9.11)$$

where k_m is the mass transfer coefficient (m.s⁻¹), C_s is water vapor concentration on the product surface (kg.m⁻³), and C_b is water vapor concentration in bulk air (kg.m⁻³).

Similar to heat transfer coefficients, mass transfer coefficients may also be estimated through dimensional analysis (Singh and Heldman, 2001).

9.3.3 INTERNAL EVAPORATION AND CONDENSATION

During baking, temperature in outer layers of dough and batter increases first. As a result, the partial water vapor pressure in pores in those layers increases. Due to vapor pressure difference, water vapor moves toward the center through pores in inner layers. In the inner layers where temperature is low, the water vapor becomes condensed. Therefore, the liquid water content in the inner layers is increased, and a liquid water gradient is built up. The liquid water starts moving toward the surface, but it is much slower than the water vapor movement toward the center (Thorvaldsson and Janestad, 1999). Thus, the moisture content of the crumb at the center is higher than the moisture content of dough. De Vries et al. (1989) found an increase of water content by 3.5 g water/100 g bread in the loaf center immediately after baking, but eventually the water content in the loaf center was decreased to the same level as that in dough. In the case of biscuits where surface layers dry out quickly and therefore the evaporation front moves very fast toward the center, inner layers are heated up quickly. Condensation in the central area might not be significant. However, boiling (evaporation) and drying could become dominating. Furthermore, condensation might be found on the surface at the beginning of baking when steam is introduced into the oven and dough surface temperature is still low (Savoye et al., 1992).

9.4   COMBINED TRANSPORT PHENOMENA DURING BAKING

During baking, transport phenomena can be grouped based on two levels: within the baking oven and within the product. For example, in an indirect-heating baking oven, heat is supplied by hot air through ducts. After the duct surface is heated up, it generates radiant heat to dough. In addition, natural convective heat transfer takes place from air in the oven chamber to the dough surface, while forced convective heat transfer can occur using convective fans. In addition to the hot duct surfaces, all metal surfaces including oven walls and ceiling can generate radiant heat. Not only heat supply should be accounted for, but heat loss also remains as a top concern. Although baking ovens have been built with good insulation, an appreciable amount of heat loss through the walls is still observed for most of the ovens, mainly by natural convective heat transfer.

For an industrial traveling-tray baking oven, Therdthai et al. (2004a) summarized the various states of heat transfer as follows:

- Convective heat transfer from hot duct surfaces to air in the oven chamber can be calculated by

$$q_a = h_a \, A \, (T_{duct} - T_{air\ inside\ oven}) \tag{9.12}$$

  with its initial condition as $T_{air\ inside\ oven} = T_{a0}$ at $t = 0$, where T_{a0} is the initial oven air temperature. Heat transfer coefficient h_a can be calculated through dimensional analysis according to the flow status inside the oven chamber.
- At the same time, radiant heat came from all hot metal parts in the oven, traveled straight through the space, and caused localized temperature differentials. This radiant heat was calculated by

$$q_b = \sigma \, \varepsilon \, A \, [(T_A+273)^4 - (T_B+273)^4] \tag{9.13}$$

  where T_A is the temperature of the heat source (°C) and T_B is the temperature of the heat sink (°C). The corresponding initial condition was $T_B = T_{B0} = 40°C$ at $t = 0$.
- Heat loss through the oven walls which were insulated with fiberglass could be very small. It was calculated by

$$q_c = h_c \, A \, (T_{oven\ inner\ wall} - T_{air\ outside\ oven}) \tag{9.14}$$

  with its initial condition as $T_{oven\ inner\ wall} = T_{c0}$ at $t = 0$, where T_{c0} is the initial inner wall temperature of the oven. The overall heat transfer coefficient h_c from the combined conduction and convection was approximately 0.3 W·m^{-2}·°C^{-1}.

The above set of equations was solved by using computational fluid dynamics (CFD) techniques for the whole oven during the entire baking period.

For biscuits baked in a natural gas indirect-fired pilot oven, Fahloul et al. (1995) described the heat balance on the product as follows:

$$ev\rho\, C_p\, \frac{\partial T}{\partial x} = \frac{(q_{cd} + q_{cv} + q_r - q_{evaporation})}{A} \tag{9.15}$$

where e is product thickness (m); v is conveyor speed (m·s^{-1}); δ is density (kg·m^{-3}); C_p is specific heat (J·kg^{-1}·K^{-1}); q_{cd}, q_{cv}, and q_r are heat transfer rates by conduction, convection, and radiation, respectively; and $q_{evaporation}$ is the energy rate required for evaporation.

C_p was estimated by the following model (Baik et al., 2001):

$$C_p = WC_{p(water)} + (1 - W) C_{p(drysolid)} \tag{9.16}$$

where W is the mass fraction of water (kg water/kg dry matter). During baking, temperature changes continuously and thereby affects the value of specific heat. The following equations can be used for estimation (Fahloul et al., 1994):

$$C_{p(drysolid)} = 5T + 25 \tag{9.17}$$

$$C_{p(water)} = 1000 \cdot \left(5.207 - 73.17 \times 10^{-4} \cdot T + 1.35 \times 10^{-5} \cdot T^2\right) \tag{9.18}$$

Mass balance could be expressed as follows (Fahloul et al., 1994):

$$m_b \cdot \frac{dX_b}{dx} = -m_{evaporation} \tag{9.19}$$

where m_b is the rate of biscuit throughput (kg.s^{-1}), X_b is moisture content of the biscuit (%), and $m_{evaporation}$ is the rate of water evaporated (kg.s^{-1}).

Zanoni et al. (1994) attempted to describe various temperature profiles within a cylindrical bread loaf at different states of baking. The dough or bread was placed in a cylindrical mold and stood on one end. At the upper surface that was exposed directly to air, temperature gradient was a result of convection between air and the dough surface and conductive heat transfer toward the inside of the dough. Moreover, there was a convective water vapor transfer between the dough surface and air. Thus, after discretizing the solution space into grids, temperature change at the upper surface could be described by the following equation:

$$\frac{dT_s}{dt} = \left(\frac{\left(\frac{h(T_a - T_s(I,J))}{\Delta x} \right) + \frac{kd^2T}{dx^2} + \frac{kd^2T}{dr^2} + \frac{kdT}{dr}\frac{1}{r}}{\rho C_p} \right) -$$

$$\frac{k_m}{\rho C_p \Delta x} \left(P_s[T,W] - P_a[T,W] \right) \Delta H[T,W]$$

$$(9.20)$$

where I, J are grid coordinators; T is temperature (K); t is time (s); x is height (m); r is radius (m); h is heat transfer coefficient (W·m^{-2}·K^{-1}); k is thermal conductivity (W·m^{-1}·K^{-1}); Δx is infinitesimal height interval (m); k_m is mass transfer coefficient (kg.s^{-1}.m^{-2}.Pa^{-1}); ρ is density (kg·m^{-3}); C_p is specific heat (J·kg^{-1}·K^{-1}); P is vapor pressure (Pa); P_s is the vapor pressure at the surface; P_a is the vapor pressure in the surrounding air; T_a is the air temperature; and T_s is the surface temperature.

Once the surface temperature reached 100°C, if there was enough liquid water on the surface, evaporation would take place at a constant temperature of 100°C. Meanwhile, the surface temperature would also remain at this constant temperature.

Due to evaporation, the moisture content at the surface continuously decreased. Crust was subsequently formed, and the surface temperature would increase toward the oven temperature. At this stage, the temperature profile at the upper surface could be expressed as follows:

$$\frac{dT_s}{dt} = \left(\frac{\left(\frac{h(T_a - T_s(I,J))}{\Delta x} \right) + \frac{kd^2T}{dx^2} + \frac{kd^2T}{dr^2} + \frac{kdT}{dr}\frac{1}{r}}{\rho C_p} \right)$$

$$(9.21)$$

where I, J are grid coordinators; T is temperature (K); t is time (s); x is height (m); r is radius (m); h is heat transfer coefficient (W·m^{-2}·K^{-1}); k is thermal conductivity (W·m^{-1}·K^{-1}); Δx is infinitesimal height interval (m); ρ is density (kg·m^{-3}); C_p is specific heat (J·kg^{-1}·K^{-1}); T_a is the air temperature; and T_s is the surface temperature.

Meanwhile, the interior of the dough was heated by conductive heat transfer in accordance with Fourier's law:

$$\frac{dT}{dt} = \frac{k}{\rho C_p} \left[\frac{d^2T}{dx^2} + \frac{d^2T}{dr^2} + \frac{dT}{dr}\frac{1}{r} \right] \qquad (9.22)$$

where T is temperature (K), t is time (s), x is height (m), r is radius (m), k is thermal conductivity (W·m^{-1}·K^{-1}), ρ is density (kg·m^{-3}), and C_p is specific heat (J·kg^{-1}·K^{-1}).

Similarly to the dough surface, when temperature at an inner dough position reached 100°C, evaporation would take place at constant temperature. Therefore, the temperature at that inner position would not change any more until the moisture at the position totally dried out.

Regarding moisture change during baking, Zanoni et al. (1994) described it according to temperature profiles as follows.

At the beginning, mass transfer at the upper surface of the dough included a convective mass transfer between air and the dough surface and a mass diffusion from the inner layers toward the surface. The moisture profile could be described by

$$\frac{dW_s}{dt} = \frac{Dd^2W}{dx^2} + \frac{Dd^2W}{dr^2} + \frac{DdW}{dr}\frac{1}{r} - \frac{k_m}{\rho \Delta x}\left(P_s\left[T,W\right] - P_a\left[T,W\right]\right) \quad (9.23)$$

where W is absolute moisture (kg water/kg dry matter), W_s is the absolute moisture at the surface, t is time (s), x is height (m), r is radius (m), D is diffusivity (m^2·s^{-1}), Δx is infinitesimal height interval (m), k_m is mass transfer coefficient (kg·s^{-1}·m^{-2}·Pa^{-1}), ρ is density (kg·m^{-3}), P is vapor pressure (Pa), P_s is the vapor pressure at the surface, and P_a is the vapor pressure in the surrounding air.

When surface temperature reached 100°C, evaporation took place at constant temperature. The moisture content could then be described by

$$\frac{dW_s}{dt} = \frac{Dd^2W}{dx^2} + \frac{Dd^2W}{dr^2} + \frac{DdW}{dr}\frac{1}{r} - \left(\frac{\dfrac{\left(h(T_a - T_s(I,J))\right)}{\Delta x} + \dfrac{kd^2T}{dx^2} + \dfrac{kd^2T}{dr^2} + \dfrac{kT}{dr}\dfrac{1}{r}}{\rho \Delta H\left[T,W\right]}\right)$$

$$(9.24)$$

where W is absolute moisture (kg water/kg dry matter), W_s is the absolute moisture at the surface, T is temperature (K), t is time (s), x is height (m), r is radius (m), D is diffusivity (m^2·s^{-1}), h is heat transfer coefficient (W·m^{-2}·K^{-1}), k is thermal conductivity (W·m^{-1}·K^{-1}), Δx is infinitesimal height interval (m), δ is density (kg·m^{-3}), ΔH is latent heat of evaporation (J.kg^{-1}), T_a is the air temperature, and T_s is the surface temperature.

When the evaporation front moved toward the inside (i.e., a dried surface), the surface temperature increased toward the oven temperature. At this stage, the moisture content at the surface could be described by

$$\frac{dW_s}{dt} = \frac{Dd^2W}{dx^2} + \frac{Dd^2W}{dr^2} + \frac{DdW}{dr}\frac{1}{r} \quad (9.25)$$

where W is absolute moisture (kg water/kg dry matter), W_s is the absolute moisture at the surface, t is time (s), x is height (m), r is radius (m), and D is diffusivity (m^2·s^{-1}).

For the interior of the dough, moisture content was normally determined according to Fick's law:

$$\frac{dW}{dt} = \frac{Dd^2W}{dx^2} + \frac{Dd^2W}{dr^2} + \frac{DdW}{dr}\frac{1}{r} \quad (9.26)$$

However, when temperature at an inner dough position reached 100°C so that evaporation at constant temperature happened, moisture content at that position could be described by

$$\frac{dW}{dt} = \left(\frac{Dd^2W}{dx^2} + \frac{Dd^2W}{dr^2} + \frac{DdW}{dr}\frac{1}{r} \right) - \frac{k}{\rho\Delta H[T,W]}\left[\frac{d^2T}{dx^2} + \frac{d^2T}{dr^2} + \frac{dT}{dr}\frac{1}{r} \right] \quad (9.27)$$

where W is absolute moisture (kg water/kg dry matter), t is time (s), x is height (m), r is radius (m), D is diffusivity (m^2.s^{-1}), k is thermal conductivity (W·m^{-1}·K^{-1}), ρ is density (kg·m^{-3}), ΔH is latent heat of evaporation (J.kg^{-1}), and T is temperature (K).

It is worth noting that the internal evaporation and condensation mechanism described in Section 9.3.3 was not considered in the above equations.

9.5   IMPACT OF HEAT AND MASS TRANSFER DURING BAKING ON PRODUCT CHARACTERISTICS

Among the whole production procedure of sweet goods, baking is the key step that develops the product characteristics, including color, texture, and flavor. For bread, baking is a process to transform dough into crumb and crust structures. For cakes, baking is a process to transform batter which contains a foam structure into a sponge structure (Cauvain and Young, 2001). The development involves several mechanisms including nonenzymatic browning reactions, starch gelatinization, and protein denaturation. Previous studies have focused on nonenzymatic browning reactions to develop color and flavor, as well as starch gelatinization and protein denaturation to develop structure and texture.

9.5.1   IMPACT ON VOLUME EXPANSION

During bread baking, volume expansion is mainly due to two mechanisms: yeast and water vapor. At temperatures below 55°C, yeast converts sugar into carbon dioxide and thereby volume expands. Due to heat transfer, dough temperature is increased. At temperatures above 55°C, yeast is inactivated. However, dough volume still increases because of increased water vapor pressure. For dough that is heated quickly at the beginning of baking, its volume expansion can be stopped at an early stage. Indeed, if crust is formed too early, it will block mass transfer from inside layers to outer layers. The resultant volume can be smaller than that desired. In addition, crust formation seems to play an important role in the formation of crumb structure in terms of porosity. With a restricted total volume confined by the crust, dough may continuously expand locally to the detriment of mechanically weak areas; as a result, the porosity in those areas was decreased (Zhang et al., 2007). Therefore, the rate of heat transfer to dough at an early baking stage should be of great concern in regard to volume expansion. However, it is worth mentioning that volume expansion during baking is generally smaller than that during proofing.

In cake baking, batter is converted to a product with desired eating characteristics. For cakes containing raising agents, when batter enters the oven and is heated, air cells in the batter begin to expand as carbon dioxide is released to inflate the cells. This phenomenon happens at first in the outer layers and extends to the inner batter. The rising of volume continues until the structure is set by starch gelatinization. Cell expansion and starch gelatinization transform the batter into cake crumb structure containing interconnected cells. For the skin, its formation rate is dependent on baking temperature. The skin of cake can be quickly formed in the presence of a high baking temperature. However, the early formed crust is not strong enough to prevent volume expansion in the case of cakes, particularly when a large amount of baking powder is used (Cauvain and Young, 2000).

9.5.2 IMPACT ON CRUMB AND CRUST DEVELOPMENT AND BAKING LOSS

Due to heat and mass transfer at the dough surface, surface moisture content could be significantly reduced. As a result, a relatively hard layer is formed which becomes crust or skin. Generally 30 min baking of a small loaf may create a 3 mm thickness of crust. If increasing the baking time to 50 min for a big loaf, crust thickness may increase to 5 mm. The crust formation rate seems to be linear with respect to time (Wiggins, 1999). When a lower temperature is applied, a longer baking time is required, which causes formation of a thicker crust.

After crust is formed, heat is still transferred to the dough which can increase the evaporation rate. However, moisture from inner layers is hardly removed. This is due to the fact that the crust acts as a barrier to block mass transfer from inner layers to outer layers. Therefore, condensation at the center of the loaf is observed, and moisture content in the inner layers is much higher than that in the crust. According to a study on water diffusion in bread during baking (Thorvaldsson and Skjoldebrand, 1998), the moisture content of crust was found to be 0 ± 2.1 g water/100 g dough, whereas the moisture content of center crumb was 49.6 ± 5.4 g water/100 g dough. Moisture loss during bread baking was mainly from the crust part including top crust, side crust, and bottom crust, as shown in Figure 9.2. For a typical bread loaf of 800 g, weight loss during baking was found to be in the range of 50 to 55 g. For some types of bread which have thicker crust, higher weight loss can be expected (Wiggins, 1999).

In addition to baking time, increasing baking temperature tends to increase the rate of moisture loss in cakes (Cauvain and Young, 2001). Similarly, Fahloul et al. (1994) demonstrated that the moisture content in bis-

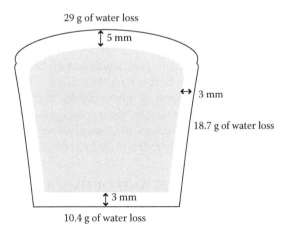

29 g of water loss

5 mm

3 mm

18.7 g of water loss

3 mm

10.4 g of water loss

FIGURE 9.2 Moisture loss during baking. (Data from Thorvaldsson, K. and Skjoldebrand, C., *Lebensmittel Wissenschaft und Technologie*, 31, 658–663, 1998.

cuits decreased from 0.24 g/g dry matter to 0.10, 0.05, and nearly 0 g/g dry matter after baking at 150°C, 200°C, and 250°C, respectively, for 6 to 8 min.

9.5.3 IMPACT ON CRUST COLOR, GLOSS, AND FLAVOR DEVELOPMENT

Crust color was developed through nonenzymatic reactions including Maillard reactions and caramelization. Both mechanisms are involved in thermal processes. Maillard reactions were between amino acids and reducing sugars under a suitable temperature that normally needs to be above 50°C (Villamiel, 2006). It can be stimulated when moisture is decreased. That means crust color can be observed as soon as evaporation at the dough surface is completed and the surface temperature increases toward the oven air temperature. When baking time is longer, the moisture content at the surface is getting lower, and the surface temperature is getting higher. Thus, crust color intensity can be increased. However, the increase in degree of browning is not linear with respect to increase in temperature due to radiant heat from metal walls in the oven. When dough gets browner, its emissivity becomes higher. Subsequently, the surface temperature is increased more quickly. Not only does higher surface temperature stimulate Maillard reactions, but it also causes caramelization. Therefore, at the late baking stage, color development rate is enhanced. There is also a similar problem as to volume expansion when a rapid heating rate is applied at the beginning of baking. A rapid heating to dough can increase the mass transfer rate and thereby moisture loss. As a result, Maillard reactions start early, which may yield the color of baked goods being too dark.

For cakes, using a lower baking temperature might cause a longer setting time, and accordingly, the texture becomes dry. Drying out can concentrate sucrose in the unset portion of a cake, resulting in caramelization. As a result, colored crust becomes thicker. In addition, the inner crumb layer that is not in contact with the oven steam becomes dry and discolored (Cauvain and Young, 2000).

To develop a gloss crust, steam is required during the first few seconds of the baking process. Gloss development needs vapor condensation on the crust surface to form a starch paste. With a minimum oven temperature of 74°C for sufficient time (optional condition is at 77°C for 10 min or 99°C for 15 sec), the starch paste can gelatinize, form dextrin, and start caramelization (Wiggins, 1999). Therefore, both color and gloss can be found on the surface.

Flavor in the form of *n*-heterocycles is developed through Maillard reactions. Major compounds found in wheat crust are 2-acetyl-1-pyroline and 2-acetyltetrahydropyridine. Flavor compounds are absorbed into the pore structure of crumb and are blocked from dispersing to air by crust (Zhou and Therdthai, 2007).

9.6 MODELING AND OPTIMIZATION OF HEAT AND MASS TRANSFER BAKING OF SWEET GOODS

Baking is a complex process that transforms dough or batter into rigid products. Many studies have been conducted to develop mathematical models in order to simulate and better understand the various phenomena during baking. The studied phenomena during baking could be broadly divided into two categories: inside the product (micro) and inside the oven (macro). This section reviews the previous studies involving baking models.

9.6.1 MODELING OF HEAT AND MASS TRANSFER AT PRODUCT LEVEL

Heat and mass transfer within bakery products have been modeled mainly to explain crust and crumb development. The progress of crust layer toward the inner region could be successfully simulated. When heat is applied to dough, water at the warmer side of a grain (pore) that absorbs latent heat of vaporization could be evaporated. Some vapor migrates to a cooler area within the loaf and becomes condensate, while some vapor escapes to the oven chamber (De Vries et al., 1989). Zanoni et al. (1993) set 100°C as the evaporation front temperature. As a result, unbound water at the surface would be evaporated through water boiling phenomenon. This mechanism could be accelerated, when water vapor pressure in the oven air is far from saturation (Eliasson and Larsson, 1993). Then crust, which is a dried layer, will be formed and separated from crumb that is moist.

Mathematical models could also be established to simulate temperature and moisture changes during baking, such as those described in Section 9.4. For mass transfer, mathematical models have been established to simulate water migration within the dough (Thorvaldsson and Janestad, 1999; Thorvaldsson and Skjoldebrand, 1998; Tong and Lund, 1993; Zanoni et al., 1994; Zhou, 2005). It was found that moisture loss was mainly from the top surface which is exposed directly to oven air. Moisture loss through the side was similar to that through the bottom (Thorvaldsson and Skjoldebrand, 1998).

9.6.2 MODELING OF HEAT AND MASS TRANSFER IN OVEN CHAMBER

Within a baking oven chamber, fluids including gas and water vapor circulate and transfer heat to dough and batter. However, different oven designs have different dominating heat transfer modes and patterns. Typical baking ovens significantly rely on radiation heat transfer from all metal pieces in the chamber (Velthuis et al., 1993). In ovens for Indian flat bread, conduction seems to be the most important heat transfer mode (Gupta, 2001). Mathematical models at the chamber level have been established for batch ovens (Tong and Lund, 1993) and continuous ovens (Fahloul et al., 1995; Gupta, 2001; Savoye et al., 1992). Due to the complexity of baking ovens, CFD has been used to simulate flow phenomena during baking (De Vries et al., 1995; Therdthai et al., 2003, 2004a; Wong et al., 2007a). In addition to CFD models, neural network models, which are blackbox models, could also be used (Kim and Cho, 1997).

To obtain a high-quality bakery product, optimization of heat and mass transfer is necessary. The impacts of process parameters on quality attributes have been studied and modeled, including crust color development (Tan and Zhou, 2003; Zanoni et al., 1995a; Zhou and Tan, 2005), starch gelatinization (Therdthai et al., 2004b; Zanoni et al., 1995b), crust thickness (Zanoni et al., 1994), volume expansion (Fan et al., 1999), and moisture loss (Thorvaldsson and Janestad, 1999; Thorvaldsson and Skjoldebrand, 1998). Using mathematical tools, optimizing temperature profiles and designing the corresponding optimum operating condition could be achieved

(Therdthai et al., 2002, 2004a). To maintain the oven operating condition at the optimum level, a good process control system should be in place to ensure the consistency of product quality (Trystram, 1997; Wong et al., 2007b).

9.7   CONCLUSIONS

During baking, heat and mass transfer play the most important role in developing physical and chemical changes in the bakery products. Dough is transformed to a product consisting of crumb and crust or skin, due to heat and mass transfer within dough as well as within the oven chamber.

Design of different baking ovens provides various combinations of heat transfer modes, including convection, radiation, and conduction, and thereby various temperature profiles in the product during baking. The interaction between rate of mass transfer and rate of heat transfer depends on many factors including mode of heat transfer, type of product, and type of oven. Due to variation in temperature profiles and mass transfer rates, the obtained baked products exhibit different characteristics including crust color, crust thickness, volume, structure, texture, and flavor, resulting in varieties of sweet goods.

The modeling of heat and mass transfer within dough and oven chamber has been studied by many researchers where various phenomena during baking have been simulated. With the knowledge of the impact of heat and mass transfer on the characteristics of products, mathematical models have been used for optimizing oven operating conditions in order to obtain good product quality and high process efficiency.

REFERENCES

Baik, O.D., M. Marcotte, S.S. Sablani, and F. Castaigne. 2001. Thermal and physical properties of bakery products. *Critical Reviews in Food Science and Nutrition* 41:321–352.
Cauvain, S.P. and L.S. Young. 2000. *Bakery Food Manufacture and Quality: Water Control and Effects*. Blackwell Science Ltd., Oxford.
Cauvain, S.P. and L.S. Young. 2001. *Baking Problems Solved*. Woodhead Publishing Limited, England.
Conforti, F.D. 2006. Cake manufacture. In *Bakery Products Science and Technology*, Ed. Y.H. Hui, 393–410. Blackwell, Oxford.
De Vries, U., H. Velthuis, and K. Koster. 1995. Baking ovens and product quality—A computer model. *Food Science and Technology Today* 9:232–234.
De Vries, U., P. Sluimer, and A.H. Bloksma. 1989. A quantitative model for heat transport in dough and crumb during baking. In *Cereal Science and Technology* Proceedings from an International Symposium, June 13–16, 1988. Ysad, Sweden. Ed. N.-G. Asp, 174–188. Stockholm, Sweden..
Eliasson, A-C. and K. Larsson. 1993. *Cereals in Breadmaking: A Molecular Colloidal Approach*. Marcel Dekker, USA.
Fahloul, D., G. Trystram, A. Duquenoy, and I. Barbotteau. 1994. Modelling heat and mass transfer in band oven biscuit baking. *Lebensmittel Wissenschaft und Technologie* 27:119–124.
Fahloul, D., G. Trystram, I. McFarlane, and A. Duquenoy. 1995. Measurement and predictive modeling of heat fluxes in continuous baking ovens. *Journal of Food Engineering* 26:469–479.

Fan, T., J.R. Mitchell, and J.M.V. Blanshard. 1999. A model for the oven rise of dough during baking. *Journal of Food Engineering* 41:69–77.

Gupta, T.R. 2001. Individual heat transfer modes during contact baking of Indian unleavened flat bread (*chapati*) in a continuous oven. *Journal of Food Engineering* 47:313–319.

Kim, S. and S.I. Cho. 1997. Neural network modeling and fuzzy control simulation for bread-baking process. *Transactions of the ASAE* 40:671–676.

Sato, H., T. Matsumura, and S. Shibukawa. 1987. Apparent heat transfer in a forced convection oven and properties of baked food. *Journal of Food Science* 52:185–193.

Savoye, I., G. Trystram, A. Duquenoy, P. Brunet, and F. Marchin. 1992. Heat and mass transfer dynamic modeling of an indirect biscuit baking tunnel-oven. Part I: Modelling principles. *Journal of Food Engineering* 16:173–196.

Singh, R.P. and D.R. Heldman. 2001. *Introduction to Food Engineering*, 3rd ed., Food Science and Technology. Academic Press, New York.

Sumnu, G., A.K. Datta, S. Sahin, S.O. Keskin, and V. Rakesh. 2007. Transport and related properties of breads baked using various heating modes. *Journal of Food Engineering* 78:1382–1387.

Sweat, V.E. 1986. Thermal properties of foods. In *Engineering Properties of Foods*, Eds. M.A. Rao and S.S.H. Rizvi, 49–87. Marcel Dekker, New York.

Tan, A. and W. Zhou. 2003. Color development of bread during baking. In *Proceedings of the 8th ASEAN Food Conference*, Hanoi.

Therdthai, N., W. Zhou, and T. Adamczak. 2002. Optimization of temperature profile in bread baking. *Journal of Food Engineering* 55:41–48.

Therdthai, N., W. Zhou, and T. Adamczak. 2003. Two-dimensional CFD modeling and simulation of an industrial continuous bread baking oven. *Journal of Food Engineering* 60:211–217.

Therdthai, N., W. Zhou, and T. Adamczak. 2004a. Three-dimensional CFD modeling and simulation of the temperature profiles and airflow patterns during a continuous industrial baking process. *Journal of Food Engineering* 65:599–608.

Therdthai, N., W. Zhou, and T. Adamczak. 2004b. Simulation of starch gelatinization during baking in a travelling-tray oven by integrating a three-dimensional CFD model with a kinetic model. *Journal of Food Engineering* 65:543–550.

Thorvaldsson, K. and C. Skjoldebrand. 1998. Water diffusion in bread during baking. *Lebensmittel Wissenschaft und Technologie* 31:658–663.

Thorvaldsson, K. and H. Janestad. 1999. A model for simultaneous heat, water and vapour diffusion. *Journal of Food Engineering*. 40:167–172.

Tong, C.H. and D.B. Lund. 1993. Microwave heating of baked dough products with simultaneous heat and moisture transfer. *Journal of Food Engineering* 19:319–339.

Trystram, G. 1997. Computerized process control for the bakery/cereal industry. In *Computerized Control Systems in the Food Industry*, Ed. G.S. Mittal, 491. Marcel Dekker, New York. 491–512.

Velthuis, H., A. Dalhuijsen, and U. De Vries. 1993. Baking ovens and product. In *Food Technology International Europe*, Ed. A. Turner, 61–66. Sterling Publications, London.

Villamiel, M. 2006. Nonenzymatic browning for cookies, crackers, and biscuits, In *Bakery Products Science and Technology*, Ed. Y.H. Hui, 433–442. Blackwell, Oxford.

Watson, E.L. and J.C. Harper. 1988. *Elements of Food Engineering*, 2nd ed. AVI, New York.

Wiggins, C. 1999. Proving, baking and cooling. In *Technology of Breadmaking*, Ed. S.P. Cauvain and L.S. Young, 120–148. Aspen, Gaithersburg, MD.

Wong, S.-Y., W. Zhou, and J. Hua. 2007a. CFD modeling of an industrial continuous bread-baking process involving U-movement. *Journal of Food Engineering* 78:888–896.

Wong, S.-Y., W. Zhou, and J. Hua, 2007b. Designing process controller for a continuous bread baking process based on CFD modeling. *Journal of Food Engineering*. 81:523–534.

Yoneya, T. and W-K. Nip. 2006. Cracker manufacture. In *Bakery Products Science and Technology*, Ed. Y.H. Hui, 411–432. Blackwell, Oxford.

Zanoni, B., C. Peri, and S. Pierucci. 1993. A study of the bread-baking process. I. A phenomenological model. *Journal of Food Engineering* 19:389–398.

Zanoni, B., C. Peri, and D. Bruno. 1995a. Modeling of browning kinetics of bread crust during baking. *Lebensmittel-Wissenschaft und-Technologie* 28:604–609.

Zanoni, B., C. Peri, and D. Bruno. 1995b. Modeling of starch gelatinization kinetics of bread crumb during baking. *Lebensmittel-Wissenschaft und-Technologie* 28:314–318.

Zanoni, B., S. Pierucci, and C. Peri. 1994. Study of bread baking process. II. Mathematical modeling. *Journal of Food Engineering* 23:321–336.

Zhang, L., T. Lucas, C. Doursat, D. Flick, and M. Wagner. 2007. Effects of crust constraints on bread expansion and CO_2 release. *Journal of Food Engineering* 80:1302–1311.

Zhou, W. 2005. Application of FDM and FEM to solving the simultaneous heat and moisture transfer inside bread during baking. *International Journal of Computational Fluid Dynamics* 19:73–77.

Zhou, W. and M.Y. Tan. 2005. Prediction of color development during bread baking, In *Proceedings of the 2nd International Conference on Innovations in Food Processing Technology and Engineering*, Bangkok.

Zhou, W. and N. Therdthai. 2007. Three-dimensional CFD modeling of a continuous industrial baking process. In *Computational Fluid Dynamics in Food Processing*, Ed. D.-W. Sun, 287–312. Taylor & Francis, London.

10 Physical and Thermal Properties of Sweet Goods

Shyam S. Sablani

CONTENTS

10.1   INTRODUCTION

In the process of baking, heat is transferred primarily by convection from the heating medium and by radiation from oven walls to the product surface. This is followed by conduction to the geometric center. At the same time, moisture diffuses outward to the product surface. The temperature and moisture distribution within the porous structure can be predicted using appropriate diffusion equations of heat and moisture transport. In order to predict the temperature and moisture distribution in the product during baking, knowledge of the product properties is needed, including physical, thermal, and moisture transport as a function of processing conditions (Rask, 1989; Sablani et al., 1998). These properties are apparent density (or specific volume), specific heat, thermal conductivity, thermal diffusivity, and moisture diffusivity.

Mathematical modeling and computer simulation based on numerical analysis have become the main tools for understanding and predicting processing phenomena. With the advent of high-speed computers and inexpensive memory, physical, thermal, and moisture transport properties of products can be treated as time- or temperature-dependent variables instead of average values for the whole process (McFarlane, 2006; Sablani et al., 1998). In addition, during the baking process, a series of physical, chemical, and biochemical changes occur in a product. These changes include volume expansion, evaporation of water, formation of a porous structure, denaturation of protein, gelatinization of starch, formation of crust, and browning reaction. A fundamental understanding of such physical, chemical, and biochemical changes will be particularly useful in the development of complete mathematical models of the baking process.

The composition of food properties affects physical, thermal, and moisture transport properties of dough and batter. In recent years, there has been worldwide interest in foods containing substances with biological activity related to disease prevention and health promotion. Food industries have been launching products with functional components. Galdeano and Grossmann (2006) demonstrated that oat hulls modified by treatment with alkaline hydrogen peroxide associated with extrusion can be used in the preparation of cookies, without damage to sensory quality. Some researchers attempted to use extruded orange pulp and soapwort extract to improve nutritional values and replace egg white (Larrea et al., 2005; Celik et al., 2007). Ronda et al. (2005) used polyols and nondigestible oligosaccharides in place of sucrose to reduce the calorie content of sponge cake. Several studies have been carried out showing the potential use of hydrocolloids in bakery products to provide dietary fiber and to impart specific functional properties such as retarding starch degradation and enhancing texture and moisture retention (Gomez et al., 2007). These ingredients also influence the physical, thermal, and moisture transport properties of dough and batter. The prediction of such properties as a function of the chemical composition and process conditions can be very useful in mathematical analyses of heat and moisture transport.

Bakery products include varieties of breads such as white bread, tortilla, tandoori roti, chapati, and sweet products such as cakes, muffins, biscuits, doughnuts, and cookies. Information about the thermophysical properties of dough and bakery products during baking is scarce when compared to that available for fruits, vegetables, and meat products. Moreover, information at various temperature, moisture, and density levels is not always readily available. Rask (1989) reviewed data on thermal properties of bakery products and prediction models, and Lind (1991) presented measurement techniques and models of thermal properties of dough during freezing and thawing. Baik et al. (2001) presented a comprehensive review of the measurement techniques, prediction models, and data on thermophysical properties of bread and nonbread products.

This chapter focuses on measurement techniques and prediction models on physical, thermal, and moisture transport properties of sweet bakery products such as cakes, biscuits, muffins, and cookies. Data on physical and thermal properties of sweet baked products at different temperatures and baking conditions are also given.

10.2   MEASUREMENT TECHNIQUES

The measurements of thermophysical properties require basic knowledge of heat and moisture transport. Mohsenin (1980), Ohlsson (1983), Murakami and Okos (1989), Rahman (1995), and recently Nasvada (2005) presented a good description of methods used for the measurement of thermal properties of food samples. Baik et al. (2001) reviewed measurement techniques applicable to bakery products. The following section discusses only those used for sweet bakery products such as cakes, biscuits, muffins, and cookies.

10.2.1 SPECIFIC HEAT

Most commonly used methods for the determination of specific heat (C_p) of sweet bakery products are the mixing method or differential scanning calorimeter. The selection of the proper measurement technique depends upon the nature and size of samples and the temperature range (Baik et al., 2001).

10.2.1.1   Mixing Method

In this method, a sample and water, of known masses, are mixed in a calorimeter at predetermined temperatures. Once the sample and water reach equilibrium temperature (T_e), the C_p is calculated from the following heat balance equation:

$$m_{cal}\, C_{p,cal}\left(T_{i,cal} - T_e\right) + m_{samp} C_{p,samp}\left(T_{i,samp} - T_e\right) = m_w C_{p,w}\left(T_e - T_{i,w}\right) + Q_{loss} \tag{10.1}$$

where m is the mass; C_p is specific heat; and subscripts *cal*, *samp*, and *w* are calorimeter, sample, and water, respectively. T_i and T_e are the initial and equilibrium temperatures, and Q_{loss} is the energy loss from or to the surroundings, which may be positive or negative. Thermal insulation is generally provided to reduce heat loss from or to the surroundings. If the sample is dissolved in water, the enthalpy of solution should be taken into account in the heat balance equation. In this case, an indirect method can be used to avoid direct contact with water. The main advantages of this method are that application is simple and that large samples can be used. The method is particularly suitable for heterogeneous foods. The measuring times can, however, be long. This method gives a mean value of C_p over a temperature range. Therefore, the technique is not suitable for the measurement of temperature-dependent specific heat. The experimental uncertainty in the measured specific heat of cookie dough using this method was in the order of 5%, which includes the uncertainties in the temperature measurements, the mass of dough, the mass of water, and the heat loss (Kulacki and Kennedy, 1978).

10.2.1.2 Differential Scanning Calorimeter (DSC)

The specific heat of homogeneous materials at specific temperatures over a wide temperature range can be determined using DSC. Nonhomogeneous samples require several replications due to the small sample size (mg) used. The amount of energy needed to change the temperature of the sample is compared with the energy needed to change the temperature of a reference material at the same rate (Baik et al., 2001; Rahman, 1995). DSC can be used either for scanning over a wide temperature range or for stepwise (isothermal) measurements where the temperature is altered in small steps. Recently, a modulated DSC (MDSC) technique was used to measure values of C_p more accurately. In MDSC, a material is exposed to a linear heating process that has a superimposed sinusoidal oscillation (temperature modulation), resulting in a cyclic heating profile. Temperature modulation (sinusoidal oscillation) of the MDSC separates total heat flow into its reversing (specific heat related) and nonreversing (kinetic) components. Thus, specific heat values obtained by MDSC are more precise than those obtained with conventional DSC. Heating rate, modulation period, and amplitude are the main variables in the measurement of C_p with MDSC (Baik et al., 1999, 2001).

10.2.2 THERMAL CONDUCTIVITY

The measurement methods of thermal conductivity (k) can be classified as steady-state and transient methods. Steady-state methods are not suitable for the assessment of thermal conductivity of bakery products due to their longer test times, which can result in moisture migration, and property changes due to long exposure to high temperatures. Transient techniques are more accepted and appropriate because testing is very fast and yields accurate results.

10.2.2.1 Steady-State Techniques

10.2.2.1.1 Guarded Hot Plate Method
In this method, the sample is placed between a heat source and a heat sink. A time-independent heat flow is generated (ASTM, 1955). This system is mathematically simple to process and easy to control experimentally. The k value can be calculated using Fourier's heat conduction equation. The estimated value of k is taken as a mean value measured over the temperature interval used in the experiment. Kulacki and Kennedy (1978) reported an experimental uncertainty of <7.4% in the measured thermal conductivity of biscuit dough. They attributed this deviation to instrumentation errors, geometrical uncertainties, and deviations from the assumed one-dimensional nature of heat flow.

10.2.2.1.2   DSC Attachment Method

Buhari and Singh (1993) used an attachment to a DSC for the determination of thermal conductivity. The main advantages of this method were relatively short duration of measurement (10 to 15 min), small sample size, and no moisture loss in the samples. The thermocouple probe was used for the measurement of the sample temperature. The DSC heating pan temperature was kept at 40°C. After 5 min, the initial temperature of the sample was recorded. The pan temperature was then immediately increased by 10°C. After 10 to 15 min, a new steady state existed, and the final sample temperature was recorded. Then k was obtained using the following equation, which is based on Fourier's heat conduction equation:

$$k = \frac{L \Delta Q}{A(\Delta T_2 - \Delta T_1)} \tag{10.2}$$

where L is the sample length, ΔQ is the difference of energy required to maintain pan temperature, A is the sample area perpendicular to heat flow, ΔT_2 is the final temperature difference between DSC heating pan and sample, ΔT_1 is the initial temperature difference between DSC heating pan and sample (Baik et al., 2001).

10.2.2.1.3   Capped Column Test Device

Zhou et al. (1994) built a capped column test device to measure thermal conductivity. A constant steady heat flux was applied on the sample during the experiment. The sample was placed in the metal cylinder capped from both sides, and there was no moisture loss. Steady heat flux was provided by circulating hot and cold water at constant temperatures at the two ends of the cylindrical test sample (diameter 3 cm, height 5 cm, and 2.5 cm). The cylinder containing sample was enclosed in polystyrene foam to minimize heat loss to the surroundings. The capped column test device was kept horizontal during the experiment to eliminate the gravity-induced migration of moisture within the sample. The steady flux resulted in temperature and moisture gradients in the sample. Once the steady state was reached, temperatures at several locations along the height of the test sample and water stream were measured. Then the sample was cut into several sections of equal heights to determine the moisture content distribution by measuring the moisture content of each section. The k value can be determined by applying simultaneous heat and moisture transfer equations to the experimental temperature and moisture gradient data. The determination is fast (within several minutes). There is no concern about experimental deviation due to a moisture gradient, because the device and data analysis are designed to evaluate thermal and mass transfer properties simultaneously (Baik et al., 2001).

10.2.2.2   Transient Techniques

In the transient methods, the sample is subjected to a time-dependent heat flow. Temperature is measured at one or more points within the sample or at its surface. The method recommended for most food applications, including bakery products, is the line heat source probe. The technique is simple and fast (i.e., 3 to 600 s) and requires relatively small samples. However, it does require a data acquisition system. The probe (0.66 mm outside diameter) consists of a constantan heater wire and chromel-constantan thermocouple wire.

The basic theory behind the use of the line heat source probe was previously discussed by several authors (Nasvada, 2005; Rahman, 1995). A line heat source probe is embedded in the sample (regarded as an infinite body), which is initially at a uniform temperature, resulting in a radial temperature distribution. After equilibrium is reached, the probe heater is energized. Heating and temperatures are monitored simultaneously. The rate of temperature rise of the heater is directly related to the sample's conductivity. The slope of the linear portion of each data set was used to determine effective thermal conductivity by Equation 10.3:

$$k = \frac{Q}{4\pi}\left[\frac{\ln(t_2 / t_1)}{T_2 - T_1}\right] \tag{10.3}$$

where k is the thermal conductivity of the sample; t_1 is the initial time when the probe heater was energized; t_2 is the final time since probe heater was energized; T_1 is the temperature of the probe thermocouple at time t_1; T_2 is the temperature of the probe thermocouple at time t_2; and Q is the heat flux generated by the probe heater.

In order to obtain correct results in a line heat source probe method, probe size and sample size should be carefully selected. Inserting a probe into an unstable structure, as done in dough and partly baked products, can also introduce errors. Rupture of the structure close to the probe can give rise to false values. Thus, the use of a linear movement probe holder is strongly recommended for k measurement using this technique. Other transient methods such as temperature history and transient hot strip methods have been used to measure thermal conductivity of tortilla and bread dough (Baik et al., 2001).

10.2.3 THERMAL DIFFUSIVITY

10.2.3.1 Indirect or Calculation Method

Thermal diffusivity (α) can be estimated from experimentally measured values of thermal conductivity, specific heat, and density (ρ). This is a preferred way of determining thermal diffusivity. In this scheme, the estimation deviation will depend mainly on the measurement of thermal conductivity, specific heat, and density (Baik et al., 2001).

$$\alpha = \frac{k}{\rho C_p} \tag{10.4}$$

10.2.3.2 Temperature History

This is the most widely used experimental system for measurement of thermal diffusivity of bakery products. The experimental apparatus and analytical solution of transient heat transfer were described by Dickerson (1965) and Rahman (1995).

In this technique, transient temperatures are collected at the surface and center of standard cylindrical geometry. Thermal diffusivity is calculated using the following solution of the transient heat transfer equation:

$$\alpha = \frac{\Omega R^2}{4(T_s - T_c)} \tag{10.5}$$

where Ω is the constant rate of temperature rise at all points in the cylinder (dT/dt, °C/s), R is the sample radius and $T_s - T_c$ is the maximum temperature difference or the establishment of steady-state conditions when the temperature gradient at any location in the sample is no longer time dependent.

10.2.3.3   Probe Method

Nix et al. (1967) mentioned that the line heat source probe method can also be used to measure thermal conductivity and diffusivity simultaneously. It can be done by adding an extra temperature sensor some distance away from the heater. Its location should be in the following range:

$$0.32\sqrt{\alpha\tau} < r_d < 6.2\sqrt{\alpha\tau} \tag{10.6}$$

where r_d is the distance of the thermal diffusivity sensor or temperature sensor from the probe heater, and τ is the time test duration.

This method is suitable for liquids, such as cake batter, or wet ingredients, and nonporous soft solids such as biscuit or bread dough. However, it is not suitable for porous structure samples in which significant volume expansion occurs, such as bread, cake, and muffin.

10.2.4 Density

10.2.4.1   Specific Gravity Bottle/Pycnometer

The density (ρ) of a sample in a liquid or semisolid state can be measured easily by specific gravity bottle or pycnometer. The sample is placed in a container whose volume is already known, and then the mass of the sample is determined (Rahman, 1995). Density can be calculated from mass and volume data.

10.2.4.2   Solid Displacement Method

The volume of irregular (dry and baked) sample can be estimated using the solid displacement method. In this method, a sample of known mass (M_s) is covered with seeds in a container, and the whole is weighed (M_{c+sd+s}). The bulk density (ρ_s) of a fine seed is determined prior to the test. The mass of the container (M_c) is known; thus, the volume of the sample (V_s) can be evaluated by the following equation:

$$V_s = V_{container} - \frac{\left[M_{c+sd+s} - M_c - M_s\right]}{\rho_s} \tag{10.7}$$

This technique is very common for density measurement of most final bakery products. Rapeseed (Rubio and Sweat, 1990) and amaranth seed (Moreira et al., 1995) have been assessed.

10.2.4.3   Geometry Cutting Technique

Final baked goods can be cut into the shape of regular geometries. The volume is calculated from the dimensions of the sample. Typically, baked samples are frozen immediately after the process. Frozen samples are then used to cut into even sides of regular geometry, such as cube or rectangular shapes (Baik et al., 1999). The mass of the pieces can be easily obtained, and then the density can be calculated from the mass divided by the volume.

10.2.5   MOISTURE DIFFUSIVITY

During baking, as the batter temperature increases, volume expansion occurs due to the chemical leavening agent, the incorporation of air, and water vaporization. The volume increases up to around two thirds to three quarters of the baking time. Then, it decreases to some extent (Baik and Marcotte, 2003). The volume change has to be taken into account in the determination of the moisture diffusivity.

Baik and Marcotte (2003) used an analytical solution of an infinite slab based on the constant volume and moisture diffusivity. The solution, however, incorporated the effect of volume change as suggested by Crank (1975) and Gekas and Lamberg (1991). They estimated moisture diffusivity through the first falling rate period from plotting the dimensionless moisture ratio against time on a semilog scale. They used an Arrhenius type of equation to model the effect of temperature and moisture content on the moisture diffusivity.

10.3   DATA COMPILATION AND PREDICTION MODELS

The physical, thermal, and moisture diffusion property data and prediction models of sweet baked products are presented in Table 10.1 and Table 10.2. The property data were classified by product, moisture, and temperature.

10.3.1   SPECIFIC HEAT

Specific heat of commercial biscuit dough was measured using the mixture method at three different temperatures (29.8, 35, and 37.9°C for AACC dough; 30, 36.5, and 39°C for hard sweet [HS] dough) by Kulacki and Kennedy (1978). The specific heat of both types of dough increased with increase in temperature. The indirect mixing method was used by Hwang and Hayakawa (1979) to measure the specific heat of biscuit and cracker. The technique allowed them to measure the specific heat at temperatures above 100°C. Samples were collected from two locations in a multizone band oven. For a moisture content varying from 3.15 to 3.87%, specific heat for biscuit ranged from 1875 to 1942.7 J/kgK.

Christenson et al. (1989) and Baik et al. (1999) used DSC to measure the specific heat of commercial muffin, biscuit, and cupcake (Table 10.1 and Table 10.2). Chris-

TABLE 10.1
Physical and Thermal Properties of Sweet Bakery Products

Product	Temperature (°C)	Moisture Content (% w.b.)	Density (kg/m³)	Specific Heat (kJ/kgK)	Thermal Conductivity (W/mK)	Thermal Diffusivity (m²/s)×10⁸	Technique	Ref.
Biscuit		3.15		1875			C_p: indirect mixing method	Hwang and Hayakawa (1979)
		3.53		1943				
		3.87		1934				
Biscuit dough AACC dough	29.8–37.9[a] 24.5–45.1[b]	4.1	1252 ± 18	2835–3128	0.405 ± 0.22	8–12	C_p: mixing method k: single plate method α: calculation method	Kulacki and Kennedy (1978)
HS (hard sweet dough)	30–39[a] 24.9–35.6[b]	8.5	1287 ± 9	2420–3182	0.39 ± 0.037	8–12		
Biscuit 2 mm from bottom 5 mm from bottom					0.07 0.16		k: hot plate steady state	Standing (1974)
Yellow cake batter		41.5	694		0.223	10.9	k: line heat source probe	Sweat (1973)
Edge		34–40	285–815	2950	0.239–0.119	8.6–15.0		
Center		35.5–39	300–815	2800	0.228–0.121	8.6–14.3		

TABLE 10.1
Physical and Thermal Properties of Sweet Bakery Products

Product	Temperature (°C)	Moisture Content (% w.b.)	Density (kg/m³)	Specific Heat (kJ/kgK)	Thermal Conductivity (W/mK)	Thermal Diffusivity (m²/s)×10⁸	Technique	Ref.
Cupcake batter	20 ± 1.2 20 ± 1.2	34.6 ± 1.93	803 ± 12.0	2517 ± 112	0.206 ± 0.007	10.2 + 0.96	C_p: MDSC k: line heat source probe α: calculation method	Baik et al. (1999)
		33.9 ± 1.69	915 ± 18.6		0.216 ± 0.011			
4 min	54 ± 7.0 20 ± 1.2	33.9 ± 1.69	662 ± 0.9	2599 ± 49.3	0.187 ± 0.005	10.8 + 0.49		
		33.5 ± 1.16	1120 ± 20		0.283 ± 0.017			
6 min	68 ± 5.7 20 ± 1.2	33.5 ± 1.16	558 ± 9.0	2629 ± 27.8	0.196 ± 0.014	13.3 ± 1.31		
		30.0 ± 1.46	236 ± 23.4		0.0703 ± 0.003			
13 min	102 ± 1 20 ± 1.2	30.0 ± 1.46	236 ± 23.4	2658 ± 106	0.106 ± 0.012	17.0 + 4.43		
15 min	103 ± 2 20 ± 1.2	30.3 ± 1.73	282 ± 17.8		0.0683 ± 0.005			
		30.3 ± 1.73	281 ± 17.8	2658 ± 46	0.116 ± 0.012	15.4 + 2.86		
19.5 min	104 ± 2	27.6 ± 2.06	287 ± 14.6		0.0683 ± 0.004			
		27.6 ± 2.06	287 ± 14.6	2613 ± 53.9	0.120 ± 0.007	16.0 + 2.06		
Muffin		17	441	2779	0.70	57.6	C_p: calculated from prediction model k: calculated as αρ C_p α: temperature history	Tou and Tadano (1991)
White layer cake								
3.25 min		1220						
4.1 min		MW at 50% power 1111					ρ: solid displacement method	Sumnu et al. (2005)
9.0 min		IR at 50% power 1020						
10.5 min		IR at 70% power 785						
7.0 min		MW (50%) + IR 950						
8.6 min		(50%) MW 785						
3 min		(50%) + IR 860						
6 min		(70%) 700						
3 min		830						
6 min		740						

TABLE 10.1
Physical and Thermal Properties of Sweet Bakery Products

Product	Temperature (°C)	Moisture Content (% w.b.)	Density (kg/m³)	Specific Heat (kJ/kgK)	Thermal Conductivity (W/mK)	Thermal Diffusivity (m²/s)×10⁸	Technique	Ref.
Sponge cake	Porosity = 0.39		880	1950	0.15	Mass permeability = $4.9 \times 10^{-13}\,\mathrm{m}^2$		Lostie et al. (2004)
Yellow layer cake								
Batter								
Control			1020					
Alginate			1055					
Carrageenan			1127					
Locust bean			1063					
Guar			1127					
HPMC			1115					
Pectin			1055					
Xanthan			1800				ρ: measuring cylinder	Gomez et al. (2007)
Cake								
Control			435					
Alginate			448					
Carrageenan			405					
Locust bean			385					
Guar			426					
HPMC			411					
Pectin			412					
Xanthan			381					

TABLE 10.1
Physical and Thermal Properties of Sweet Bakery Products

Product	Temperature (°C)	Moisture Content (% w.b.)	Density (kg/m³)	Specific Heat (kJ/kgK)	Thermal Conductivity (W/mK)	Thermal Diffusivity (m²/s)×10⁸	Technique	Ref.
Sponge cake			710					
Batter Control SE25			740					
SE50 SE75 W25			750					
W50 W75 Cake			770					Celik et al. (2007)
Control SE25 SE50			810 880 850					
SE75 W25 W50			232 248 253					
W75			254 299 297					
			293					
Cookies Control 5%								
orange pulp 15%	48 54 57 78		645 741 763					Larrea et al. (2005)
orange pulp 25%			813					
orange pulp								
Sponge cake Control								
Maltitol Mannitol			242 292 356					
Sorbitol Xylitol			255 262 286				ρ: solid displacement method	Ronda et al. (2005)
Isomaltose			256 290					
Oligofructose								
Polydextrose								
Angel food cake	Oven temperature							
LEW/WPI 100/0		190 190 180 180	186 286 209					
100/0 100/0 100/0	Air pressure (bar) 0 0.5 0.05 0.1 0.5	160 160 190 180	231 277 348				ρ: solid displacement method	Morr et al. (2003)
100/0 100/0 75/25	1.0 0.05 0.05 0.1	180 190 190 190	206 208 220					
75/25 75/25 50/50	0.05 0.5 1.0 1.5	190	329 481 497					
50/50 50/50 50/50			605					

TABLE 10.1
Physical and Thermal Properties of Sweet Bakery Products

Product	Temperature (°C)	Moisture Content (% w.b.)	Density (kg/m³)	Specific Heat (kJ/kgK)	Thermal Conductivity (W/mK)	Thermal Diffusivity (m²/s)×10⁸	Technique	Ref.
Cake								
Wheat control			400				ρ: solid displacement method	Ragaee and Abdel-Aal (2006)
15% Barley			426					
15% Millet			417					
15% Rye			456					
15% Sorghum			426					
Cookies								Galdeano et al. (2006)
Untreated oat hulls			505					
Treated oat hulls[c]			450					
Cake	41.5 35.5							Rask (1989)
Batter			700	2950	0.22	11		
Crumb			300	2800	0.12	14		
Biscuit								
Rotation speed[d]							ρ: solid displacement method	Edoura-Gaena et al. (2007)
250		6.1	247					
350		5.7	257					
500		6.0	246					
1000		6.5	271					
Duration[d]		6.5	227					
1 min		6.1	241					
5 min		5.8	258					
20 min		5.7	270					
40 min		6.5	279					
80 min								

[a] Measurement temperature for C_p.
[b] Measurement temperature for k.
[c] SE25, SE50, and SE75 are egg white protein replaced with soapwort extract by 25, 50, and 75%, respectively. W25, W50, and W75 are egg white protein replaced by water by 25, 50, and 75%, respectively. LEW is liquid egg white; WPI is whey protein isolates.
[d] 15 g WPI/100 g solution.
[e] 11 g WPI/100 g solution.
[f] Chemically (alkaline hydrogen peroxide) and physically (extrusion) treated.
[g] During the aeration step.

TABLE 10.2

Prediction Models for Physical and Thermal Properties of Sweet Bakery Products

Product	Equation	Reference
Muffin	$C_p = [0.40 + 0.0039 \text{ T}] \times 103, 298\text{–}358\text{K}; R^2 = 0.96$	Christenson et al. (1989)
Biscuit	$C_p = [0.80 + 0.0030 \text{ T}] \times 103, 331\text{–}358\text{K}; R^2 = 0.95$	
	$C_p = [1.17 + 0.0030 \text{ T}] \times 103, 303\text{–}331\text{K}$	
	$\ln k = -48.0 - 10.9 \text{ m} + 0.272\text{T} + 0.053 \text{ m T} - 4.1 \times 10^{-4} \text{ T}^2; R^2 = 0.917$	
Muffin	$\ln k = -7.79 - 7.80 \text{ m} + 0.015 \text{ T} + 0.043 \text{ m T};$ $R^2 = 0.904$	
	$\ln k = -15.8 - 7.90 \text{ m} + 0.072\text{T} + 0.038 \text{ m T} - 9.7 \times 10^{-5} \text{ T}^2; R^2 = 0.819$	
Biscuit	$\ln k = -5.95 - 8.61 \text{ m} + 0.0098 \text{ T} + 0.041 \text{ m T};$ $R^2 = 0.820$	
Yellow layer cake	$C_p = [1.0 - 0.5 (1\text{-m})] C_{pw}$	Sweat (1973)
Cake	$k = 0.0844 + 0.0000892 \, \rho;$ standard deviation: 0.0073	Rubio and Sweat (1990)
	$C_p = 7107 \text{ m} + 18.7 \text{ T} - 45.3 \text{ m T}; R^2 = 0.999$	
	$k = 0.00263 \text{ T} - 0.831 \text{ m} - 0.00091 \, \rho + 0.00422 \text{ m} \rho;$ $R^2 = 0.991$	Baik et al. (1999)
Cupcake	$\alpha = 2.55 \times 10^{-8} \text{ m} - 1.75 \times 10^{-10}$ $\rho - 3.95 \times 10^{-10} \text{T} + 2.42 \times 10^{-7}; R^2 = 0.971$	
	$D = 29.6 \, \varepsilon \exp(-8020/\text{T})$	Baik and Marcotte (2003)

Note: Where m is moisture content, T is the temperature, ε is the porosity, and C_{pw} is the specific heat of water.

tenson et al. (1989) determined the specific heat at a temperature range of 20 to 85°C for the samples having moisture content between 0 and 60%. To get various moisture content levels, wet samples were either air dried or microwave oven dried for appropriate times. Baik et al. (1999) baked samples in a pilot-scale electric oven at a temperature/time cycle (177°C/19.5 min) that produced physical properties similar to a cake baked in an industrial oven. During baking, the samples were removed at specific baking times (0, 4, 6, 13, 15, and 19.5 min). Sample specific heats were determined using MDSC. Then, 10 to 15 mg of samples were scanned at a heating rate of 3°C/min over the temperature range of 20 to 110°C. An 80-s modulation period (single cycle) and 1°C amplitude were used. They observed that the specific heat of cake batter increased from 2516.8 J/kgK to 2658 J/kgK after 13 min baking time, and then decreased to 2613 J/kgK by the end of baking. Specific heat increased with rising temperature and decreased with reducing moisture. After 13 min of baking, the temperature of the product remained nearly constant, but the moisture content continued to decrease. As a result, a small decrease in the specific heat of the cake was observed. The traditional mass fraction model was applied to their experimental data, but it did not fit their experimental data well. Using a PROC GLM, a model for specific heat was developed as a function of moisture content and temperature. The specific heat of a cupcake was affected significantly ($p < 0.001$) by moisture content and temperature. Interaction effects of the two variables were also significant ($p < 0.001$). Sweat (1973) estimated the specific heat of yellow layer cake by assuming that C_p for the nonwater fraction was half of that of water. During baking, the specific heat decreased slightly from 2950 J/kgK of the batter to 2800 J/kgK of the fully baked cake.

10.3.2 THERMAL CONDUCTIVITY

Standing (1974) estimated the thermal conductivity in biscuits using a guarded hot plate steady-state process. The temperatures were recorded at two locations, 2 and 5 mm, from the base. During these measurements, the biscuits were heated only by conduction from a hot plate. The plate temperature was varied from 159 to 208°C, and the thermal conductivity was calculated from a heat balance over the heated plate and biscuit. Thus, the calculated conductivity included the resistance to heat transfer between the plate and the surface of the product.

A significant difference in thermal conductivity between two positions was seen (5 mm from the base: 0.16 W/mK, 2 mm: 0.07 W/mK). Standing (1974) also found the thermal conductivity to be lower at higher plate temperature. Thermal conductivity of commercial biscuit dough was determined by a single-plate method based on a steady-state method and was measured at several temperatures between 24 and 64°C (Kulacki and Kennedy, 1978). The thermal conductivity of two doughs first increased with temperature (AACC: <31.9°C, HS: <30.4°C), then dropped to relatively constant values at higher temperatures. The increase of the thermal conductivity of water with temperature accounts for the initial increase in the thermal conductivity of the dough. However, the decrease in the thermal conductivity at higher temperatures could not be explained.

Sweat (1973) used a line heat source probe to measure the thermal conductivity of yellow layer cake baked at different temperatures and at different locations. The samples were cooled to about 28°C before measurement. During baking, the thermal conductivity decreased from 0.223 W/mK in the batter to 0.121 W/mK at the end of baking. Halfway through baking, the thermal conductivity at the cake edge showed lower values than at the center. Rubio and Sweat (1990) measured the thermal conductivity of three types of cake at room temperature using a line source probe. The water content (34.64 to 41.13%) did not influence the thermal conductivity, but they found a positive correlation between thermal conductivity (about 0.058 to 0.14 W/mK) and density (about 85 to 540 kg/m³). The model developed based on experimental data yielded better predictions than theoretical models such as the parallel model.

Baik et al. (1999) measured the thermal conductivity of a cupcake at specific baking times. Moisture content, density, temperature, and the interaction between moisture content and density had significant effects ($p < 0.002$) on thermal conductivity of cupcakes during baking. The prediction accuracy of their model was higher than that of models reported for muffin (Christenson et al., 1989) and cake (Rubio and Sweat, 1990).

Using a line heat source probe, Christenson et al. (1989) measured the thermal conductivity of muffin and biscuit at a temperature range of 20 to 85°C and a moisture content range of 0 to 60%. Thermal conductivity data for biscuit had greater variation than for muffin. The variation was caused by nonhomogeneity of biscuit. Tou and Tadano (1991) estimated thermal conductivity of muffin from thermal diffusivity, density, and specific heat. The calculated thermal conductivity of muffin was 0.706 W/mK.

The models developed for the prediction of thermal conductivity are mostly product specific (Table 10.2). A neural-network-based model was developed by Sablani et al. (2002) for calculating thermal conductivity of a variety of bakery products under a wide range of conditions of moisture content, temperature, and apparent density. The developed model (Appendix A) can easily be incorporated in the numerical analysis of heat and moisture transfer during baking.

10.3.3 Thermal Diffusivity

Kulacki and Kennedy (1978) used an indirect method to calculate the thermal diffusivity of commercial biscuit dough. The thermal diffusivity varied from 8.0 to 12.0 $\times 10^{-8}$ m^2/s in the range of moisture content 4.1 to 8.5%, density 1252.3 to 1286.6 kg/m^3. The total uncertainty in the thermal diffusivity of two biscuit doughs was at maximum 13.4% (AACC) and 15% (HS). Sweat (1973) estimated the thermal diffusivity of yellow layer cake from bulk density, thermal conductivity, and specific heat. It increased from 1.09×10^{-7} to 1.43×10^{-7} m^2/s for the center of the cake during baking. Changes were more significant at the edge of the cake than at the center. Baik et al. (1999) also estimated the thermal diffusivity of a cupcake using an indirect method. The initial thermal diffusivity of batter at room temperature was 1.02×10^{-7} m^2/s. The thermal diffusivity increased to 1.696×10^{-7} m^2/s by the end of baking with a maximum of 1.698×10^{-7} m^2/s after 13 min. These changes were similar to those reported by Sweat (1973).

Baik et al. (1999) also developed a prediction model of thermal diffusivity during simulated baking as a function of density, temperature, and moisture. The temperature history method was used to obtain the thermal diffusivity of muffin (Tou and Tadano, 1991). Samples were put into a cylindrical container (diameter 80 mm, height 30 mm) and baked at 220°C, and the temperature of the center of dough was measured. The thermal diffusivity value obtained was 5.76×10^{-7} m^2/s.

10.3.4 Density/Specific Volume

The changes in density follow the volumetric change in the product during baking. Sweat (1973) reported that during baking, the volume increased up to 70% due to formation of gas that occurred from the action of chemical leavening. This resulted in a density decrease of about 75% at the end of the process.

Christenson et al. (1989) measured the density of interior portions of samples by the geometry cutting technique. Baik et al. (1999) used pycnometer and geometry cutting to estimate the density of cake batter/semisolid batter and baked cake, respectively. They used frozen sample and a sharp knife to prepare regular shapes. The samples were weighed, and after thawing, each dimension was measured to calculate volume. They reported that the batter density decreased sharply from 803 to 236 kg/m^3 during 13 min of baking. It then increased to 281 kg/m^3 after 15 min due to collapse. After that, it remained nearly constant to the end of baking.

Several researchers have used the solid (seed) displacement method to measure the volume/density of baked products (Gomez et al., 2007; Morr et al., 2003; Ragaee and Abdel-Aal, 2006; Ronda et al., 2005; Sumnu et al., 2005). Specific volume/density was used as a quality index to study the influence of different ingredients and operating conditions on baked products. Morr et al. (2003) studied the influence of applied air pressure in the oven to improve baking properties of whey protein isolate (WPI) in angel food cake. Cakes baked with 75% liquid egg white and 25% WPI with variable air pressure exhibited improved physical, textural, and sensory properties compared to those baked at atmospheric pressure or constant air pressure.

Edoura-Gaena et al. (2007) demonstrated that the speed and aeration duration had a significant effect on the density of biscuits. The density increased from 247 to 271 kg/m³ and 227 to 279 kg/m³ as the speed of rotation increased from 250 to 1000 rpm and duration from 1 to 80 min, respectively. Ronda et al. (2005) assessed the effect of polyols and oligosaccharides on various quality attributes including density and specific volume of sugar-free sponge cakes. They found that xylitol and maltitol produced sponge cakes more similar to the control which was manufactured with sucrose. The functionality of different hydrocolloids on the quality of yellow layer cake was studied by Gomez et al. (2007). They reported that incorporation of hydrocolloid led to yellow layer cakes with higher volume than the control, except when alginate was used. Cakes with the highest volume were obtained with xanthan gum followed by locust bean gum.

An experimental study by Celik et al. (2007) showed that egg white protein can be partially replaced with soapwort extract in the sponge cake formulation with only a small change in the density of the final baked product. Larrea et al. (2005) partially replaced wheat flour with extruded orange pulp and produced cookies with accepted flavor and textural quality. The specific volume varied from 1.23 to 1.55 as the amount of orange pulp decreased from 25 to 5%. Galdeano and Grossmann (2006) demonstrated that physically and chemically modified oat hulls can be used to partially replace the wheat flour in cookies in order to improve the fiber content without modification of physical and sensory properties. Sumnu et al. (2005) used microwave and infrared (IR) alone and in combination for the baking of cakes. The cake specific volume was low when microwave and infrared alone were used for baking. They showed that improved quality cakes could be obtained when IR heating was combined with microwave heating. By using IR–microwave combination baking, both the time-saving advantages of microwave and the surface browning advantage of IR can be obtained. The specific volume also improved significantly by using microwave and infrared in combination.

10.3.5 MOISTURE DIFFUSIVITY

Baik and Marcotte (2003) estimated the moisture diffusivity of cupcake for different initial moisture contents and oven temperatures. The moisture diffusivity ranged from 9.0×10^{-11} to 4.4×10^{-8} m²/s for the industrial cake batter during baking. Temperature and porosity strongly affected the effective moisture diffusivity ($p < 0.0001$). As porosity and temperature of the batter increased, the moisture diffusivity increased.

10.4   THEORETICAL MODELS

Several theoretical models have been proposed to predict physical and thermal properties of foods at desired conditions. These models are mostly based on the chemical composition as follows:

$$C_p = \sum X_i^w C_{pi}$$

(10.8)

$$k = \sum X_i^v k_i \quad \text{(Parallel model)}$$

(10.9)

$$\frac{1}{k} = \sum \frac{X_i^v}{k_i} \quad \text{(Perpendicular model)}$$

(10.10)

$$\rho = \frac{1}{\sum \dfrac{X_i^w}{\rho_i}}$$

(10.11)

where X_i^v and X_i^w are volume and mass fraction of a unit component i. These models have been applied successfully for different food materials, and the models of specific heat and density are also valid in porous foods, including bakery products. Parallel and perpendicular models have been found to provide the upper and lower limits of thermal conductivity, respectively, of most food materials. However, applications of such models to bakery products have been limited.

10.5   CONCLUSIONS

Both the DSC and mixing methods have been favored for the measurement of the specific heat of bakery products. Because specific heat is independent of density, the mass fraction model is usually employed for prediction. For the direct measurement of thermal conductivity of bakery products, transient techniques were more predominant. Among these techniques, the line heat source probe was the most popular; second was the temperature history method. As the porosity affects the thermal conductivity significantly, the mass fraction model was not suitable for the estimation of thermal conductivity of bakery products. The favored structure model for air-containing bakery products was the parallel model. This was based on the volume fraction of each component of the sample food. During processing, there are great structural changes. Chemical and physical reactions (e.g., phase transition, distillation heat transfer) occur as well as complicated temperature history and moisture content. Volume is increasing. Very often, these changes are interacting.

Thus, a regression model based on experimental data is applied more than a structural model.

The indirect method of calculation from C_p, ρ, and k and the temperature history method are the most common for the measurement of thermal diffusivity. The thermal diffusivity increased during baking while thermal conductivity decreased and specific heat changed slightly. This was attributed to the migration of vapor, structural changes, and the water-holding capacity. Solid (seed) displacement and geometry cutting methods have been used for the measurement of volume and density of baked products, and the pycnometer method has been used for the density measurement of batter. Moisture-diffusivity-related measurements have been very limited. Efforts are needed to develop more generic correlation to predict physical and thermal properties of bakery products under a wide variation of conditions.

10.6 ACKNOWLEDGMENT

The author would like to acknowledge the assistance of Mattheus F.A. Goosen, New York Institute of Technology, Amman, Jordan, for providing valuable comments on the manuscript.

REFERENCES

ASTM. 1955. Standard method of test for thermal conductivity of materials by means of the guarded hot plate. *ASTM Standards* Part 3, 1084.

Baik, O.D. and M. Marcotte. 2003. Modeling the moisture diffusivity in a baking cake. *Journal of Food Engineering* 56: 27–36.

Baik, O.D., M. Marcotte, S.S. Sablani, and F. Castaigne. 2001. Thermal and physical properties of bakery products. *Critical Reviews in Food Science and Nutrition* 41(5): 321–352.

Baik, O.D., S.S. Sablani, M. Marcotte, and F. Castaigne. 1999. Modeling the thermal properties of a cup cake during baking. *Journal of Food Science* 64: 295–299.

Buhari, A.B. and R.P. Singh. 1993. Measurement of food thermal conductivity using differential scanning calorimetry. *Journal of Food Science* 58: 1145–1147.

Celik, I., Y. Yilmaz, F. Isik, and O. Ustun. 2007. Effect of soapwort extract on physical and sensory properties of sponge cakes and rheological properties of sponge cake batters. *Food Chemistry* 101: 907–911.

Choi, Y. and M.R. Okos. 1986. Effects of temperature and composition on thermal properties of foods, in *Food Engineering and Process Applications: Transport Phenomena*, LeMaguer, M. and P. Jelen (eds.), Vol. 1, Elsevier: London, 93–101.

Christenson, M.E., C.H. Tong, and D.B. Lund. 1989. Physical properties of baked products as functions of moisture and temperature. *Journal of Food Processing and Preservation* 13: 201–217.

Crank, J. 1975. *Mathematics of Diffusion*. Oxford University Press, London.

Dickerson, R.W. 1965. An apparatus for measurement of thermal diffusivity of foods. *Food Technology* 19(5): 198–200.

Edoura-Gaena, R.-B., I. Allais, G. Trystram, and J.B. Gros. 2007. Influence of aeration conditions on physical and sensory properties of aerated cake batter and biscuits. *Journal of Food Engineering* 79: 1020–1032.

Galdeano, M.C. and M.V.E. Grossmann. 2006. Oat hulls treated with alkaline hydrogen peroxide associated with extrusion as fiber source in cookies, *Ciência e Tecnologia de Alimentos, Campinas* 26(1): 123–126.

Gekas, V. and I. Lamberg. 1991. Determination of diffusion coefficients in volume-changing systems—Application in the case of potato drying. *Journal of Food Engineering* **14:** 317–326.

Gomez, M., F. Ronda, P. Caballero, C.A. Blanco, and C.M. Rosell. 2007. Functionality of different hydrocolloids on the quality and shelf-life of yellow layer cakes. *Food Hydrocolloids* 21: 167–173.

Hwang, M.P. and K.I. Hayakawa. 1979. A specific heat calorimeter for foods. *Journal of Food Science* 44: 435–438, 448.

Kulacki, F.A. and S.C. Kennedy. 1978. Measurement of the thermo-physical properties of common cookie dough. *Journal of Food Science* 43: 380–384.

Larrea, M.A., Y.K. Chang, and F. Martinez-Bustos. 2005. Some functional properties of extruded orange pulp and its effect on the quality of cookies, *LWT–Food Science and Technology* 38: 213–220.

Lind, I. 1991. The measurement and prediction of thermal properties of food during freezing and thawing—A review with particular reference to meat and dough. *Journal of Food Engineering* 13: 285–319.

Lostie, M., R. Peczalski, and J. Andrieu. 2004. Lumped model for sponge cake baking during the "crust and crumb" period. *Journal of Food Engineering* 65: 281–286.

McFarlane, I. 2006. Control of final moisture content of food products baked in continuous tunnel ovens. *Measurement Science and Technology* 17: 241–248.

Mohsenin, N.N. 1980. *Thermal Properties of Foods and Agricultural Materials. Can. J. Tech.* 31: 57–69.

Moreira, R.G., J. Palau, V.E. Sweat, and X. Sun. 1995. Thermal and physical properties of tortilla chips as a function of frying time. *Journal of Food Processing and Preservation* 19: 175–189.

Morr, C.V., W. Hoffmann, and W. Buchheim. 2003. Use of applied pressure to improve the baking properties of whey protein isolates in angel food cakes. *Lebensmittel Wissenschaft und Technologie* 36: 83–90.

Murakami, E.G. and M.R. Okos. 1989. Measurement and prediction of thermal properties of foods. In: *Food Properties and Computer-Aided Engineering of Food Processing Systems*, R.P. Singh and A.G. Medina (Eds.), Kluwer Academic, Norwell, MA, pp. 3–48.

Nasvada, P. 2005. Thermal properties of unfrozen foods. In: *Engineering Properties of Foods*, 3rd ed., M.A. Rao, S.S.H. Rizvi, and A.K. Datta (Eds.), CRC Press, Boca Raton, FL, pp. 149–173.

Nix, G.H., G.W. Lowery, R.I. Vachon, and G.E. Tanger. 1967. Direct determination of thermal diffusivity and conductivity with a refined line-source technique. In: *Progress in Astronautics*, G.R. Heller (Ed.), Academic Press, New York, 20, pp. 865–878.

Ohlsson, T. 1983. The measurement of thermal properties. In: *Physical Properties of Foods*, Vol. 1, R. Jowitt et al. (Eds.), Applied Science Publishers, London, pp. 313–353.

Ragaee, S. and E.S.M. Abdel-Aal. 2006. Pasting properties of starch and protein in selected cereals and quality of their food products. *Food Chemistry* 95: 9–18.

Rahman, M.S. 1995. *Food Properties Handbook*. CRC Press, Boca Raton, FL.

Rask, C. 1989. Thermal properties of dough and bakery products: A review of published data. *Journal of Food Engineering* 9: 167–193.

Ronda, F., M. Gomez, C.A. Blanco, and P.A. Caballero. 2005. Effects of polyols and nondigestible oligosaccharides on the quality of sugar-free sponge cakes. *Food Chemistry* 90: 549–555.

Rubio, A.R.I. and V.E. Sweat. 1990. and Measurement and modeling thermal conductivity of baked products. Presented at Annual Meeting of ASAE, Chicago, IL, December 18–21.

Sablani, S.S., O.D. Baik, and M. Marcotte. 2002. Neural networks for predicting thermal conductivity of bakery products. *Journal of Food Engineering* 52: 299–304.

Sablani, S.S., M. Marcotte, O.D. Baik, and F. Castaigne. 1998. Modeling of simultaneous heat and water transport in the baking process: A review. *Lebensmittel Wissenschaft und Technologie* 31: 201–209.

Standing, C.N. 1974. Individual heat transfer modes in band oven biscuit baking. *Journal of Food Science* 39: 267–271.

Sumnu, G., S. Sahin, and M. Sevimli. 2005. Microwave, infrared and infrared-microwave combination baking of cakes. *Journal of Food Engineering* 71: 150–155.

Sweat, V.E. 1973. Experimental measurement of the thermal conductivity of a yellow cake. *Proceedings of 13th International Conference on Thermal Conductivity*, 213–216, Lake Ozark, MO, November 5–7.

Tou, K. and T. Tadano. 1991. Study on effective thermal diffusivity of muffin during baking. *Bulletin of the College of Agriculture and Veterinary Medicine, Nihon University* 48: 199–204.

Vos, B.H. 1955. Measurements of thermal conductivity by a nonsteady state method. *Applied Scientific Research* A5: 425–438.

Zhou, L., V.M. Puri, and R.C. Anantheswaran. 1994. Measurement of coefficients for simultaneous heat and mass transfer in food products. *Drying Technology* 12: 607–627.

Appendix A

Neural-network-based equations for estimation of thermal conductivity, k (W/mK) of dough and bakery products as a function of temperature (T, °C), moisture content (M, %, wet basis), and apparent density (ρ, kg/m^3) (adapted from Sablani et al., 2002):

$$X2 = T * (0.00856) + (-0.627)$$

$$X3 = M * (0.0423) + (-1.059)$$

$$X4 = \rho * (0.00193) + (-1.31)$$

$$X5 = \tanh[(-3.66) + (5.64) * X2 + (0.298) * X3 + (-1.08) * X4]$$

$$X6 = \tanh[(-0.812) + (-1.33) * X2 + (-1.385) * X3 + (1.43) * X4]$$

$$X7 = \tanh[(-0.157) + (0.0634) * X2 + (0.122) * X3 + (0.22) * X4 + (-0.686) * X5 + (-0.403) * X6]$$

$$k = X7 * (0.802) + (0.549)$$

11 Alternative Baking Technologies

Dilek Kocer, Mukund V. Karwe,
Servet Gülüm Sumnu

CONTENTS

11.1 INTRODUCTION

Baking technology is continuously changing to increase energy efficiency and savings, and to improve product quality. Today's ovens have advanced from earlier simple wood baking stoves to sophisticated microchip-controlled devices. Earlier baking ovens were natural convection ovens which were followed by forced convection and gas-fired ovens. Then, microwave ovens and jet impingement ovens were introduced. Microwave ovens and jet impingement ovens provide noticeable improvements in baking technology and have been studied by researchers as alternative baking technologies. These studies focus on the following:

1. Understanding the effect of using alternative technologies on the physico-chemical changes occurring during baking, and the physical characteristic of the finished baked product
2. Improving product formulations and oven designs to overcome any problems associated with the use of alternative baking technologies

3. Defining optimum processing conditions to obtain products with high-quality parameters

The first part of this book chapter covers these alternative baking technologies and their applications. The studies on baking show that any single mode of baking (microwave, natural convection, and impingement) has its limitations as well as its advantages. These limitations have been encouraging researchers to study combination baking. The combination of alternative baking technologies has the advantage of producing a high-quality product with shorter and more efficient processes. The last part of this chapter focuses on these combination oven technologies as microwave and impingement combination heating, and microwave and infrared combination heating.

11.2   JET IMPINGEMENT OVEN TECHNOLOGY

Natural convection baking is a slow and inefficient process that results in variations in product quality due to nonuniform heat transfer over the product surface (Henke, 1985; Walker, 1987). Increasing the air movement within the oven with the use of fans and blowers improves the heat transfer rate but still is not enough to achieve product uniformity (Henke, 1985; Walker, 1987). In order to provide more uniform distribution of air over the product surface compared to natural convection, jet impingement technology was introduced. Jet impingement ovens, first designed by Donald Smith (1975), are a special class of forced convection ovens in which high-velocity (10 to 50 m/s) jets of hot air (100 to 250°C) impinge vertically on a food product (Figure 11.1).

The impingement of high-velocity air vertically onto the product surface results in a higher rate of heat transfer so that the products with internal and external characters similar to conventionally baked ones can be produced at lower temperatures and in shorter times (Li and Walker, 1996; Walker, 1987). This rapid heat transfer technology has been successfully introduced in small fast-food ovens as well as commercial tunnel ovens (Walker, 1987). Jet impingement ovens have been used in the food industry for the baking of tortilla, potato chips, pizza crust, pretzels, crackers, cookies, breads, and cakes and to toast ready-to-eat cereals (Li and Walker, 1996; Walker, 1987; Walker and Sparman, 1989).

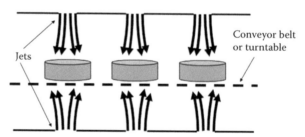

FIGURE 11.1   A multiple jet impingement oven.

The heat transfer from an impingement jet can be expressed by Newton's equation (1):

$$Q = hA\Delta T \tag{11.1}$$

where Q is the heat transfer rate, h is the heat transfer coefficient, A is the heat transfer area, and ΔT is the temperature difference (K) between the product surface and the jet medium. The heat transfer coefficient, h (W/m²K), that is affected by the air velocity can be determined experimentally and predicted numerically (Walker, 1987). The heat transfer coefficients associated with air impingement ovens are 5 to 20 times higher than those associated with natural convection (Kocer and Karwe, 2005; Marcroft and Karwe, 1999; Marcroft et al., 1999; Nitin and Karwe, 2001, 2004; Nitin et al., 2006; Walker, 1987). Typical heat transfer coefficient values are 6 to 12 W/m²K for natural convection, 13 to 30 W/m²K for forced convection to flat surfaces, and 40 to 200 W/m²K for air impingement.

11.2.1 Engineering and Design Aspects of Jet Impingement Ovens

In jet impingement ovens, the heated air is directed to the food through the nozzles that can be simple holes, short nozzles, or long nozzle tubes. The important factors that should be considered in jet impingement oven design are the distance between the impingement nozzle and the product surface, nozzle diameter and width, and spacing between nozzles (Ovadia and Walker, 1998). Jets used in jet impingement ovens for heating and baking of food products are submerged turbulent jets where the jet fluid is the same as the surrounding medium.

The flow field of an impinging jet has been divided into three regions: the free jet region, the stagnation region, and the lateral spread region (Figure 11.2) (Gardon and Akfirat, 1965; Sarkar et al., 2004). The free jet region is further divided into three subregions: the potential core region, the developing flow region, and the developed flow region (Figure 11.2).

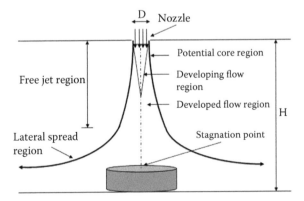

FIGURE 11.2 Regions of impinging jet flow. (Modified from Sarkar, A., Nitin, N., Karwe, M.V., and Singh, R.P., *Journal of Food Science* 69(4): 113–122, 2004. With permission from Blackwell Publishing.)

In the stagnation region, the axial velocity decreases rapidly, and in the lateral spread region, the radial velocity rapidly increases near the stagnation region and later decreases (Figure 11.3) (Marcroft and Karwe, 1999; Marcroft et al., 1999; Nitin and Karwe, 2004). In general, for a single jet impinging on a flat surface, the heat transfer coefficient is high at the stagnation region and it decreases along the radial direction due to a growing boundary layer (De Bonis and Ruocco, 2005; Gardon and Akfirat, 1965; Martin, 1977; Nitin and Karwe, 2004; Nitin et al., 2006; Olsson et al., 2004). The value of the heat transfer coefficient is known to be a function of nozzle-to-plate spacing as well as Reynolds number. Jets are generally located 2 to 5 times the nozzle diameter above the product (Walker, 1987). A recent study of numerical simulation of fluid flow and heat transfer for an impingement flow on a model cookie showed that the maximum heat transfer coefficient is observed at the stagnation point (Figure 11.4a), and the local maximum shifts away from the stagnation point for high z/d (z is nozzle-to-plate distance, and d is hydraulic diameter of the jet at inlet) ratios at higher jet velocities (30 and 40 m/s) as seen in Figure 11.4b (Nitin and Karwe, 2004). Another interesting study that incorporated the water vapor transport model showed the nonuniformity of heat and mass transfer along the exposed surface with evaporation and depletion of liquid water at the stagnation region and rapid removal of surface vapor away from the stagnation point, inducing more diffusion within the food (De Bonis and Ruocco, 2005).

Multiple jets have characteristics similar to those of a single jet; however, the possible interactions between surrounding jets may disturb the stagnation region and lead to a reduction in the rate of heat transfer. The interactions between the jets can lead to strong reverse flows or "upward jet fountains" that can result in secondary heat transfer maxima on the impinging surface due to high levels of turbulence (Goldstein and Timmers, 1982; Huber and Viskanta, 1994; Olsson et al., 2005; Saripalli, 1983).

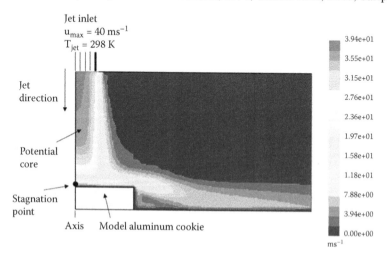

FIGURE 11.3 Contour plot of total velocity in a turbulent impinging jet on a model cookie at z/d = 3. (From Nitin, N. and Karwe, M.V., *Journal of Food Science* 69(2), 59–65, 2004. With permission from Blackwell Publishing.) (See color insert after p. 158.)

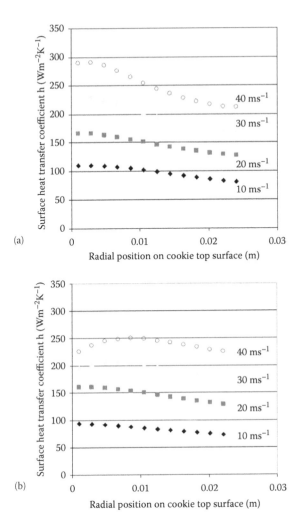

FIGURE 11.4 Variation of local surface heat transfer coefficient on the top surface of the model cookie as a function of position and jet velocity at jet temperature of 450 K for (a) z/d = 2, (b) z/d = 5. (From Nitin, N. and Karwe, M.V., *Journal of Food Science* 69(2), 59–65, 2004. With permission from Blackwell Publishing.) (See color insert after p. 158.)

Heat transfer during multiple jet impingement is affected by the Reynolds number, nozzle-to-plate spacing, jet–jet mixing, nozzle-to-nozzle spacing, and nozzle array geometry. A recent study showed that spacing between the jets should be high enough but not too high to attain high heat transfer rate (Olsson et al., 2005). If the distance between the jets is too small, the entrainment of air between the jets is suppressed and almost no recirculation zone is observed, whereas if the distance between the jets is too large, a large recirculation zone is created between the jets, leading to more kinetic energy loss and lower rate of heat transfer (Olsson et al., 2005).

Different techniques have been used to study the fluid flow and heat transfer associated with jet impingement flows. The use of the smoke wire method is a qualitative approach to study the flow pattern of impinging jets in food applications (e.g., baking or roasting of hot dogs) (Cornaro et al., 1999; Popiel and Trass, 1991; Viskanta, 1993). The quantitative experimental measurements, like hot wire anemometry (Gardon and Akfirat, 1965; Sugiyama and Usami, 1979) and pitot tube (Kocer and Karwe, 2005; Stoy and Ben-Haim, 1973) which measure point velocity of impinging jets, are invasive techniques and are likely to induce changes in the flow field. On the other hand, laser doppler anemometry (LDA) (Durst et al., 1981; Marcroft and Karwe, 1999; Marcroft et al., 1999) and particle imaging velocimetry (PIV) (Adrian, 1991) are noninvasive techniques that have been developed to quantitatively measure the flow field, although these techniques are expensive.

The lumped capacitance technique has been used to determine average heat transfer coefficient during jet impingement processing (Kocer and Karwe, 2005; Nitin and Karwe, 2001). Liquid crystals (Baughn, 1995; Goldstein and Timmers, 1982; Lee and Lee, 2000; Mesbah et al., 1996), two-dimensional infrared radiometer (Pan et al., 1992; Pan and Webb, 1995), and naphthalene film-indirect approach (Angioletti et al., 2003; Sparrow and Lovell, 1980) have been used to measure the spatial variation of surface heat transfer for a model object.

11.2.2 BAKING IN JET IMPINGEMENT OVENS

Baking is a process where heat and mass transfer occur simultaneously within the food system. Heat transfer causes temperature rise; mass transfer causes migration and loss of moisture in the food. This temperature rise, moisture migration, and evaporation cause starch gelatinization, protein denaturation, crust formation, color development, and flavor formation. The extent of all these physical and chemical changes determines the final product quality, which is defined by texture, color, flavor, and shelf life.

During baking, while moisture migrates to the product surface, it encounters a heavy, cool, moist, and stagnant layer around the product surface, which acts as an insulator and greatly slows heat transfer (Walker and Sparman, 1989). Natural convection is not enough to move this cold layer around the product surface (Henke, 1985; Walker and Sparman, 1989). Blowing air horizontally around the product surface by forced convection only decreases this cold layer (Henke, 1985). On the other hand, impinging high-velocity hot air jets onto the product surface removes this cold boundary layer and replaces it with hotter and drier air. This results in increased heat and mass transfer rate at the product surface which helps to achieve the desired uniform baking (Henke, 1985; Walker, 1987; Walker and Sparman, 1989; Walker and Li, 1993; Yin and Walker, 1995).

Although the moisture loss rate is higher in jet impingement ovens, the net moisture loss in the products baked in jet impingement ovens is less than that of the ones baked in conventional ovens due to shorter baking times (Olsson et al., 2005; Walker and Li, 1993; Walker and Sparman, 1989). Therefore, the foods baked in jet impingement ovens have a better yield and higher moisture content, and their shelf life and texture are improved (Walker and Li, 1993; Walker and Sparman, 1989; Yin and Walker, 1995).

The other reason for high moisture retention in products baked in jet impingement ovens is the quick crust formation. A higher rate of surface moisture removal under hot air jet impingement results in quick crust formation. Because the crust has lower moisture diffusivity, the product retains more moisture inside, which can enhance the perceived quality of the processed food (Walker, 1987), including an increase in shelf life as well as retention of some key health-promoting nutraceutical compounds such as Omega-3 fatty acids (Borquez et al., 1999). In addition, greater moisture retention results in improved flavor retention (Henke, 1985).

The crust and color formation are key parameters that identify the acceptability of the baked products by consumers. The formation of crust and increase in surface temperature beyond the evaporation temperature result in color development. Browning reactions, mainly caramelization, have been known to be responsible for the development of the crust color, and they need higher temperatures (>100°C) to occur. The air temperature is the most significant parameter affecting the color development (Olsson et al., 2005). When baking at the same temperature, the rate of color development in jet impingement ovens is faster than that with conventional ovens due to faster surface temperature rise (Olsson et al., 2005; Wahlby et al., 2000). The crust thickness is also affected by air temperature and velocity and process time. Shorter heating time results in a thinner crust, whereas the crust thickness increases with increase in air temperatures (Olsson et al., 2005). In addition, the center temperature increases faster during impingement baking, leading to faster settling of the crumb (Wahlby et al., 2000).

Table 11.1 shows the comparison of baking times and air temperatures for products baked in conventional versus jet impingement ovens (Walker and Sparman, 1989). It is important to note that processing times and temperatures may vary with

TABLE 11.1

Comparison of Baking Times and Air Temperatures for Products Baked in Conventional versus Jet Impingement Ovens

Product	Conventional Oven		Jet Impingement Oven	
	Time (min)	Air Temperature (°C)	Time (min)	Air Temperature (°C)
Muffins	26	174	12	154
Layer cake	26	159	16	149
Pound cake	75	134	55	124
Croissant	18	171.5	12	154
Puff pastry	22	166.5	13	159
Apple danish	20	171.5	10	149
Cherry turnovers	28	171.5	14	154
Raisin oatmeal cookies	15	166.5	12	159
Raisin nut oatmeal cookies	16	169	12	154

Source: Walker, C.E. and Sparman, A.B., *AIB Research Department, Technical Bulletin*, XI, 11, November, 1989. With permission.

the oven type, product formulation, sample weight, and pan characteristics (Walker and Sparman, 1989). The jet impingement oven used in this study was a Jet Sweep® air impingement oven set with the upper jet fingers located 8 cm above the product surface. Air velocity was 15 m/s at 4 cm away from the orifice and 6 to 13 m/s at the product surface. The study showed that both external and internal appearances of the finished foods were similar (Walker and Sparman, 1989); however, the baking time was substantially less in the case of jet impingement ovens.

Another study investigated the baking of cakes in five different types of ovens: an electrically heated convection oven, an electrically heated conveyor-type jet impingement oven, a gas-heated conveyor-type jet impingement oven, a jet impingement oven, and a jet impingement–microwave hybrid oven. Table 11.2 shows the process times and temperatures, heat transfer coefficients, and total energy required baking a 20-cm layer cake in these ovens (Li and Walker, 1996). The study showed that the cakes baked in jet impingement ovens had slight crumb compaction and firmer texture than those baked in the conventional oven. It was also observed that increasing baking time, air velocity, and temperature resulted in increase in both

TABLE 11.2
Comparison of Process Parameters for Various Types of Commercial Ovens

	Conventional Oven (electrically heated)	Conveyor-Type Jet Impingement Oven		Jet Impingement Oven	Hybrid Oven (jet impingement and microwave)
		(electrically heated)	(direct-fired gas)		
Model	Despatch Mini-Bake (Despatch Oven Co., Minneapolis, MN)	Middleby-Marshall model PS200T (Middleby-Marshall Inc., Elgin, IL)	Blodgett Mastertherm® model MT70PH (Blodgett Oven Co., Burlington, VT)	Windshear Jet Sweep® (Enersyst, Dallas, TX)	Enersyst Food Finisher III® (Enersyst, Dallas, TX)
Optimum baking time	30 min	18 min	18 min	14 min	6 min
Optimum baking temperature	177°C	149°C	166°C	149°C	227°C
Apparent convective heat transfer coefficient		**Upper target plate**			
	23.3 W/m²K	83.8 W/m²K	66.4 W/m²K	84.8 W/m²K	49.1 W/m²K
		Lower target plate			
	17.4 W/m²K	110.9 W/m²K	91.4 W/m²K	105.0 W/m²K	98.2 W/m²K
Total energy required	174 kJ	158 kJ	175 kJ	109 kJ	144 kJ

Source: Li, A. and Walker, C.E., *Journal of Food Science* 61(1): 188–191, 197, 1996. With permission.

crust color and cake firmness. The cakes baked in conveyor-type ovens had lower volumes than the ones baked in conventional ovens, and they showed stripes where they had passed too closely beneath the nozzles.

To summarize, the benefits of air impingement over conventional baking are decreased processing time, lower process temperatures, energy efficiency, uniform heating, and reduced moisture loss (Henke, 1985; Olsson et al., 2005; Walker and Li, 1993). The products baked in jet impingement ovens have improved texture, uniformly baked surface, and internal structure (Henke, 1985).

In some baking applications, jet impingement is not required during the whole baking process. Instead, different baking zones with different processing times, air velocities, and temperatures are applied to achieve products with desired baking qualities. The use of high velocities of air at low temperatures during the early stages of baking gives a good oven spring with minimum crust development. This permits rapid heat movement to the center of the product, which then is held in an intermediate zone where the crumb structure develops. Finally, the application of high oven temperature results in crisp and brown crust (Walker, 1987). If high temperature is applied to the product at the early stages of baking, a dry, thick crust can develop. The formation of a thick crust retards proper baking at the center because the crust acts as an insulator (Walker, 1987). Different combinations of air temperature, velocity, and processing times can be used to achieve a crust with a desired thickness (Walker and Sparman, 1989). Rapid initial baking is critical in some products with dense, moist centers, such as fruit pies, to achieve a brown, flavorful crust, with the darkening characteristic of dextrinization, caramelization, and the Maillard reactions, but overbaking should be avoided while the center heats completely (Walker, 1987; Walker and Sparman, 1989).

11.3   MICROWAVE BAKING TECHNOLOGIES

Microwave baking has the advantages over conventional baking in terms of reduction of baking time and energy. Various studies have been conducted on microwave baking of soft wheat products, and these studies showed that conventional baking time was significantly reduced in the presence of microwave heating. Table 11.3 shows the comparison of baking times in different ovens for different products.

The usage of microwave baking in the food industry is limited. In the early 1990s, APV Baker (UK) introduced a microwave–conventional baking oven for postbaking. This oven was developed as an alternative for radio frequency (RF) heating (Bengtsson, 2001). Because microwave equipment is more compact, flexible, and considered as advanced technology, there is recently a tendency toward hybrid ovens instead of RF in baking and postbaking processes. Microwave ovens working at 896 MHz have been reported in Britain for bread baking and production of bread crumbs (Bengtsson, 2001).

11.3.1   PRINCIPLES OF MICROWAVE BAKING

There are two microwave heating mechanisms of foods: dipolar rotation and ionic conduction. In dipolar rotation, polar molecules placed in an alternating electric field

TABLE 11.3

Comparison of Process Parameters for Microwave and Conventional Ovens

Product	Oven Type and Baking Condition		Ref.
	Microwave Oven	Conventional Oven	
Madeira cake	100% power, 40 s (Full power: 900 W)	200°C, 600 s	Megahey et al., 2005
Model layer cake	100% power, 6 min (Full power by IMPI test: 600 W)	180°C, 25 min	Sumnu et al., 2000
High-ratio white layer cake	100% power, 5.5 min (Full power: 650 W)	190°C, 25 min	Martin and Tsen, 1981
White layer cake	50% power, 4 min (Full power by IMPI test: 706 W)	175°C, 24 min	Sumnu et al., 2005
Canned biscuit dough	50 s, 100% (Full power: 900 W)	218°C, 10 min	Pan and Castell-Perez, 1997

experience a rotational force, which orients them in the direction of the field. As molecules try to rotate in the direction of the field, they collide randomly with their neighbors. As the field changes its direction, molecules try to line up with the direction of the field, and further collisions take place. This results in heating.

Ionic conduction is observed in materials containing ions such as salts. Salt is composed of positive sodium and negative chloride ions in dissociated form. Positive charged particles will be accelerated toward the direction of the electric field, and negative charged particles will be accelerated in the reverse direction. An accelerating particle collides with its neighbor and sets its neighbor into more agitation. Thus, the temperature of the particle increases. This heat is then transferred to other parts of the food by conduction.

In microwave baking, heat is generated inside the food by the energy equation (Equation 11.2):

$$\frac{\partial T}{\partial t} = \alpha \nabla^2 T + \frac{Q}{\rho C_P}$$

(11.2)

where T is temperature (K), t is time (s), α is thermal diffusivity (m²/s), ρ is density (kg/m³), C_p is specific heat capacity of the material (J/kg.K), and Q is the heat generated per unit volume of material (W/m³) which represents the conversion of electromagnetic energy.

The relationship between Q and electric field intensity (E) at that location can be derived from Maxwell's equations of electromagnetic waves as shown by Metaxas and Meredith (1983):

$$Q = 2\Pi\varepsilon_0\varepsilon'' fE^2$$

(11.3)

where ε_0 is the dielectric constant of free space, ε'' is the dielectric loss factor of the food, f is the frequency of the oven, and E is the root-mean-squared value of the electric field intensity.

After heat is generated inside the food, the rest of the food is heated by conduction. In conventional baking, food is heated by conduction, convection, and radiation. The air inside the microwave oven is not heated as in conventional ovens. Therefore, foods baked by microwaves are not heated but are cooled by convection at the surface.

Dielectric properties are the physical properties of a food that affect microwave heating. Dielectric properties include the dielectric constant and the dielectric loss factor of foods. The dielectric constant and the loss factor represent the ability of a food material to store electrical energy and convert electrical energy into heat, respectively. The dielectric properties of food depend on temperature, moisture content, salt content, and composition of the product. There are limited studies in the literature on the dielectric properties of baked products.

Kim and Cornillon (2001) studied the effects of mixing time on dielectric properties of wheat dough. As mixing time increased, the dielectric constant and the loss factor of wheat dough decreased due to a low amount of mobile water in the sample after mixing. The increase in temperature was shown to increase the loss factor of dough which could be due to the increased amount of dissolved ions.

The dielectric properties of starch slurries have recently been studied by Motwani et al. (2007). The dielectric constant of starch slurry decreased, but its loss factor increased with increasing starch concentration. The dielectric constant decreased with increasing temperature for all frequencies. The variation of the loss factor with temperature was a function of frequency. It increased with temperature between 15 MHz and 450 MHz and then decreased with increase in temperature between frequencies of 450 MHz and 3 GHz. The dielectric constant of 20% starch slurry was found to be significantly correlated with gelatinization (Motwani et al., 2007).

The first study on the variation of dielectric properties of baked products during microwave–infrared (MIR) and microwave–jet impingement (MJET) was performed on bread by Sumnu et al. (2007). The dielectric properties of breads were shown to decrease sharply during the initial stages of baking and then remained constant (Figure 11.5). The sharp decrease in dielectric properties with baking time was explained by the increase in the porosity during baking. The dielectric properties of crust were shown to be significantly higher than the crumb portion, because crust was less porous than the crumb portion.

Dielectric constant and loss factor of cake samples were shown to be dependent on formulation, baking time, and temperature (Sakiyan et al., 2007). The increase in baking time and temperature decreased dielectric constant and loss factor of all formulations. Fat content was shown to increase dielectric constant and loss factor of cakes. Variation of dielectric properties of cakes during baking was explained by porosity and moisture content.

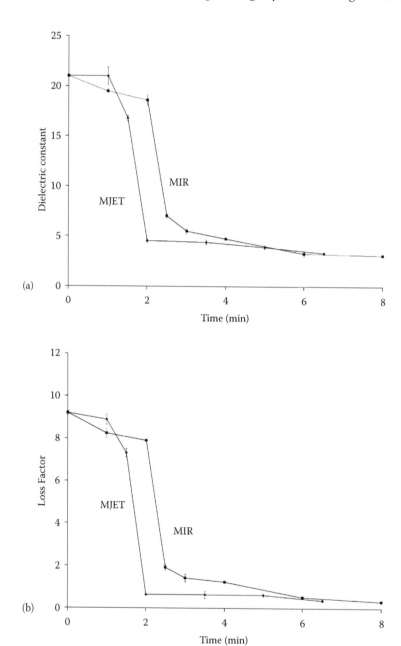

FIGURE 11.5 Transient dielectric constant (a) and loss factor (b) of breads measured at the central region of crumb during baking in different heating modes. (MJET: microwave–jet impingement, MIR: microwave–infrared.) (From Sumnu, G., Datta, A.K., Sahin, S., Keskin, S.O., and Rakesh, V., *Journal of Food Engineering* 78(4), 1382–1387, 2007. With permission from Elsevier.)

11.3.2 QUALITY DEFECTS IN MICROWAVE-BAKED PRODUCTS

Microwave-baked soft wheat products have various quality problems such as low volume, firm or tough texture, and high moisture loss (Sumnu, 2001). The low volume and firm or tough texture can be explained by the insufficient starch gelatinization. Due to the short time of microwave heating, there is not enough time for starch to complete its gelatinization. It is assumed that the specific interaction of microwave with gluten is responsible for the toughness of baked products. The exact mechanism of interaction of microwaves with gluten is not known. Campana et al. (1993) showed that drying of wheat by microwaves did not affect total protein content but changed the functionality of gluten. Rogers et al. (1990) showed that the microwave toughening effect was not the result of cross-linking by disulfide bond formation.

Another reason for the hard texture of baked products in microwave ovens is the high moisture loss in these products. It was shown by various researchers that microwave-baked products were drier than conventionally baked ones (Keskin et al., 2004; Lambert et al., 1992; Seyhun et al., 2003; Sumnu et al., 1999, 2005). This can be explained by the difference in the mechanisms of microwave heating and conventional heating. As explained in Section 11.3.1, there is internal heat generation in microwave heating. This creates significant pressure and concentration gradients, which increase the flow of moisture through the food to the boundary (Datta, 1990). In addition, the crust formed in conventional products cannot be obtained in microwave-baked products. Thus, there is no crust on the surface of these products to restrict moisture loss.

Crust and surface color cannot be formed in microwave-baked products. Microwaves are generated inside the foods and the air inside the microwave oven is not heated. Therefore, the surface temperature of the products cannot reach temperatures necessary for Maillard and caramelization reactions. In addition, the moisture removed from microwave-baked products condenses when it comes in contact with the air at ambient temperature at the surface. Because the moisture content at the surface of the product is high, a dry crust cannot be formed.

Because Maillard and caramelization reactions are not observed in microwave-baked products, flavors generated as a result of these reactions are also absent in these products, and the aroma profile of a microwave-baked cake is similar to that of batter. Whorton and Reineccius (1990) showed that aromas (nutty, brown, and caramel type) observed in a conventional cake were absent in microwave-baked cakes. In addition, unwanted flavors such as flour or egg-like flavors develop during the microwave baking of cakes. It is possible to mask these undesired flavors and to obtain a similar flavor profile with conventionally baked cakes by adding flavoring agents to the cake recipe (Sumnu and Sahin, 2005).

Rapid staling is another problem in microwave-baked cakes. The staling mechanism of microwave-baked cakes is still unclear. High moisture loss in microwave-baked products was thought to be one of the reasons for the staling of microwave-baked products. It was shown that shelf life of bread was increased by increasing its moisture content by 2% (Stauffer, 2000). A high amount of amylose that is leached during microwave baking may be another reason for the rapid staling of microwave reheated breads (Higo and Noguchi, 1987) and microwave-baked

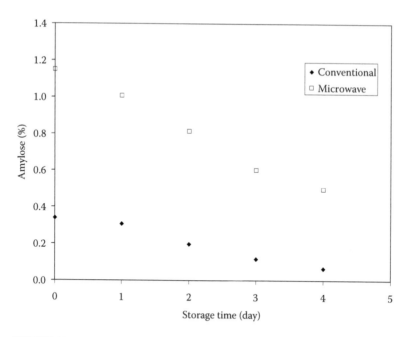

FIGURE 11.6   Variation of amylose content of cakes baked in different ovens. (From Sey-hun, N., Sumnu, G., and Sahin, S., *Food and Bioproducts Processing* 83, 1–5, 2005. With permission from Institution of Chemical Engineers.)

cakes (Seyhun et al., 2005). When amylose contents of microwave and convention-ally baked cakes were compared just after baking and during storage, it was seen that more amylose was leached out during microwave baking than conventional bak-ing (Figure 11.6).

11.3.3 Starch Gelatinization in Microwave Baking

Starch is one of the main components of soft wheat products, so it is necessary to discuss what happens to a starch granule when it is heated by microwaves. Studying starch gelatinization in a microwave oven will be helpful to improve the quality of microwave-baked products.

Gelatinization of wheat starch has been studied by Goebel et al. (1984). In this study, microwave-heated starch samples were found to be nonuniform as compared to conventionally heated ones. Zylema et al. (1985) showed that the distribution of swollen granules and the degree of swelling depended on heating method (micro-wave or conduction heating). Starch granules heated by conduction were swollen to the same extent as those heated by microwaves in limited water systems (1:1 and 1:2) but were less swollen for systems containing higher amounts of water.

Sakonidou et al. (2003) showed that starch gelatinization after microwave heating was incomplete as compared to conventional heating when maize starch suspensions at different concentrations were heated by microwaves. Although the required tem-perature was reached during microwave heating, gelatinization was not complete due to the limited starch–water interaction during the short time of microwave heating.

TABLE 11.4

Differential Scanning Calorimetry Values of Native and Microwave-Heated Cereal Starches

Starch Type	Native			Microwaved		
	T_0 (°C)	T_p (°C)	ΔH (J/g)	T_0 (°C)	T_p (°C)	ΔH (J/g)
Wheat	53.6	59.5	11.5	67.4	72.0	3.2
Corn	61.0	69.5	13.8	72.1	76.1	7.3
Waxy corn	60.4	68.6	14.7	66.4	75.1	13.6

Source: Reprinted from Lewandowicz, G., Janowski, T., and Fornal, J., *Carbohydrate Polymers* 42(2): 2000. With permission from Elsevier.

Lewandowicz et al. (2000) showed that microwave heating reduced crystallinity, solubility, and swelling characteristics of wheat and corn starches. However, microwave heating did not affect those characteristics of waxy corn starch. Starch gelatinization of microwave-treated samples occurred at higher temperatures. Table 11.4 shows the gelatinization temperatures and gelatinization enthalpy of native and microwave-heated cereal starches. As can be seen in Table 11.4, wheat and corn starches were partially gelatinized in the microwave oven. However, gelatinization of waxy corn in microwave heating was found to be insignificant because gelatinization enthalpy values of native and microwave-treated samples were almost the same.

Palav and Seetharaman (2006) investigated whether the starch gelatinization steps in the microwave oven were different from those in conventional heating. They found that swelling of starch granules heated by microwaves did not occur prior to the loss of birefringence. However, in starch granules heated by conduction, swelling and loss of birefringence occurred simultaneously. The loss of crystalline arrangement in microwave-heated samples occurred at a lower temperature compared to that observed for conduction-heated samples. Due to the rotational motion of polar molecules, the crystalline lamella of amylopectin was affected, crystal arrangement was destroyed, and no swelling occurred. Granule swelling was observed at temperatures greater than 65°C in microwave heating. Granule swelling followed loss of birefringence. This is in contrast to conduction heating in which the swelling of starch granules and melting of crystallites are semicooperative processes. In conduction heating, even at 90°C, the starch granule maintains its integrity; however, in microwave heating, granular residues were observed. This could be explained by the rupture of granules during microwave heating which is due to the mechanism of dipolar rotation.

11.3.4 MICROWAVE-BAKED CAKES

In the literature, there are various studies about microwave-baked cakes. These studies are about improving the quality of these products either by using different formulations or different baking conditions.

Hydration level had a significant effect on heating rate of batter for both rice- and wheat-starch-formulated cakes (Sumnu et al., 1999). As the hydration level increased, more water molecules became available for the absorption of microwave energy, which increased batter temperature. When cakes were formulated with wheat, rice, or corn starch, it was observed that wheat-starch-containing cakes had the higher volume as compared to others (Sumnu et al., 2000). This was explained by the failure of rice- and corn-starch-containing cakes to set their structure during the short microwave baking period. The microwave power level was found to be the most effective independent variable in affecting all quality parameters such as volume and texture of cakes.

The effects of two different flours on quality of microwave-baked cakes were compared (Bilgen et al., 2004). Flour A was straight grade flour, and flour B was whole wheat flour. Gluten content and moisture content of flour A were higher than flour B. The baking loss from cakes made with flour A was greater than that of cakes made from flour B. The specific volumes of cakes made from flour B were higher than flour A and similar to conventionally baked ones. Ozmutlu et al. (2001) also showed that microwave-baked breads with lower gluten flour had higher volume than the ones formulated with higher gluten flour. Combining conventional heating with microwave heating produced cakes with qualities similar to those that were conventionally baked (Bilgen et al., 2004).

Cake quality was found to be a function of concentration of monocalcium phosphate monohydrate, which is present in the baking powder. As the concentration of monocalcium phosphate monohydrate increased, the specific volume of cakes decreased, and crumb firmness increased (Martin and Tsen, 1981). When cell structures of microwave and conventionally baked cells were investigated by scanning electron microscope, it was found that cell structures of microwave-baked cakes were coarser than those of conventionally baked cakes. Cells in microwave-baked cakes were irregular and had thicker cell walls than did those in conventionally baked cakes.

The effects of sucrose addition in crystalline form or in liquid form (solubilized in water) on the quality of microwave and conventionally baked cakes were compared (Baker et al., 1990a). It was found that cake structure of conventionally baked cakes showed more variation as a function of formulation compared to microwave-baked cakes.

Pan type was found to be a significant factor in affecting the heating profile of microwave-baked cakes (Baker et al., 1990b). When cakes were baked in glass pans, the edge temperature of cake batter was higher than the center temperature during baking (Baker et al., 1990b; Sumnu et al., 1999). On the other hand, the edge temperature of cake batter baked in a metal pan was lower than the center temperature during microwave baking (Baker et al., 1990b).

There are various patents about microwaveable cakes which focus on obtaining high-quality cakes in the microwave oven. A formulation is given in the U.S. patent of 4,396,635 to obtain soft and moist cakes (Roudebush and Palumbo, 1983). The cake formulation contains leavening, a sugar-to-flour ratio of 1.4:1 to 2:1, 0 to 16% shortening, and 2 to 10% emulsifier. In another study, sponge cakes were prepared by using mesophase gels, which rose and formed an acceptable cake with high volume

(McPherson et al., 2002). Mesophase gel is formed with emulsifiers and an aqueous phase. It contains either a mixture of high and medium emulsifiers with HLB values of 11 to 25 and 6 to 10, respectively, or a mixture of high, medium, and low emulsifiers with HLB values of 11 to 25, 6 to 10, and 2 to 10, respectively. Mesophase gel is added 5 to 15% to cake batter. The name of the emulsifiers that can be used in cake batter and formulation of the cake batter are given in detail in the patent.

Seyhun et al. (2005) added different types of starches (corn, potato, waxy corn, amylomaize, and pregelatinized) to cake formulations to reduce staling of microwave-baked cakes. The control cake formulation contained no additional starch. Starches except amylomaize significantly reduced firmness of cakes as compared to control cakes. Pregelatinized starch, the most effective starch on the retardation of staling, can be recommended for cakes to be baked in the microwave oven. The use of emulsifiers and gums was shown to retard the staling of microwave-baked cakes (Seyhun et al., 2003).

Characteristics of cake batter during baking were studied by Megahey et al. (2005). In microwave baking, batter expanded rapidly during the initial 30 to 50 sec, but during conventional baking it was within 420 sec. This was followed by a period of slight shrinkage of cakes. Cakes baked at 250 W power in a microwave oven showed improved springiness, firmness, and moisture content as compared to cake baked in a conventional oven.

11.3.5 MICROWAVE-BAKED COOKIES

In the baking industry, checking is a failure in cookies, which is due to uneven expansion or contraction of moisture due to nonuniform distribution of moisture within the product. Microwave baking was found to significantly reduce checking to 5% compared to 61% in a conventionally baked cookie (Ahmad et al., 2001). Within 24 h after baking, biscuits had an average of 18% checking, and microwaved biscuits had an average of 1% checking. This showed that more uniform internal moisture profiles can be obtained in microwave baking. The postbaking of biscuits can be done by microwaves to reduce checking and to improve product quality (Ahmad et al., 2001).

The rate of weight loss of microwave-baked biscuits was significantly higher than conventional baking (Pan and Castell-Perez, 1997). When final baking in conventional gas and microwave ovens was compared, higher moisture loss and minimal color change occurred in microwave heating (Sosa-Morales et al., 2004).

11.4   HYBRID TECHNOLOGIES

Because there are quality problems in microwave-baked products, it was recently realized that microwaves should be combined with other heating methods in order to obtain high-quality products. These hybrid technologies are microwave–jet impingement ovens and microwave–infrared combination ovens.

11.4.1 Hybrid Jet Impingement and Microwave Ovens

Although microwave ovens provide more uniform heating of the interior regions of a product in a shorter time, the lack of crust formation and surface color development are their drawbacks (Walker and Li, 1993). Intensive microwave heating causes high internal pressure that pushes more water to the product surface which should be removed by the addition of hot air or infrared heating (Ni et al., 1999). A study investigating the temperature and moisture profiles for infrared and hot air assisted heating showed that for foods with larger infrared penetration depth, infrared heating can actually increase the surface moisture that has to be removed by convection heating to obtain crust (Datta and Ni, 2002). Another study investigating the effect of air flow on heat and mass transfer in a microwave oven showed that convection improves the heat transfer and reduces moisture accumulation inside the oven (Verboven et al., 2003).

Considering the advantages of air impingement and microwave ovens, rapid microwave baking of the interior of the product matches quite well with rapid impingement baking which leads to quick crust formation and color development (Datta et al., 2005; Walker and Li, 1993; Yin and Walker, 1995). During baking in a microwave oven, when the water vapor leaves the product, it comes in contact with cool air at the surface, which causes condensation of the vapor at the product surface and leads to a soggy surface texture. In addition, this cool ambient air inside the microwave oven causes surface cooling, and the low surface temperature prevents Maillard reactions and caramelization from taking place. Impinging high-velocity hot air to the product surface replaces this cool stagnant air with hot and dry air, increases the heat transfer rate, and removes the surface moisture, which leads to quick crust formation and color development. Combining microwave with high-velocity impingement heating would further decrease the processing time and form a surface crust rapidly, which would lock the moisture inside and thus prevent excessive drying of the product.

Browning reactions, mainly caramelization, have been known to be responsible for the development of the crust color, and they need higher temperatures (>100°C) to occur. With only microwave heating, even though we could reduce the moisture content to the equilibrium moisture content, the temperature at the surface could not exceed

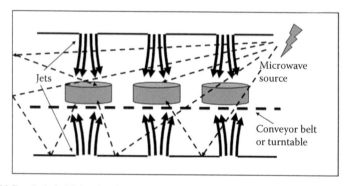

FIGURE 11.7 A hybrid jet impingement–microwave oven showing the two modes of energy transfer.

the evaporation temperature due to lack of convective heat transfer at the surface, and hence, the browning could not take place (Shukla and Anantheswaran, 2001).

A schematic diagram of a hybrid jet impingement–microwave oven is shown in Figure 11.7. For cooking and baking purposes, hybrid ovens combining microwave and impingement (Sharp R90-GC, Thermador JetDirect, Jenn-Air Accellis 5XP, Fujimak SuperJet) have been developed. Thermador JetDirect was designed for use in the home, and Fujimak SuperJet was designed for food service use by Enersyst Development Center (Babyak, 2000).

The combination of microwave heating and impingement heating was studied (Li and Walker, 1996; Ovadia and Walker, 1998; Sumnu et al., 2007; Walker and Li, 1993; Yin and Walker, 1995). A study showed that more compact crumb structure, firmer texture, surface cracking, and low volume were observed in cakes baked in hybrid jet impingement and microwave ovens compared to conventional convection and jet impingement ovens (Li and Walker, 1996). The shorter baking time was the key problem in hybrid jet impingement and microwave baking due to incomplete functioning of leavening acids before the structure began to set. Incorporating more air into the batter by intense creaming and use of liquid shortening, replacing the conventional baking powder with a rapid-acting one, and adjusting water and emulsifier contents appeared to reduce this problem (Li and Walker, 1996). As shown in Table 11.2, higher operating temperatures were used during baking in hybrid jet

TABLE 11.5
Process Parameters for Products Baked in Three Different Ovens

Product	Oven Type	Oven Temperature (°C)	Baking Time (min)
Apple danish	Conventional	190	12
	Impingement	193	5
	Hybrid	193	2.5
Puff pastry	Conventional	204	20
	Impingement	204	12
	Hybrid	204	6.5
Chocolate chip cookies	Conventional	190	12
	Impingement	204	4
	Hybrid	204	2
Blueberry muffin	Conventional	204	12
	Impingement	204	7
	Hybrid	204	3.5
Corn muffin	Conventional	204	14
	Impingement	204	7
	Hybrid	204	3.8

Source: Modified from Walker, C.E. and Li, A., *AIB Research Department Technical Bulletin*, XV, 9, September, 1993. With permission.

impingement and microwave ovens to provide the desired surface color at much shorter baking times (Li and Walker, 1996).

Another study compared the baking of different products in a conventional, impingement, and hybrid jet impingement and microwave oven (Table 11.5). The study showed that although the baking time was the shortest in a hybrid oven, overall appearance and quality of the product were similar for the three ovens (Walker and Li, 1993). In the same study, some formulation adjustments were made in layer cakes by substituting a very rapid leavening acid with the conventional double-acting baking powder to achieve the desired volume in the finished product (Walker and Li, 1993).

It can be visualized that the moisture distributions and crust thickness of the final product would depend on the path followed by the baking process (Figure 11.8). By studying the various combinations of power levels, air temperature, and velocity, an optimum combination can be found to produce a particular product with a desired moisture content, crust thickness, and color.

A numerical study was conducted to understand the effect of sequencing power levels, air temperature, and velocity on moisture distribution (Kocer, 2005). Two different combinations of microwave power level, air velocity, and air temperature were studied as listed in Table 11.6. For all processes, the total process time was chosen as 1200 sec, in which 180 sec of microwave heating was applied with 50% microwave power. In case 1, the process was started with only impingement heating, followed by hybrid impingement and microwave heating, followed by only impingement heating at the end. In the second case, on the other hand, the process was started with hybrid jet impingement and microwave heating, followed by only jet impingement heating. Figure 11.9 shows the moisture profiles after 1200 sec of baking. As seen from the

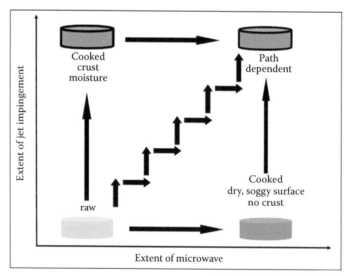

FIGURE 11.8 The effect of jet impingement and microwave sequencing during hybrid baking. (From Kocer D., PhD dissertation, Rutgers, The State University of New Jersey, New Brunswick, 2005.)

TABLE 11.6

Cases Used to Explore the Effect of Sequencing

Stages	Process Variables	Case 1	Case 2
1. Stage	Process time (seconds)	900	180
	Air velocity (m/s)	10	2.5
	Heat transfer coefficient (W/m²K)	40	20
	Air temperature (K)	450	450
	Microwave power	0	50
2. Stage	Process time (seconds)	180	1020
	Air velocity (m/s)	2.5	10
	Heat transfer coefficient (W/m²K)	20	40
	Air temperature (K)	450	450
	Microwave power	50	0
3. Stage	Process time (seconds)	120	NA
	Air velocity (m/s)	10	NA
	Heat transfer coefficient (W/m²K)	40	NA
	Air temperature (K)	450	NA
	Microwave power	0	NA

figures, the effect of focusing microwave energy was observed for case 1, but in case 2 it was not observed. From these results, it can be seen that increasing the product temperature at the beginning of the process by fast microwave heating, followed by only impingement heating resulted in thicker crust. In addition, the focusing effect driven by microwave heating was prevented. The study showed that by sequencing energy modes, we can obtain a product with a desired moisture profile. Microwaves, with their ability to penetrate deep within the product, should be used in the first zones, rapidly raising the internal temperature to a point just below the gelatinization of starch (Walker, 1989).

11.4.2 MICROWAVE–INFRARED COMBINATION OVENS

A microwave–halogen lamp combination oven was produced in 1999. It was sold by the name of Advantium® by General Electric Company (Louisville, KY). This oven includes three halogen lamps in addition to a classical microwave oven design. Two 1500-W halogen lamps are located at the top, and one 1500-W halogen lamp is at the bottom. After this oven was introduced into the market, the effects of this oven on the quality of baked products such as cakes, cookies, and breads were studied (Keskin et al., 2004, 2005; Sevimli et al., 2005; Sumnu et al., 2005; Turabi et al., 2008).

Halogen lamp heating provides near-infrared radiation, and it has lower penetration depth than the other infrared radiation categories. Near-infrared radiation

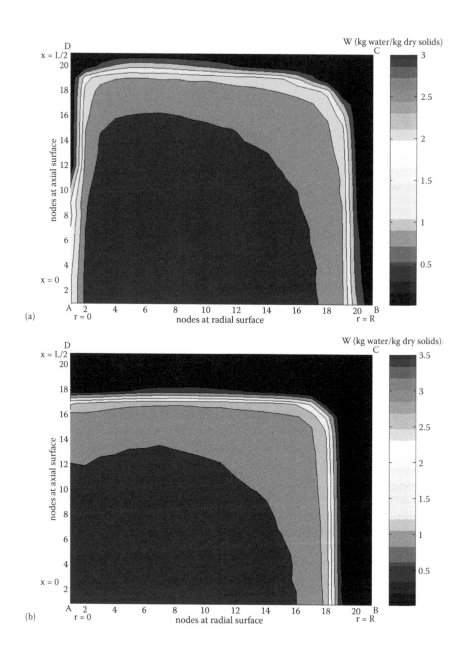

FIGURE 11.9   Moisture contours in potato for (a) case 1, (b) case 2. (From Kocer, D., PhD dissertation. Rutgers, The State University of New Jersey, New Brunswick, 2005.) (See color insert after p. 158.)

mainly affects the surface of the product. As discussed before, heat is generated inside the product when microwaves are used. It is known that a microwave–halogen lamp oven provides the browning and crisping advantages of near-infrared heating with the time-saving advantages of microwave heating.

In the study by Sevimli et al. (2005), baking conditions in the halogen lamp–microwave combination oven were optimized for cakes. It was found that 5 min of baking at 70% upper halogen lamp power, 60% lower halogen lamp, and 30% microwave power resulted in cakes having quality comparable with conventionally baked cakes. The firmness and weight loss of cakes increased as upper halogen lamp power increased. The lower halogen lamp power was found to be insignificant in affecting volume, weight loss, and color of cakes. By using the halogen lamp–microwave combination oven, the conventional baking time of cakes was reduced by 79%.

In the microwave–near infrared combination oven, it was shown that the increase in upper halogen lamp power and baking time increased the color change (ΔE value) of cakes (Figure 11.10) (Sumnu et al., 2005). The higher halogen lamp power might increase the surface temperature of cakes, which might affect the surface color formation. It was possible to obtain a similar color value with conventionally baked cakes in microwave–near infrared combination baking at 70% halogen lamp powers and 50% microwave power after 5 min.

Keskin et al. (2005) investigated the possibility of using a microwave–near-infrared combination oven for the baking of cookies. It was possible to obtain cookies having similar characteristics as conventional ones when 70% halogen lamp,

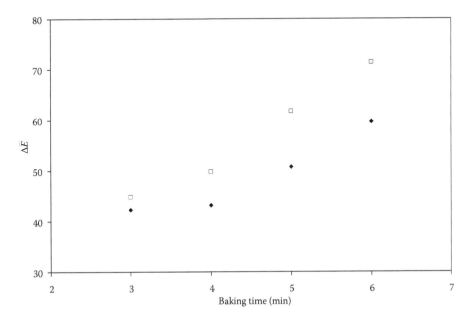

FIGURE 11.10 Variation of color change (ΔE value) in cakes during infrared (IR)–microwave combination baking at 50% microwave power and at different IR powers (◆): 50%, (□): 70%. (From Sumnu, G., Sahin, S., and Sevimli, M., *Journal of Food Engineering* 71, 150–155, 2005. With permission from Elsevier.)

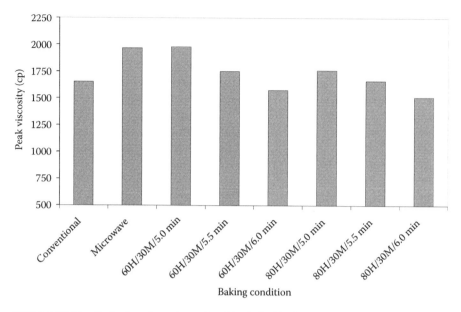

FIGURE 11.11   Peak viscosity values of cookies baked in different ovens. (H: halogen lamp power, M: microwave power.) (Data from Table 11.2, Keskin, S.O., Ozturk, S., Sahin, S., Koksel, H., and Sumnu, G., *European Food Research & Technology*, 220, 546–551, 2005. With permission from Springer Science and Business Media.)

20% microwave power, and 5.5 min baking time were used. The increase in halogen lamp increased the hardness of cookies. Higher spread ratio can be obtained in microwave–near-infrared combination baked products. A rapid viscoanalyzer (RVA) was used to compare starch gelatinization in conventional, microwave, and microwave–near-infrared baked cookies. According to peak viscosity results obtained by RVA, the increase in baking time reduced the peak viscosity, meaning that more starch was gelatinized (Figure 11.11). Microwave-baked samples had higher peak viscosity than conventionally baked ones due to a smaller degree of gelatinization. As near-infrared power and baking time increased, peak viscosity approached that of conventionally baked cookies (Figure 11.11).

Recently, the possibility of using a microwave–near-infrared combination oven was studied for the baking of rice cakes (Turabi et al., 2008). Rice cakes containing no gluten are critical for patients with celiac disease, because these patients cannot consume any gluten-containing products. It was possible to produce high-quality cakes in the oven when rice cakes were combined with different gums and the emulsifier Purawave® (Puratos, Belgium). Xanthan gum and the emulsifier Purawave were recommended to be used in rice cakes to achieve high volume and soft texture.

11.5   CONCLUSIONS

Jet impingement ovens are a unique type of convection oven in which high-velocity jets of hot air impinge on a food product. It has the advantages of a higher rate of surface heat transfer and rapid moisture removal, which result in quick crust formation and color development. Microwave ovens use electromagnetic energy to heat the product by molecular absorption of energy, mainly by water and salt molecules in the food product. Microwave ovens provide faster processing by accelerated heat transfer and moisture migration. Even though baking operation in a microwave oven requires considerably less time as compared to a conventional convection oven, the lack of crust and desirable color formation are its main limitation. Because microwave heating offers many opportunities, researchers have been investigating ways to have the ease of fast processing while maintaining quality. Changing the formulation, using susceptors, and adding flavors are some of the ways to reduce quality problems in microwave baking. Other efforts include changing the design of the ovens. Finally, the introduction of hybrid technologies such as microwave and jet impingement, and microwave and near-infrared are promising improvements to overcome the quality problems associated with microwave ovens. These combination ovens offer the advantages of energy efficiency due to a faster rate of heat transfer; energy savings due to operation at lower temperature and less processing times; and quality improvement with crust formation, surface browning, and flavor development, and at the same time retention of moisture inside the baked product. More studies on heating mechanism, physicochemical changes during baking, physical and electrical properties of the products, product quality improvement, and process optimization are needed to effectively use these alternative baking technologies.

REFERENCES

Adrian, R.J. 1991. Particle-imaging techniques for experimental fluid mechanics. *Annual Review of Fluid Mechanics* 23: 261–304.

Ahmad, S.S, M.T. Morgan, and M.R. Okos. 2001. Effects of microwave on the drying, checking and mechanical strength of baked biscuits. *Journal of Food Engineering* 50: 63–75.

Angioletti, M., R.M. Di Tommaso, E. Nino, and G. Ruocco. 2003. Simultaneous visualization of flow field and evaluation of local heat transfer by transitional impinging jets. *International Journal of Heat and Mass Transfer* 46: 1703–1713.

Babyak, R. 2000. Racing Recipes. Appliance design. www.appliancedesign.com/CDA/Archives/e3cd0c5404938010VgnVCM100000f932a8c0 (accessed December 12, 2006).

Baker, B.A., E.A. Davis, and J. Gordon. 1990a. The influence of sugar and emulsifier type during microwave and conventional heating of a lean formula cake batter. *Cereal Chemistry* 67: 451–457.

Baker, B.A., E.A. Davis, and J. Gordon. 1990b. Glass and metal pans for use in microwave and conventionally heated cakes. *Cereal Chemistry* 67: 448–451.

Baughn, J.W. 1995. Liquid crystal methods for studying turbulent heat transfer. *International Journal of Heat and Fluid Flow* 16: 365–375.

Bengtsson, N. 2001. Development of industrial microwave heating of foods in Europe. *Journal of Microwave Power and Electromagnetic Energy* 36: 227–240.

Bilgen, S., Y. Coskuner, and E. Karababa. 2004. Effects of baking parameters on the white layer cake quality by combined use of conventional and microwave ovens. *Journal of Food Processing and Preservation* 28: 89–102.

Borquez, R., W. Wolf, W.D. Koller, and W.E.L. Spie. 1999. Impinging jet drying of pressed fish cake. *Journal of Food Engineering* 40: 113–120.

Campana, L.E., Sempe, M.E., and R.R. Filgueira. 1993. Physical, chemical and baking properties of wheat dried with microwave energy. *Cereal Chemistry* 70: 760–762.

Cornaro, C., A.S. Fleischer, and R.J. Goldstein. 1999. Flow visualization of a round jet impinging on cylindrical surfaces. *Experimental Thermal and Fluid Science* 20: 66–78.

Datta, A.K. 1990. Heat and mass transfer in the microwave processing of food. *Chemical Engineering Progress* 86: 47–53.

Datta, A.K. and H. Ni. 2002. Infrared and hot-air-assisted microwave heating of foods for control of surface moisture. *Journal of Food Engineering* 51: 355–364.

Datta A.K., S.S.R. Geedipalli, and M.F. Almedia. 2005. Microwave combination heating. *Food Technology* 59: 36–40.

De Bonis, M.V. and G. Ruocco. 2005. Modeling local heat and mass transfer in food slabs due to air impingement. *Journal of Food Engineering* 78: 230–237.

Durst, F., B. Lehmann, and C. Tropea. 1981. Laser Doppler system for rapid scanning of flow fields. *Review of Scientific Instruments* 52: 1676–1681.

Gardon, R. and J.C. Akfirat. 1965. The role of turbulence in determining the heat transfer characteristics of impinging jets. *International Journal of Heat and Mass Transfer* 8: 1261–1272.

Goebel, N.K., J. Grider, E.A. Davis, and J. Gordon. 1984. The effects of microwave energy and conventional heating on wheat starch granule transformations. *Food Microstructure* 3: 73–82.

Goldstein, R.J. and J.F. Timmers. 1982. Visualization of heat transfer from arrays of impinging jets. *International Journal of Heat and Mass Transfer* 25: 1857–1868.

Henke, M.C. 1985. Air impingement baking—Technology for the future. *The Consultant* XVIII (2, Spring): 28–33.

Higo, A. and S. Noguchi. 1987. Comparative studies on food treated with microwave and conductive heating. I. Process of bread hardening by microwave heating. *Journal of the Japanese Society for Food Science and Technology* 34: 781–787.

Huber, A.M. and R. Viskanta. 1994. Impingement heat transfer with a single rosette nozzle. *Experimental Thermal and Fluid Science* 9: 320–329.

Keskin, S.O., G. Sumnu, and S. Sahin. 2004. Bread baking in halogen lamp-microwave combination baking. *Food Research International* 37: 489–495.

Keskin, S.O., S. Ozturk, S. Sahin, H. Koksel, and G. Sumnu. 2005. Halogen lamp-microwave combination baking of cookies. *European Food Research and Technology* 220: 546–551.

Kim, Y.R. and P. Cornillon. 2001. Effects of temperature and mixing time on molecular mobility in wheat dough. *Lebensmittel Wissenschaft und Technologie* 34: 417–423.

Kocer, D. 2005. Numerical simulation and experimental investigation of the baking process in a hybrid jet impingement and microwave (JIM) oven. PhD dissertation. Rutgers, The State University of New Jersey.

Kocer, D. and M.V. Karwe. 2005. Thermal transport in a multiple jet impingement oven. *Journal of Food Process Engineering* 8: 378–396.

Lambert, L.L.P., J. Gordon, and E.A. Davis. 1992. Water loss and structure development in model cake systems heated by microwave and convection methods. *Cereal Chemistry* 69: 303–309.

Lee, J. and S.J. Lee. 2000. Effect of nozzle configuration on stagnation region heat transfer of axisymmetric jet impingement. *International Journal of Heat and Mass Transfer* 43: 3497–3509.

Lewandowicz, G., T. Janowski, and J. Fornal. 2000. Effect of microwave radiation on physico-chemical properties and structure of cereal starches. *Carbohydrate Polymers* 42: 193–199.

Li, A. and C.E. Walker. 1996. Cake baking in conventional, impingement and hybrid ovens. *Journal of Food Science* 61: 188–191, 197.

Marcroft, H.E. and M.V. Karwe. 1999. Flow field in a hot air jet impingement oven—Part I: A single impinging jet. *Journal of Food Processing and Preservation* 23: 217–233.

Marcroft, H.E., M. Chandrasekaran, and M.W. Karwe. 1999. Flow field in a hot air jet impingement oven—Part II: multiple impingement jets. *Journal of Food Processing and Preservation* 23: 235–248.

Martin, D.J. and C.C. Tsen. 1981. Baking high ratio white layer cakes with microwave energy. *Journal of Food Science* 46: 1507–1513.

Martin, H. 1977. Heat and mass transfer between impinging gas jets and solid surfaces. *Advanced Heat Transfer* 13: 1–60.

McPherson, A.E., W. Chen, A. Akashe, and M. Miller. 2002. Microwaveable sponge cake. U.S. Patent 6,410,073.

Megahey, E.K., W.A.M. McMinn, and T.R.A. Magee. 2005. Experimental study of microwave baking of madeira cake batter. *Food and Bioproducts Processing* 83: 1–11.

Mesbah, M., J.W. Baughn, and C.W. Yap. June 25–28, 1996. The effect of curvature on the local heat transfer to an impinging jet on a hemispherically concave surface. In: *Proceedings of the Ninth International Symposium on Transport Phenomena in Thermal-Fluids*, Singapore.

Metaxas, A.C. and R.J. Meredith. 1983. *Industrial Microwave Heating.* Peter Peregrinus, London.

Motwani, T., K. Seetharaman, and R.C. Anantheswaran. 2007. Dielectric properties of starch slurries as influenced by starch concentration and gelatinization. *Carbohydrate Polymers* 67: 73–79.

Ni, H., A.K. Datta, and K.E. Torrance. 1999. Moisture transport in intensive microwave heating of biomaterials: A multiphase porous media model. *International Journal of Heat and Mass Transfer* 42: 1501–1512.

Nitin, N. and M.V. Karwe. 2001. Measurement of heat transfer coefficient for cookie-shaped object in a hot air jet impingement oven. *Journal of Food Process Engineering* 24: 51–69.

Nitin, N. and M.V. Karwe. 2004. Numerical simulation and experimental investigation of conjugate heat transfer between a turbulent hot air jet impinging on a cookie. *Journal of Food Science* 69: 59–65.

Nitin, N., R.P. Gadiraju, and M.V. Karwe. 2006. Numerical simulation and experimental investigation of conjugate heat transfer between turbulent hot air jet impinging on a hot-dog-shaped objects. *Journal of Food Process Engineering* 29: 386–399.

Olsson, E.E.M., L.M. Ahrne, and A.C. Tragardh. 2004. Heat transfer from a slot air jet impinging on a circular cylinder. *Journal of Food Engineering* 63: 393–401.

Olsson, E.E.M., L.M. Ahrne, and A.C. Tragardh. 2005. Flow and heat transfer from multiple slot air jets impinging on circular cylinders. *Journal of Food Engineering* 67: 273–280

Ovadia, D.Z. and C.E. Walker. 1998. Directing jets of fluid such as air against the surface of food provides advantages in heating, drying, cooling, and freezing. *Food Technology* 52: 46–50.

Ozmutlu, O., G. Sumnu, and S. Sahin. 2001. Effects of different formulations on the quality of microwave baked bread. *European Food Research und Technologie* 213: 38–42.

Palav, T. and K. Seetharaman. 2006. Mechanism of starch gelatinization and polymer leaching during microwave heating. *Carbohydrate Polymers* 65: 364–370.

Pan, B. and M.E. Castell-Perez. 1997. Textural and viscoelastic changes of canned biscuit dough during microwave and conventional baking. *Journal of Food Process Engineering* 20: 383–399.

Pan, Y. and B.W. Webb. 1995. Heat transfer characteristics of arrays of free-surface liquid jets. *Transactions of the ASME, Journal of Heat Transfer* 117: 878–883.

Pan, Y., J. Stevens, and B.W. Webb. 1992. Effect of nozzle configuration on transport in the stagnation zone of axisymmetric, impinging free-surface liquid jets—Part 2: Local heat transfer. *Transactions of the ASME, Journal of Heat Transfer* 114: 880–886.

Popiel, C.O. and O. Trass. 1991. Visualization of a free and impinging round jet. *Experimental Thermal and Fluid Science* 4: 253–264.

Rogers, D.E., L.C. Doescher, and R.C. Hoseney. 1990. Texture characteristics of microwave reheated bread. *Cereal Chemistry* 67:188–191.

Roudebush, R.M., and P.D. Palumbo. 1983. Microwave cake mix. U.S. Patent 4,396,635.

Sakiyan, O., G. Sumnu, S. Sahin, and V. Meda. 2007. Investigation of dielectric properties of different cake formulations during microwave and infrared-microwave combination baking. *Journal of Food Science* 72(4): E205–E213.

Sakonidou, E.P., T.D. Karapantsios, and S.N. Raphaelides. 2003. Mass transfer limitations during starch gelatinization. *Carbohydrate Polymers* 53: 53–61.

Saripalli, K.R. 1983. Visualization of multijet impingement flow. *American Institute of Aeronautics and Astronautics Journal* 21: 483–484.

Sarkar, A., N. Nitin, M.V. Karwe, and R.P. Singh. 2004. Fluid flow and heat transfer in air jet impingement in food processing. *Journal of Food Science* 69: 113–122.

Sevimli, M., G. Sumnu, and S. Sahin. 2005. Optimization of halogen lamp-microwave combination baking of cakes: A response surface study. *European Food Research and Technology* 221: 61–68.

Seyhun, N., G. Sumnu, and S. Sahin. 2003. Effects of different emulsifiers, gums and fat contents on retardation of staling of microwave baked cakes. *Nahrung-Food* 47: 248–251.

Seyhun, N., G. Sumnu, and S. Sahin. 2005. Effects of different starch types on retardation of staling of microwave-baked cakes. *Food and Bioproducts Processing* 83: 1–5.

Shukla, T.P. and R.C. Anantheswaran. 2001. Ingredient interactions and product development for microwave heating. In: *Handbook of Microwave Technology for Food Applications*, Ed. A.K. Datta and R.C. Anantheswaran, Chapter 11, 355–395. New York: Marcel Dekker.

Smith, D.P. 1975. Cooking Apparatus. U.S. Patent 3,884,213.

Sosa-Morales, M.E., G. Guerrero-Cruz, H. Gonzale-Loo, and J.F. Velez-Ruiz. 2004. Modeling of heat and mass transfer during baking of biscuits. *Journal of Food Processing and Preservation* 28: 417–432.

Sparrow, E.M. and B.J. Lovell. 1980. Heat transfer characteristics of an obliquely impinging circular jet. *Journal of Heat Transfer* 102: 202–209.

Stauffer, C.E. 2000. Emulsifiers as antistaling agents. *Cereal Foods World* 45(3): 106–110.

Stoy, R.L. and Y. Ben-Haim. 1973. Turbulent jets in a confined crossflow. *Journal of Fluids Engineering-Transactions of the ASME* 95(4): 551–556.

Sugiyama, Y. and Y. Usami. 1979. Experiments on flow in and around jets directed normal to a cross flow. *Bulletin of Japan Society of Mechanical Engineers* 22: 1736–1745.

Sumnu, G. 2001. A review on microwave baking of foods. *International Journal of Food Science and Technology* 36: 117–127.

Sumnu, G. and S. Sahin. 2005. Baking using microwave processing. In: *Microwave Processing of Foods*, Ed. M. Reiger and H. Schubert, 119–142. Cambridge: Woodhead.

Sumnu, G., M.K. Ndife, and L. Bayindirli. 1999. Temperature and weight loss profiles of model cakes baked in the microwave oven. *Journal of Microwave Power and Electromagnetic Energy* 34: 221–226.

Sumnu, G., M.K. Ndife, and L. Bayindirli. 2000. Optimization of microwave baking of model layer cakes. *European Food Research and Technology* 211: 169–174.

Sumnu, G., S. Sahin, and M. Sevimli. 2005. Microwave, infrared, infrared-microwave combination baking of cakes. *Journal of Food Engineering* 71: 150–155.

Sumnu, G., A.K. Datta, S. Sahin, S.O. Keskin, and V. Rakesh. 2007. Transport and related properties of breads baked using various heating modes. *Journal of Food Engineering* 78: 1382–1387.

Turabi, E., G. Sumnu, and S. Sahin. 2008. Rheological properties and quality of rice cakes formulated with different gums and an emulsifier blend. *Food Hydrocolloids* 22: 305–312.

Verboven, P., A.K. Datta, N.T. Anh, N. Scheerlinck, and B.M. Nicolai. 2003. Computation of airflow effects on heat and mass transfer in a microwave oven. *Journal of Food Engineering* 59: 181–190.

Viskanta, R. 1993. Heat transfer to impinging isothermal gas and flame jets. *Experimental Thermal and Fluid Science* 6: 111–134.

Wahlby, U., C. Skjoldebrand, and E. Junker. 2000. Impact of impingement on cooking time and food quality. *Journal of Food Engineering* 43: 179–187.

Walker, C.E. 1989. Impingement oven technology—Part I: Principles. *AIB Research Department Technical Bulletin* XI, 11, November.

Walker, C.E. and A. Li. 1993. Impingement oven technology—Part III: Combining impingement with microwave (hybrid oven). *AIB Research Department Technical Bulletin* XV, 9: 1–6.

Walker, C.E. and A.B Sparman. 1989. Impingement oven technology—Part II: Applications and Future. *AIB, Research Department, Technical Bulletin* XI, 11, November.

Whorton, C. and G. Reineccius. 1990. Current developments in microwave flavours. *Cereal Foods World* 35: 553–559.

Yin, Y. and C.E. Walker. 1995. A quality comparison of breads baked by conventional versus nonconventional ovens: A review. *Journal of the Science of Food and Agriculture* 67: 283–291.

Zylema, B.J., J.A. Grider, J. Gordon, and E.A. Davis. 1985. Model wheat starch systems heated by microwave irradiation and conduction with equalized heating times. *Cereal Chemistry* 62: 447–453.

12 Low-Sugar and Low-Fat Sweet Goods

Manuel Gómez

CONTENTS

12.1   INTRODUCTION

There are two seemingly contradictory trends spanning Western societies. One dietary trend is the increasing consumption of sugars, fats, and calories recorded in recent decades; a contrasting trend is the growing awareness shown by consumers of anything related to personal physical appearance, fitness, and health issues. Of the top ten mortality rates in the ranking of leading causes of death, five are diet related (heart disease, cancer, stroke, diabetes, and atherosclerosis). As a result, and thanks also to the dietary recommendations made by health-care professionals and much information presented by the media—albeit not always reliable—a marked increase in the consumption of low-calorie, low-fat, and sugar-free products has taken place (Figure 12.1). Cakes, cookies, and a number of yeast-raised bakery products account for a substantial proportion of the fat and sugar intake. Hence, the development of special dietary foods (low-calorie, low-fat, and sugar-free foods) with good organoleptic properties should enable manufacturers to diversify their production and make inroads into an emergent sector of the market. Development of such special products can help implement and maintain specific diets, particularly if available to consumers when eating out away from home (Sigman-Grant, 1997).

Although for this kind of product a reduction in calories can be obtained by substituting fibers, or raw materials rich in fiber, for flour (Kaack and Pedersen, 2005), most low-calorie baked goods are produced by lowering fat and sugar contents. Products rich in sugars and fats are usually placed on the upper levels of the food pyramid, and hence the informed advice is that they should be eaten sparingly. However, the challenge of developing special dietary products first requires the sifting of truth from myth both on the adverse effects of the consumption of fats and sugars, and on the virtues of a fat- and sugar-free diet. In fact, consumers of low-fat products show reduced intakes of total fat, saturated fats, and cholesterol, but their diets may include insufficient amounts of some other nutrients (Peterson et al., 1999).

Products rich in fats and sugars are usually associated with pleasant sensations, and they are highly regarded organoleptically; it is therefore no mean task to match their sensory properties and their widespread acceptability. To be successful in the substitution of fats and sugars, the food industry must resort to fat and sugar replacers capable of mimicking their roles, but these replacers seldom perform all the functions attributable to fats and sugars, and consequently, the new products are bound to lead to less favorable sensory evaluation results. The development of special dietary products thus demands a thorough understanding of the roles played by fats and sugars in each formulation as well as a detailed knowledge of the options available as potential replacers. By and large, consumers find these products not as pleasant as

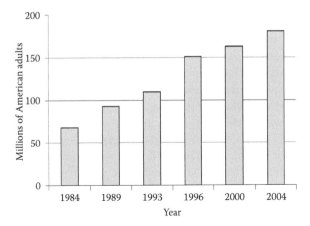

FIGURE 12. Consumer consumption of low-calorie, sugar-free foods and beverages in the United States. (From Calorie Control Council, Trends and Statistics, 2006, www.caloriecontrol.org/lcchart.html.

the regular products, although they readily admit they are "better for you" (Tuorila et al., 1997). When compared with the regular products, the purchase intent of these special dietary products is higher than the "overall degree of liking" (Guinard, et al., 1996). The challenge that the bakery industry faces is the development of modified products capable of receiving high hedonic ratings. Furthermore, in many cases, the modified product is required to be as similar as possible to the regular product it is trying to replace.

12.2 NUTRITIONAL PROBLEMS OF THE CONSUMPTION OF FATS AND SUGARS

12.2.1 SUGARS

12.2.1.1 Sugar and Dental Caries

It has been shown that the consumption of sugar-rich foods promotes the development of dental caries (Makinen and Isokangas, 1988; Sreebny, 1982). Dental caries is a complex process occurring, in part, through the action of microorganisms capable of fermenting certain carbohydrates in the mouth. The end products of this fermentation process are lactic acid and some polysaccharides that adhere to the surface of the teeth and form plaques. This process often leads to cavitation of the teeth. Sugars are carbohydrates susceptible to fermentation by microorganisms in the mouth, and although this is not the only factor responsible for tooth decay, there is no doubt it is closely involved in its development. It is also known that there are some other factors affecting the incidence of dental caries, such as tooth structure, mouth microflora of each individual, proper oral hygiene habits, and the inclusion of fluorine in drinking water. In fact, over the last few years, a marked reduction in the incidence of dental caries has been recorded in advanced societies owing to better oral hygiene practices and the fluoridation of drinking water.

12.2.1.2   Sugar and Blood Glucose

The term glycemic index has come to be regarded as important in the determination of plasma glucose levels. This term refers to the relationship between glucose absorption and a particular food item. A high glycemic index means a rapid absorption of carbohydrates, whereas a low glycemic index means a slow absorption of carbohydrates. However, glucose absorption also depends upon the physical state of the foodstuff, the form in which it is eaten, and the individual in question. The glycemic index of foodstuffs is determined through personal tests and complex procedures, nevertheless it is quite useful for the diabetic whose paramount concern must be to avoid blood glucose increases. Although some researchers have questioned its usefulness in the planning of diets, there are signs the glycemic index is becoming a valuable concept in the development of special diets.

Overall, sugars are rapidly absorbed in the gut, producing a sharp rise in blood glucose levels, and thus rapidly increasing insulin secretion to allow cells the use of this glucose and stimulate the synthesis of triglycerides. After this stage, a decline in the blood glucose level occurs, and hunger arises. Hence, intake of simple carbohydrates produces transitory satiety. Conversely, complex carbohydrates are absorbed much more slowly, resulting in a lower increase of blood glucose and thus in a smaller secretion of insulin. Triglyceride synthesis is minimal, blood glucose levels are maintained over longer periods of time, and hunger arises much more gradually.

12.2.1.3   Sugar and Diabetes

Diabetes mellitus is a disease characterized by hyperglycemia because of disturbances in the normal insulin production mechanism or the inability of the secreted insulin to adequately perform its function. As a result, high levels of blood sugar ensue, and shock or even death might occur. There are two main categories of the disease. The first is called insulin-dependent diabetes, or type I diabetes; it is usually known as juvenile diabetes because its onset takes place at an early age. Patients with this condition require an injection of insulin daily. The second is type II diabetes or noninsulin-dependent diabetes, also known as maturity-onset diabetes; it manifests itself among adults and seems to be influenced by a genetic component. In this latter case, patients show significant amounts of insulin in their plasma; their diet must be carefully controlled, and they must exercise regularly. It has been reported that the incidence of type II diabetes is higher among obese individuals.

These patients must reduce the intake of foodstuffs with a high glycemic index and combine them with some low-index foodstuffs. Fructose is the only sugar with a metabolism independent from the presence of insulin, hence its very low glycemic index and its widespread use as glucose and sucrose substitute in the manufacture of products aimed at diabetics. Overall, there is no clear evidence that a correlation exists between sugar consumption and diabetes. On the contrary, it has been reported that diets rich in carbohydrates reduce the risk of developing diabetes (Feskens and Kromhout, 1990; Marshall et al., 1994). Nevertheless, some other authors have pointed out that diets with a high glycemic index that are low in fiber increase the risk of developing type II diabetes (Salmeron et al., 1997a, 1997b).

12.2.1.4   Sugar and Obesity

Although the belief that a diet rich in sugars promotes obesity is widely held, there is in fact no evidence to support it. Sugars, like other carbohydrates, and proteins yield 4 kcal/g, fats yield 9 kcal/g, and alcohol yields 7 kcal/g. The human body is capable of gaining weight when more calories are being consumed than are required. Consequently, sugar does not promote obesity to a greater extent than proteins or other carbohydrates do, and is well below fats and alcohol in calorie yield (Hill and Prentice, 1995). It has been reported, however, that low-sugar diets can result in weight reduction (Colditz et al., 1990), in all probability because of their lower calorie content; some researchers have suggested that obese individuals show a marked preference for sugar-rich foods (Drewnowski et al., 1991). It may very well be that all this evidence has helped keep alive the notion that sugar consumption results in weight gain.

12.2.1.5   Sugar and Cardiovascular Disease

Studies on the link between sugar consumption and the development of cardiovascular diseases show conflicting results (Howard and Wylie-Rosett, 2002). Yudkin and Evans (1972) and Yudkin (1978) asserted that an excessive consumption of sugars is correlated with cardiovascular disease, but later researchers have begged to differ from that view (Bolton-Smith and Woodward, 1994; García-Palmieri et al., 1980). A more recent study has again restated the validity of the link between consumption of foods with a high glycemic index and cardiovascular disease in women (Liu et al., 2000). At present, there is no conclusive evidence on this point, and it is to be hoped that further research will provide new clarifying data.

In addition, several studies about the influence of sugar on cholesterol and triglycerides have been carried out. Thus, it has been shown that a diet with a high sucrose content increases triglyceride levels, although this effect depends upon the sugar level and the rest of the intake nutrients (Frayn and Kingman, 1995). Similarly, it seems equally established that when sugar consumption is on the rise, there is a drop in the level of high-density lipoprotein (HDL) cholesterol (Archer et al., 1998).

12.2.1.6   Sugar and Hyperactivity

It is a quite commonly held belief that sugar consumption leads to hyperactivity in children. However, Wolraich et al. (1985) established that a decrease in the sugar levels in the diet of hyperactive children failed to show any effect on the degree of hyperactivity.

12.2.2   Fats

12.2.2.1   Fats and Obesity

An excess of dietary fat entails a high caloric density of many of the component foodstuffs which, in turn, causes the excess fat to be deposited as adipose tissue, eventually leading to obesity. When the level of fat is too high but the caloric intake is adequate, the drawback often is the lack of essential nutrients due to a restricted intake of products, such as cereals and legumes, rich in carbohydrates and proteins.

For further information on the relationship between fat consumption and obesity, there are useful reviews by Bray and Popkin (1998) and Bray et al. (2004).

12.2.2.2   Fats and Atherosclerosis

Another parameter associated with fat intake worth mentioning is the blood cholesterol level (cholesterolemia). In fact, a high blood cholesterol level (hypercholesterolemia) is considered to be the first factor of risk for cardiovascular disease. It has been established that serum cholesterol rises with high fat intake either because the rate of saturated fats is excessive or because the daily fat intake exceeds 300 mg/day. For their transport in the bloodstream, lipids must be bound to proteins, thus forming lipoproteins. Lipoproteins are classified according to their density and are commonly divided into two groups: low-density lipoproteins (LDL) and high-density lipoproteins (HDL). LDL carry endogenous cholesterol toward the cells which capture and use it through their membrane receptors. When this reception is limited or the cholestrol level is too high, the level of LDL in the blood rises; this is one of the most significant risk factors for atherosclerosis. On the other hand, HDL transport cholesterol used up by the cells toward the liver to be disposed of as bile acids. An increase in HDL levels in the blood means an improved protection against atherosclerosis.

Polyunsaturated fatty acids (seed and fish oils) lower blood cholesterol levels but do not help raise the HDL levels. Alternatively, monounsaturated fatty acids (olive oil) do not modify blood cholesterol levels but raise HDL levels and have then a positive effect. As a rule, a reduced intake of saturated fatty acids not over 10% of daily calorie intake is recommended. The recommended ratio in the fatty acid intake is 1/1/1 (saturated/monounsaturated/polyunsaturated), keeping in mind that there should be an adequate intake of essential fatty acids (no more than 1% of kilocalories consumed). Outstanding reviews about the effect of different fatty acids on cholesterol levels and atherosclerosis are available (Khosla and Sundram, 1996; Kritchevsky, 2000).

Fat intake has also been connected with a greater risk of developing coronary conditions and colon and lung cancer, but studies disagree, and it can be safely stated that no current consensus exists on these points. Rothstein (2006) recently published a historical overview dealing with these studies.

12.2.2.3   Trans Fatty Acids

A final factor to be considered when discussing fat intake is the presence of trans fatty acids in the diet. It is known that intake of this kind of fatty acids increases fat cholesterol levels and those of LDL cholesterol, whereas the HDL cholesterol level is lowered with the corresponding risk of clogged arteries (Ascherio, 2006; Ascherio and Willett, 1997; Katan et al., 1995; Lichtenstein, 2000). Moreover, intake of trans fatty acids has been associated with a greater risk of developing type II diabetes, although the evidence provided is somewhat contradictory (Odegaard and Pereira, 2006).

In conclusion, in their continuing effort to develop products with better nutritional characteristics, manufacturers of bakery products should not just focus on producing low-calorie products, but, in addition, they should meet the challenge of nutritionally improving the fats they use.

12.3   FUNCTIONS OF SUGARS AND FATS IN SWEET GOODS

12.3.1   Sugar

Not all bakery products require added sugar, but the common practice with most of these products is to include sugar in their formulations. Amounts range from less than 5% in some breads to over 25% in certain cakes. High-ratio cakes contain more sugar than flour, and the effect of the sugar content on their processing and quality is obvious. Traditionally, sucrose has been the sweetener of choice, though bakers have available to them a wide range of sugars derived from the hydrolysis of cereal starch, such as glucose and fructose syrups.

In some foods, sugar is just a sweetening agent. The substitution strategy must then rely on the use of intense sweeteners such as saccharin and aspartame. However, in bakery products, the functions of sugar are much more complex and vary depending on different types of products. Apart from its sweetening action, sugar performs a variety of functions, including fermentation substrate, formation of crust color, flavor enhancer, texture modifier, development of structure, and shelf-life improvement. Sugar replacement is surely a daunting task because sometimes not only must ingredients be replaced, but even the processes involved must be changed or replaced. The functions of sugars in bakery products have been dealt with extensively in the literature (Alexander, 1998; Ponte, 1990; Sluimer, 2005), and a detailed understanding of these functions is a necessary prerequisite to adopting successful substitution strategies.

12.3.1.1   Yeast-Raised Products

In leavened doughs, yeasts transform sugars into alcohol and carbon dioxide. If sugars are not included in the dough formula, yeasts will mainly transform maltose resulting from the enzyme hydrolysis of starch. When small amounts of sucrose (1 to 2%) are added, yeasts will first transform this sugar before affecting maltose, and a slight increase in the rate of fermentation, particularly in the early stages, will take place. When the amount of added sugar exceeds 5%, a fall in water activity and a rise in osmotic pressure occur in doughs. This brings about a reduction in yeast activity and fermentation rates. Under such conditions, it will be necessary to extend fermentation times or raise yeast doses. Another option might be the use of osmotolerant yeasts.

During baking, dough sugars are involved in two browning reactions that determine the final color of the finished product: Maillard and caramelization reactions. The Maillard reaction occurs through interactions between reducing sugars and amino acids or peptides in the dough which results in melanoidin formation. The resulting final color and aroma will depend upon the type of sugars and amino acids present in the dough. On the other hand, the caramelization reaction consists of a thermal degradation of sugars that change color from a pale yellow in the initial stages to a dark brown in the final stages of the process. As a rule, products with high levels of added sugar are baked at lower temperatures to prevent the adverse effects an excessive caramelization could bring about. Maillard and caramelization reactions require high temperatures, over 100°C, and thus they can only take place on the crust of baked products or on toasted bread.

Both Maillard and caramelization reactions, together with the fermentation process, are determinant factors in the final flavor of bakery products. Thus, caramelized sugars impart an appealing caramel-like flavor in the initial stages or a strong bitter flavor and burnt aroma in more advanced stages. The type of sugar in the dough also affects the compounds generated by the Maillard reaction and thus the aroma of the product.

In doughs where gluten development is essential during kneading, sugars compete with proteins for free water, and gluten development is delayed. Kneading times should then be extended. The presence of sugars has a relaxation effect on doughs, because their resistance and consistency are reduced. Hence, doughs including sugar in their formulation require stronger flours. It is also common to reduce the amount of water added to these types of doughs to improve their consistency. For other products, such as cakes and some cookies, gluten development is supposed to be prevented, and the presence of sugars can only be beneficial.

An additional effect of sugar is its influence on the starch swelling pattern in doughs. When significant amounts of sugar are involved, as in the case of sweet goods, the gelatinization temperature of starch is raised; the greater the amount of sugar in the dough, the higher the gelatinization temperature. As a result, the dough sets during baking, a fraction later allowing longer dough expansion times and providing products of greater volume. This effect is particularly important when baking cakes.

The water holding capacity of sugars favors a soft and tender crumb. It also inhibits staling and the growth of microorganisms, and improves the shelf life of bakery products.

12.3.1.2 Cakes

The presence of sugar in cakes not only influences their flavor, color, texture, and shelf life but also exerts a significant influence on the final volume of the product. Cake manufacture starts with batter preparation or mix of ingredients. At this stage, the aim is to trap air in the mix in the form of tiny bubbles. During subsequent baking, gas from baking powder and the water vapor released fill the bubbles to create the final porous structure. Once large numbers of minute bubbles are entrapped in the batter, the final grain of the cake crumb can be fine and uniform. However, if the entrapped air eventually forms larger bubbles, the final grain will have large and irregular pores. In addition to a correct aeration, the batter must reach an appropriate viscosity, because a batter with low viscosity would result in the escape of the entrapped gases. As temperature rises in the oven, the viscosity of the batter falls until the gelatinization of the starch in the flour is triggered and a significant amount of water is absorbed. At this point, viscosity increases until the gelatinization is completed and the structure of the cake is set, thereby arresting the expansion of the batter. Gas production should mainly occur throughout the gelatinization process, because it is then, prior to cake setting, the capability of retaining gas is higher. In the final analysis, it can be said that sugar plays a triple role in the production of cakes as it promotes the entrapment of air in the batter, increases viscosity, and delays the temperature of starch gelatinization (Bean and Yamazaki, 1978; Ngo and Taranto, 1986), all of which will contribute to the final volume of the product in a decisive manner.

Broadly speaking, cakes fall into two major groups. The first group includes all those cakes with substantial amounts of shortening in their formulations, such as yellow layer cakes, white layer cakes, or pound cakes. They can be termed shortening cakes because shortening plays a fundamental role in the incorporation of air into the batter. The second group is made up of cakes, such as sponge cakes or angel food cakes, with formulations lacking shortening. Cakes belonging to the second group are usually known as foam-type cakes, and the incorporation of air into these batters is closely related to the foaming properties of egg white.

Several methods can be followed in the preparation of cake batters or the mixing of ingredients. With shortening cakes, the most common approach is known as the creaming method, in which the process is started by beating the shortening and sugar. In this initial step, sugar helps cream air into the fat by forming numerous minute bubbles. The rest of the ingredients are added in successive steps. Once in the oven, shortening melts, and the bubbles undergo an aqueous phase expanding because of the carbon dioxide from baking powder. These cakes are characterized by a very fine and uniform grain. A second approach in the preparation of shortening cake batters is the one-stage method in which all ingredients are added at once. The use of emulsifying agents permits direct incorporation and stabilization of air bubbles in the batter. With both methods, sugar is the determining factor in the incorporation of air, so that if sugar is removed from the formulation, cake volumes will be drastically reduced.

With foam-type cakes, egg and sugar are whipped until a stable foam is obtained. Some formulations include only egg whites, but occasionally, whites and yolks are whipped separately. In foam-type cakes, sugar has an essential function as a whipping aid. After the foam is formed, flour is gently folded in without breaking the structure.

12.3.1.3 Cookies

The market offers consumers a vast array of cookies differing both in their formulations and processing methods. These are some of the major types: deposit cookies, wire-cut cookies, or rotary-molded cookies. For each type, there will be different requirements as far as the consistency and texture of cookie doughs are concerned. The effect of sugar on these characteristics must be studied. Like cakes, most cookies are chemically leavened and have high sugar and shortening contents but a rather low water content. For most cookies, the initial step consists of a creaming process in which sugar plays a central role in the incorporation of air. As was the case in cakes, sugars have a significant effect on the final structure of cookies by delaying the gelatinization temperature of starch. Furthermore, they serve as sweetening and browning agents through Maillard and caramelization reactions.

Dough consistency tends to be regarded as a very important property, particularly when processing cookies at an industrial scale. Deposit cookie doughs must retain enough fluidity to make depositing feasible; wire-cut cookie doughs must be properly extruded. In addition, doughs of rotary-molded cookies must be firm when entering the mold and cohesive enough to keep their shape once they have been released from the mold. Achieving all these different levels of consistency depends on the types and amounts of sugars and shortenings involved.

During oven baking, doughs undergo a rise in temperature which, in turn, results in the melting of the shortening and the granular sweeteners. As a result of these changes, water is liberated, leading eventually to a more fluid dough and an improved cookie spreading. Spreading is also affected by the gas released in the process, and it is considered an important quality parameter in cookies. The amount and type of sweetener used, as well as other ingredients, such as shortenings and eggs, determine the final cookie spread. After dough cooling, sucrose recrystallizes and hardens, helping to create the characteristic fracturability or snap of cookies. If a chewier and softer texture is sought, different sweeteners and ingredients capable of minimizing sucrose recrystallization must be introduced.

12.3.1.4   Fillings and Icings

The final stage in the processing of many sweet bakery products is the addition of some kind of filling or icing. Most fillings are prepared by whipping shortening and sugar into a creamy consistency. Fillings are composed of other ingredients that provide color, flavor, or consistency. Apart from its sweetening power, sugar provides bulk and structure, creating aeration nuclei when shortenings are whipped. Sugar granulometry is also important; coarse granulated sugar will yield a grainy and sandy texture. Fine crystals are preferable for a smooth and creamy texture. As for icings, the type of sugar used will determine the degree of recrystallization after cooling and so whether the final product shows a matte or glossy appearance.

12.3.2 Fats

The concentration of fats or oils in bakery products ranges widely. In some breads, fat is absent from their formulation; in others, fat is present at percentages below 5 to 6% (flour basis); finally, brioche-like doughs may have fat percentages well over 50%. Similarly, in some cakes (foam type), no fats are added, but in others, percentages may range between 40 and 60% (flour basis), and fats play an essential role. In addition, in cookie production, fat percentages may range between 10 and 55%. As mentioned above in the case of sugars, the functions of fats in bakery products are very complex and vary depending on the type of fat utilized, the amount added, and the product being produced. Several studies have dealt with this point in detail (Pyler, 1988; Sluimer, 2005; Stauffer, 1996).

12.3.2.1   Yeast-Raised Products

The inclusion of small amounts of fat in bread doughs has a significant effect on their processing and on the quality of the finished product. Such doughs are more extensible, and their machinability is markedly improved. After proofing, doughs with shortening are more stable and resistant to possible shocks when the pieces are transferred from the proofer to the oven. Increased oven-spring is also noticeable during the baking of such doughs, because shortening brings about a delay in the reactions marking the end of dough expansion, gelatinization of starch granules, and gluten denaturation. The final product exhibits larger volume and crumb of a finer grain. Fat strengthens the sidewalls of bread and minimizes the possibility of keyholing in the final product. Such breads have softer texture, less crumbliness, and greater moist mouthfeel. The combination of all these effects is known as lard effect; it occurs mainly with solid fats, provided they are evenly dispersed throughout the dough, suggesting that the presence of saturated fatty acids is important. All these effects are already noticeable with shortening percentages of 1%, but it has been established that maximum volume is obtained with 5 to 6% of shortening. In any case, volume increase changes with different proofing procedures. Thus, long proofing periods give volume increases of up to 10%, whereas short proofing periods may yield increases of up to 25%.

With formulations containing greater amounts of fat, as those of some buns and rolls or of brioche doughs, other factors must be taken into account. For example, when processing these kinds of doughs, the amount of water should be reduced and kneading times extended. Doughs with higher percentages of shortening may require that kneading be carried out in several steps. Dough consistency will be reduced and dough stickiness will increase. To minimize these effects, in addition to reducing the amount of water in the formulation, the final kneading temperature may be lowered and stronger flours may be used. Flours of higher strength are required to attain the correct gas holding capacity. There is less oven-spring in these doughs, and occasionally, they may collapse during or after baking.

In addition to the effects mentioned, all leavened doughs containing shortening show an antistaling effect. Staling increases the firmness of bread crumb as a result of a very slow process, lasting several days, during which the recrystallization of amylopectin occurs. When small amounts of shortening are included, loaf acceptability may be extended by one or two days, while with greater amounts of shortening, loaf shelf life can be prolonged for weeks. Fatty acid chains form a complex with starch molecules hindering amylopectin recrystallization and delaying crumb staling. This effect is similar to that observed when certain emulsifiers such as monoglycerides of saturated fatty acids are added.

Finally, fats have a lubricating effect, making the baked product easy to swallow without adhering to the surfaces of the mouth. They also help to produce bread that can be cleanly and evenly sliced.

12.3.2.2 Cakes

Fats play an essential role in the processing of shortening cakes, including layer cakes or similar products, such as muffins or cake doughnuts. Fats are responsible for air incorporation in the form of many small bubbles for a fine and uniform grain in the final product. When the shortening percentage on a flour basis is incorrect or the wrong kind of shortening is used, air will be entrapped abnormally, larger bubbles will be formed, and the final product will have less volume and a coarse grain.

The effect of shortening is at its peak when the ingredients are mixed through the creaming method in which shortening and sugar are mixed as air is incorporated; with the remaining methods, shortening also exerts considerable influence.

In the creaming method, shortenings must be solid, so that air does not escape, and plastic, so that bubbles can be surrounded. The presence of emulsifiers also contributes to a better dispersion of the entrapped air bubbles. In the one-stage production method, air is entrapped in the water phase and stabilized by egg proteins and flour; fats tend to have a defoaming effect. To prevent this, an emulsifier is included in the shortening. Cakes produced with oil are more tender and exhibit better moistness than those produced with fats, although their grain tends to be coarser. It has been shown that the presence of saturated fatty acids improves air absorption during cake batter mixing. In fact, hydrogenation and the incorporation of fats with higher melting points improve the creaming quality of oils. The formation of a structure incorporating small air bubbles thinly dispersed also improves the stability of the batter and prevents the escape of the bubbles toward the surface and the collapse of the dough during baking before dough setting has occurred.

Fats also improve the eating qualities of cakes as they help to create a more tender texture; they allow increases in the content of other ingredients such as sugars, milk, or eggs; and in certain instances, like butter or lard, they can act as flavoring agents. Finally, fats extend the shelf life of cakes by slowing the rate of staling through their influence in the retrogradation of amylopectin.

12.3.2.3 Cookies

As in cakes, fats have an important function in the incorporation of air in cookie doughs. They also determine the consistency and stickiness of doughs that have to meet the particular requirements of each production process; for example, they must extrude smoothly and should not stick to surfaces. Different amounts and kinds of shortening will modify cookie spread during baking, although the sugar content in the formulation will also be a determining factor. In doughs with high sugar contents (90% of the flour weight), an increase in shortening from 35 to 55% lowers spread about 10%. However, in doughs with a sugar content of 50% of flour weight, a similar increase in shortening will lead to an increase in spread of approximately 25% (Stauffer, 1996). Variation in other ingredients or in the kind of shortening used results in different spreading rates. In general, reducing shortening levels gives cookies of greater fracturability (Baltsavias et al., 1999) and hardness (Sudha et al., 2007).

12.3.2.4 Fillings

Fillings used in the production of certain sweet goods, such as sandwich cookies or sugar wafers, are mainly composed of sugar and fat. The kind and amount of fat used will influence air incorporation, consistency, stickiness, and mouthfeel. The consistency of fillings must be such that they can be readily extruded or spread over cookies, wafers, or cakes; they must be firm at ambient temperature and occasionally (cookies and wafers) sticky enough to hold onto the cookie. Fat must also melt in the mouth to avoid a waxy mouthfeel and provide a cooling effect.

12.4   GENERAL STRATEGIES FOR THE SUBSTITUTION OF SUGARS AND FATS

When manufacturers embark on the production of sugar- and fat-free products, they usually rely on the use of alternative products, although in certain situations the option of reformulating the recipes should not be completely disregarded. If substitutes are to be used, they should meet the following requirements:

1. Be safe and comply with national and international regulations
2. Have flavor characteristics similar to those of regular products
3. Be stable under processing conditions (pH, temperature, etc.)
4. Be soluble
5. Possess nutritional characteristics adequate to the needs of the product
6. Be reasonably cost effective

12.4.1 SUGAR REPLACERS

Product developers have followed three different approaches in their search for new sugar-free products. The first approach has focused on the substitution of sucrose by a different sugar capable of minimizing the adverse effects associated with sucrose consumption. In fact, these products should not be labeled as "sugar-free," because the substituting substance is another sugar. Fructose has been used mostly in the production of foods for diabetics. The second approach opts for the introduction of intense sweeteners. As these products possess greater sweetening power than sucrose, they can be used in smaller doses and thus a sharp fall in caloric intake occurs. They represent a very attractive option for the development of low-calorie carbonated beverages; in baked goods though, sugar is much more than just a flavor enhancer and a different solution must be found. When the main function of sugar is to provide volume and texture, as it is the case in many baked goods, then developers have usually applied the third approach—the introduction of bulking agents, alone or in combination with intense sweeteners. In all cases, a careful consideration of the synergistic effects of different sugar replacers is necessary to optimize the sensory characteristics of the finished product (Hanger et al., 1996; Montijano et al., 1998). Comprehensive reviews dealing with most of these points have been published over the years (Beereboom, 1979; Frye and Setser, 1993; Giese, 1993; Newsome, 1993; Olinger and Velasco, 1996; Shinsato, 1996).

12.4.1.1   Fructose

In addition to glucose, fructose is one of the constituent monosaccharides of sucrose, and, as its name indicates, it occurs in many fruits. Like sucrose and glucose, it acts as substrate, fermentable by baker's yeast, promotes the development of textural properties in baking products, and affects the final color through Maillard and caramelization reactions, though a slightly different color than that provided by sucrose results. Fructose has a sweetening power greater than sucrose (1.8 greater), so smaller amounts are adequate, and there is a significant caloric reduction in the finished product. The chief advantage of fructose is its low glycemic index; diabetics tolerate it much better than sucrose as no insulin is needed for its metabolism. However, tolerance is not complete, and diabetics are advised to refrain from an indiscriminate consumption of products containing fructose.

Although fructose is an ideal replacer of sucrose in certain products, it should always be kept in mind that, like sucrose or glucose, it is just another sugar. Therefore, any labeling of products with fructose stating "no sugar added" can be considered misleading. In certain countries, labels can be currently found declaring, "no sugar, with fructose" or "no sugar, no fructose." Another aspect to be considered is that fructose has a similar caloric value to sucrose; thus, the only way of obtaining a caloric reduction in the finished product is to lower its content in the formulation. Moreover, fructose can be a factor in the development of dental caries, and its effect on diseases such as cardiovascular disease is still under study.

12.4.1.2 Intense Sweeteners

12.4.1.2.1 Saccharin

Saccharin is a sugar substitute with a sweetening effect between 300 and 500 times greater than sucrose, but it is not metabolized in the human organism and is excreted in the urine without providing calories. Saccharin is stable under the pH, moisture, and temperature conditions undergone by baked goods, and it also shows adequate solubility. An additional desirable property is its synergistic effect with other sweeteners and bulking agents which results in an enhanced sweetening effect. The main drawback is that significant numbers of consumers detect a slightly bitter aftertaste. For this purpose, a combination of saccharin and cyclamate is ideal, as the latter can partially mask the unpleasant aftertaste.

Saccharin was widely used for over 100 years in all kinds of foods and beverages. Consumption reached its peak during the 1970s when it was the only low-calorie sweetener available. It was used as a coffee sweetener or in other products destined for diabetics or concerned overweight people. For many, it became part of their daily lives. Now, although it has been pushed aside by newly developed sweeteners with better organoleptic characteristics, its very low cost still justifies its inclusion in beverages, chewing gums, jams, and sauces. Its Acceptable Daily Intake (ADI) is 5 mg/kg.

12.4.1.2.2 Cyclamate

The sweetening effect of cyclamate is 30 times greater than sucrose, with a pleasant taste that is similar to sucrose. Cyclamate exhibits good solubility and stability over a wide range of temperatures. Unlike saccharin, it does not leave a bitter aftertaste, but the sweetness sensation, although persisting longer, has a more delayed onset than sucrose. Cyclamate is synergistic with most sweeteners, masks bitterness, and is capable of enhancing some flavors. Even though a small section of the population can metabolize cyclamate, most people are unable, and hence it is considered as a noncaloric sweetener.

Cyclamate was first marketed during the 1950s, but in the late 1960s, its safety was questioned when evidence from studies with rats became known. A ban in many countries ensued. Today, having been declared safe for human consumption, it has been reintroduced in many countries. Its ADI is 11 mg/kg. It is now used in low-calorie beverages or chewing gums.

12.4.1.2.3   *Aspartame*

Aspartame is 180 to 220 times sweeter than sucrose with a similar taste. Like other sweeteners, aspartame has synergistic effect and enhances flavors, namely acid fruit flavors. Its tendency to degrade at high temperatures constitutes its main drawback when added to baked goods. Hence, its use is inadvisable when baking processes are involved. An expensive encapsulated version capable of withstanding high temperatures has been in the market for a number of years. Aspartame can be useful in the production of creams and other products that need not undergo severe heat treatments.

Aspartame has currently become the most widely used sweetener in the food industry, even though it was not approved until the early 1980s. It is metabolized in the intestinal tract producing aspartic acid, phenylalanine, and methanol. After extensive studies confirmed its safety for all population groups, it was approved by the European Union with an ADI of 50 mg/kg, the equivalent of a daily sucrose intake of 600 g for a person weighing 60 kg. However, a small number of individuals with the disease phenylketonuria, intolerant to phenylalanine, must limit aspartame intake.

12.4.1.2.4   *Acesulfame-K*

Discovered in 1967, acesulfame-K has by now been approved in most countries. It is not metabolized and is therefore completely excreted without providing calories. In the European Union its ADI is 9 mg/kg, and in the United States it is 15 mg/kg. Acesulfame-K presents a clean taste, no aftertaste, and its sweetness does not last, though is rapidly perceived. When in combination with other sweeteners, acesulfame-K exhibits synergistic effects and a certain capacity for masking off-flavors. It is 200 times sweeter than sucrose. This compound is stable under a wide range of pH and temperature conditions, and, in particular, it can withstand baking conditions common in the processing of bakery products (Klug et al., 1992).

There are other intense sweeteners approved as sugar substitutes under the regulations of different countries or with applications for use and regulatory reviews still pending (e.g., taumatine, neohesperidine DC, sucralose, and alitame). All of these sugar substitutes face regulatory or economic difficulties.

12.4.1.3   Bulking Agents

12.4.1.3.1   *Polyols*

Polyols are obtained by the catalytic hydrogenation of different sugars. These substances are similar to sucrose for the texture characteristics and volumes they provide but without promoting dental caries, with better tolerance rates among diabetics, and with lower caloric values. Their main limitation is that their sweetness is below that of sucrose. Table 12.1 shows the sweetness of several sugar substitutes. Further, as polyols are not involved in processes such as Maillard or caramelization reactions, they also perform less decisively than sucrose in the development of crust color; finished products containing polyols usually exhibit lighter colors. But polyols surpass sucrose as inhibitors of bacteria and mold growth and have high hygroscopicity.

TABLE 12.1

Relative Sweetness of Some Sugar Substitutes

	Sugars		
Sucrose	1		
Fructose	1.2–1.8	**Polyols**	
		Xylitol	1
High-Intensity Sweeteners		Maltitol	0.85–0.95
Acesulfame-K	130–200	Sorbitol	0.55–0.7
Aspartame	180–220	Isomalt	0.45–0.65
Saccharin	200–700	Lactitol	0.35
Cyclamate	30	Erythritol	0.65
Oligosaccharides		Mannitol	0.5
Polydextrose	0	Hydrogenated starch hydrolysates	0.7–0.9
Oligofructose	0.3–0.5		

In some people, an excessive consumption of polyols leads to intestinal disorders, such as a persistent laxative effect or diarrhea, reminiscent of the problems caused by high fiber intakes. Symptoms will vary with individuals and the rest of the diet, and though they are rarely severe or lasting, a restrictive intake is advisable. Polyols are absorbed in the small intestine in a slow and incomplete manner and are converted into energy with little or no insulin being required. Each polyol has its own specific caloric value and the average value is considered to be 2.4 kcal/g.

Maltitol is one of the most widely used polyols among bakers, because its functional properties (hygroscopicity, solubility, melting point) make it similar to sucrose. Its caloric value is 3 kcal/g, and its sweetening effect is about 80 to 90% of that of sucrose without objectionable aftertaste. Considering its cooling effect, it is below sorbitol and xylitol but above sucrose. It can be optimally adapted to chocolate manufacturing because of a melting point similar to that of sucrose (Rapaille et al., 1995). Second in the ranking of polyols most widely used in baked goods is sorbitol. Its caloric value is 2.6 kcal/g, and compared with sucrose it has a sweetening effect of 60%. It does not have any difficulty in solubilizing. It is a highly hygroscopic polyol used as a humectant in the commercial production of baked goods to delay staling. From an economic standpoint, both maltitol and sorbitol are very cost-effective sugar replacers. In addition, the use of hydrogenated starch hydrolysates is widespread; this is a general term used to refer to blends of polyols. The final product is a blend of sorbitol, maltitol, and other larger hydrogenated saccharides with differing properties (sweetening effect, hygroscopicity, solubility, etc.) depending on the manufacturing process. Caloric values are below 3 kcal/g, and their sweetening effects range between 40 and 90%.

Lactitol and isomalt, having very low hygroscopicity, are widely used with low moisture products such as cookies and candies. Their energy value is 2 kcal/g, and their sweetening effects are 30 to 40% for lactitol and 45 to 65% for isomalt, always in relation to the sweetening effect of sucrose. Isomalt, like xylitol, prevents tooth decay. In fact, xylitol is commonly used in toothpaste and other oral hygiene products. Xylitol is possibly the most suitable replacer for the production of sucrose-free products, but its applications have been limited due to its high cost. In sweetness and solubility, it is similar to sucrose, and its energy value is only 2.4 kcal/g. With high hygroscopicity, it can act as a humectant, and its cooling effect is fairly high.

Maltitol is the least adaptable to the necessities of the baking industry, as it has a high laxative effect, low hygroscopicity, and a low sweetening effect (40 to 50% of sucrose). Its cooling effect is not only inferior to that of sucrose but is well below the rest of the polyols. Erythritol is not very widely used in the baking industry. Its strong laxative effect and low to medium solubility limit the number of possible applications. It has a sweetening effect ranging from 60 to 70% of that of sucrose, and its cooling effect, although higher than that of sucrose or maltitol, is lower than that of xylitol and sorbitol. In Japan, it has been in use since the 1990s but is still under review in the rest of the world. Its main advantage is its low energy value, approximately 0.2 kcal/g.

12.4.1.3.2   Oligosaccharides

Polydextrose and oligofructose belong to the oligosaccharides group. Both compounds are made up of glucose or fructose branched chains with a very slight metabolization in the human body, and thus with a negligible caloric effect. Like dietary soluble fiber, they can provide similar nutritional benefits and, with excessive intake, a laxative effect. They are not cariogenic and are better tolerated by diabetics than sucrose.

Polydextrose is a branched glucose polymer containing amounts of sorbitol and citric acid with a caloric value of 1 kcal/g and a maximum daily intake of 50 to 90 g. When first marketed, polydextrose was acid, yielded some unusual flavors and fat rancidity, but these problems have been overcome with a newly developed improved version of polydextrose. As its sweetening power is limited, polydextrose must be blended with some intense sweeteners; nevertheless, it imparts a very clean flavor, without undesirable aftertaste or incompatibilities with other sweeteners. In the development of texture and volume in baked goods, it is similar to sucrose and presents good solubility. Highly hygroscopic, it keeps freshness in foods and can be used as a humectant. In addition to a sugar replacer, polydextrose can be used as a fat replacer.

Oligofructose is a complex of short branched chains of fructose obtained from inulin, a chicory root extract. Inulin has the same composition but a higher molecular weight, and its best application is as a fat replacer. Like polydextrose, it has a low sweetening effect and no cooling effect but a clean taste, and in its ability to impart texture and volume to the final product, it is similar to sucrose. Nevertheless, its main advantages are related to nutritional benefits. Its caloric value is 1.5 kcal/g and it acts as soluble fiber, helping digested food pass easily through the intestinal tract, increasing fecal bulk, reducing constipation, and facilitating bowel movement. According to some studies, oligofructose seems to reduce cholesterol and triglyceride levels. Finally, oligofructose promotes the production of bifidobacteria in the intestinal tract (prebiotic effect) and calcium absorption.

12.4.2 Fat Replacers

To reduce the fat content of a food item, manufacturers have favored the introduction of fat replacers, ingredients combining some of the functional properties of fats with less caloric effect. Fat replacers are also known as fat mimetics, and they can be based on carbohydrates, proteins, fats, or a combination of these. A second option for manufacturers has been to use fat extenders, products optimizing fat functionality and allowing a reduction in the amount of fat included in the formulation. Comprehensive reviews of the characteristics and applications of these products can be found in the literature (ADA, 2005; Frye and Setser, 1993; Glueck et al., 1994; Lucca and Tepper, 1994). Some of the most commonly used fat replacers and their manufacturers are shown in Table 12.2.

12.4.2.1 Carbohydrate-Based Fat Mimetics

These products are capable of absorbing large amounts of water to form a gel-like matrix with some of the functional properties of fats. They are similar to fats in the properties they influence: viscosity, body, creaminess, and mouthfeel. Because they are mixed with water, their caloric value is reduced to 1 to 2 kcal/g, even though carbohydrates can provide 4 kcal/g. Fibers, cellulose, and gums have no caloric effect. Their use with fried goods is not recommended, and their high moisture content increases water activity with greater risk of promoting microbiological growth and thus reducing shelf life. This group of fat mimetics includes modified starches, fibers, cellulose, gums, maltodextrins and dextrins, polydextrose, and inulin. Both polydextrose and inulin were discussed in the previous section as bulking agents, and they can replace sugars as well as fats.

12.4.2.1.1   Starch Derivatives
These products, including modified starches, maltodextrins, and dextrins, are mixed with three parts of water, are gel-like, and provide texture and mouthfeel usually associated with fats. Their caloric value will be of 1 kcal/g instead of 9 kcal/g like fats. Modified starches are obtained through physical or chemical treatments of native starches so that they can withstand extreme conditions (temperature, acidity, shear) and alter their pasting behavior. Final characteristics will depend on the parent starch and the modifications it has been forced to undergo. It was reported that small-granule starch having a granule diameter similar to that of lipid micelles (less than 2 μm) might have potential as a fat replacer (Lucca and Tepper, 1994). Maltodextrins and dextrins are hydrolyzed starches with a dextrose equivalent of less than 20, which may result in dark colors when heat processed, but this should not prevent its use in baked products. Modified starches, maltodextrins, and dextrins can mask certain flavors and impart new ones deserving specific attention.

12.4.2.1.2   Gums and Cellulose
Neither of these products is absorbed in the intestinal tract, and they can be included in the fiber group; they provide nutritional advantages with no caloric effect. Gums are hydrocolloids with great capacity for water absorption which can impart viscosity, stabilize water systems, and inhibit synerisis. Examples of gums include xanthan, locust bean, carrageenan, and pectin. Pectin is used primarily as a gelling agent.

TABLE 12.2

Fat Replacers

Type	Brand Name	Manufacturer
	Carbohydrate-based	
Cellulose	Avicel	FMC Corp.
	Novagel	FMC Corp.
Modified starches	Stellar	A.E. Staley Mfg. Co.
	Sta-Slim	A.E. Staley Mfg. Co.
	Amalean	American Maize-Products Co
	N-lite	National Starch & Chemical Co.
Maltodextrin	Paselli (potato)	Avebe
	Oatrim (oat)	Rhone-Poulenc
	Maltrin (corn)	Grain Processing Corp.
	N-Oil (Tapioca)	National Starch & Chemical Co.
	RiceTrin (rice)	Zumbro Inc.
	C*Light	Cerestar
Fibers	WonderSlim	Natural Food Technologies
	Z-Trim	FiberGel Technologies Inc.
	Betatrim	Rhone-Poulenc
	DairyTrim	Meyhall Chemical AG
Polydextrose	Litesse	Cultor Food Science, Inc.
Inuline	Raftiline	Orafti
Pectin	Slendid	Hercules Inc.
	Grindsted	Danisco
Gums	Kelgum	Kelco
	Kelcogel	Kelco
	Nutricol	FMC Corp.
	Protein-based	
Microparticulate protein	Simplesse	The NutraSweet Co.
Whey protein concentrate	Dairy-Lo	Cultor Food Science, Inc.
Nonfat milk, gums, modified starch, and emulsifiers	N-Flate	National Starch & Chemical Co.
	Fat-based	
Altered triglycerides	Salatrim	Nabisco
	Caprenin	Procter & Gamble Co.
	Benefat	Cultor Food Science, Inc.
Sucrose polyester	Olean	Procter & Gamble Co.

Cellulose gels or microcrystalline cellulose result from the processing of nonfibrous cellulose to reduce particle size in such a way that between 60 and 70% of particles have lengths of less than 2 microns. The cellulose microfibers join other ingredients, such as carboxymethylcellulose or guar gum, to hold together as a network.

12.4.2.2   Protein-Based Fat Mimetics

There is a patented procedure to obtain microparticulated proteins. This procedure consists of the simultaneous application of two processes—pasteurization (heat treatment) and homogenization (high shear), producing spherical protein particles of less than 2 microns in diameter. Whey, milk, and egg proteins are the usual raw materials, and people with allergies should refrain from consuming these products. Because of particle size and shape, the tongue fails to detect the individual particles and instead perceives a creamy, smooth, and fluid product coating mouth surfaces in ways usually associated with fats. This coating action helps flavors reach the taste buds more gradually, but it also masks some bitter and astringent flavors characteristic of low-fat products. Microparticulated proteins, marketed in several forms under the trademark Simplesse®, tend to imbibe water in such a way that 1 g of protein-based fat mimetics can replace 3 g of fat with the resulting caloric reduction. Like carbohydrate-based fat mimetics, they are unsuitable for use with fried foods, although some versions withstand heat processing better than others.

An additional fat substitute is derived from whey protein, modified through a treatment combining heating and an acidic medium. Changes in protein concentration, temperature, and pH, and the presence of other ingredients result in a wide range of products with varying functional properties, including opacity, water absorption capacity, particle size, and emulsifying capacity. Finally, fats may also be partially replaced by products composed of a blend of animal and vegetable proteins with gums, starches, and water.

12.4.2.3   Fat-Based Replacers

Fat-based fat replacers are lipids modified or synthesized in such a way that they are not fully metabolized in the human body, and thus their caloric effect is below the 9 kcal/g level of conventional fats. In their physical properties, they are similar to fats. They can withstand high temperatures, including frying, and can replace fats totally or partially.

One of the best known is Caprenin, formed by esterification of glycerol with three fatty acids (capric, caprylic, and behenic). It is partially absorbed in the intestine and provides only 5 kcal/g. Caprenin is similar to cocoa butter and is commonly used as a replacer in soft candy and chocolate confectionery coatings. Salatrim is a similar product, obtained by esterification of monostearin with short-chain acids (acetic, propionic, and butyric). The resulting product provides just 4 kcal/g with varying plastic properties depending on the ratio of short-chain fatty acids used. It thus has the capacity to replace all-purpose shortenings as well as filler fats.

In addition, there are substances with thermal and organoleptic properties similar to fats, too large for intestinal absorption, unavailable for hydrolysis by gastric and pancreatic lipase, and providing no calories. The best known is olestra, a sucrose polyester made up of hexa, hepta, and octa esters of sucrose, esterified with long-chain fatty acids derived from edible oils, including soy, corn, and cotton. By varying the degree of esterification and the number of fatty acids used, it is possible to obtain a wide range of products with functional properties, appearance, mouthfeel, stability, shelf life, and organoleptic characteristics usually associated with conventional fats. Although at one time, in some countries, manufacturers wishing to include olestra in their products were required to add fat-soluble vitamins (A, D, E, and K) and warn on the label that the absorption of fat-soluble vitamins could be inhibited, currently, after much evidence accumulated over the years, this has come to be considered as an unnecessary precaution.

Emulsifiers can behave as fat extenders. Their caloric effect is the same as that of fats (9 kcal/g) but, as only between 25 and 75% of emulsifier is necessary to mimic the effect of conventional fats, there is a significant calorie reduction. Emulsifiers are used to replace all or part of the fat content in certain products. In others, they also help retain moisture and increase volume. The most widely used emulsifiers in the manufacturing of low-fat products are monoglycerides and polysorbates; in other applications, manufacturers can also use lecitin, sodium stearoyl lactylates (SSLs), or diacetyl tartaric esters of mono- and diacylglycerols (DATEM). Blends of emulsifiers aggregate the functional properties of the separate components.

12.5 SUBSTITUTION OF SUGAR AND FATS IN CAKES AND COOKIES

12.5.1 SUGAR SUBSTITUTION

12.5.1.1 Cakes

Strategies for the production of sugar-free cakes are usually based on the substitution of sugar by one or a combination of bulking agents capable of yielding a finished product with adequate volume, texture, and grain. It has been established that different bulking agents affect the behavior of doughs during baking in varying degrees, thereby affecting the eventual volume and structure (Ikawa, 1998). Ronda et al. (2005) compared the capacity of seven bulking agents as sugar substitutes in sponge cakes. Top performers were xylitol and maltitol, followed by sorbitol; the high price of xylitol, however, has resulted in its being superseded by maltitol as the most widely used in sugar-free cake manufacture. The same study pointed out that cakes manufactured with polyols yielded lighter cakes because they are unaffected by Maillard reactions. Conversely, those cakes manufactured with oligofructose and polydextrose resulted in darker colors, as repeatedly reported by other studies (Esteller et al., 2006; Hicsasmaz et al., 2003) and suggesting that the combination of these bulking agents with polyols as sugar substitutes may provide positive results (Penna et al., 2003). In fact, Frye and Setser (1992) optimized a sugar-free cake formulation using blends of polydextrose and different polyols with good results. Moreover, emulsifiers can be used to buffer the adverse effects of the substitution of sugar by bulking agents (Kamel and Rasper, 1988). In these instances, a correct choice of type and dosage of emulsifier is essential.

Apart from their appearance, volume, and texture, sugar-free products must also be assessed by their flavor and aroma. It is an established fact that different bulking agents exhibit varying sweetening effects as well as diverse aromatic and taste profiles. In some formulations, particularly when using bulking agents with low sweetening power, such as polydextrose, the addition of an intense sweetener, like acesulfame-K, aspartame, or any other, may be required (Attia et al., 1993; Freeman, 1989). Some sugar substitutes cause an unpleasant aftertaste, like some types of polydextrose (Frye and Setser, 1992), and have to be masked by other sweetening or flavoring agents.

Overall, it is difficult to make general recommendations for the substitution of sugar in cakes. The divergence of views reported in the literature highlights that difficulty. Thus, it is known that polydextrose raises the temperature of the gelatinization of starch (Pateras et al., 1994) and lowers batter stability (Hicsasmaz et al., 2003). This, in turn, affects air retention during baking, the expansion of the product, and the final volume. However, although Hicsasmaz et al. (2003) found that cakes manufactured with polydextrose as sucrose substitute showed fine and uniform grain, Kocer et al. (2007) reported opposite results. On the other hand, Rosenthal (1995) suggested that the rise in the starch gelatinization temperature produced by polydextrose can be used to improve the manufacture of cakes with aged eggs as a result of an increase in the egg protein denaturation temperature. All these differing results indicate that cake manufacturing depends on the interaction of many factors, for example, the type of cake sought, formulation, and raw materials. In each particular formulation, therefore, different blends of sugar substitutes should be tested, and then the results verified, including both the physicochemical and sensory properties of the finished product.

12.5.1.2 Cookies

Strategies for the production of sugar-free cookies closely resemble those already discussed for sugar-free cakes. The removal of sugar brings about drastic changes in doughs and in the finished products, and these cannot be minimized just by the introduction of intense sweeteners (Lim et al., 1989). In cookies, sugar and fat replacement can affect texture to a greater extent than flavor (Perry et al., 2003). Thus, the addition of a bulking agent or a combination of several of them is necessary to control the changes in the texture of doughs and cookies due to the removal of sugar. Zoulias et al. (2000a) compared the functionality of several bulking agents as sugar replacers in cookies. Cookies with maltitol scored the highest results in hedonic evaluations, even higher than the control, and they were also considered as similar to the original product in shape, color, and texture. However, sugar replacement with maltitol significantly impacted dough adhesiveness and cohesiveness, an important point in the commercial production of cookies. The same authors observed an improved acceptability when an intense sweetener (acesulfame-K) was added to the sugarless cookies, even with sugar replacers, like fructose, of greater sweetening effect than sucrose. Maltitol has been shown to have better characteristics than other bulking agents as sugar substitute in low-fat cookies (Zoulias et al., 2002a).

Another bulking agent commonly used, alone or combined with polyols, in the manufacture of sugar-free cookies is polydextrose. Because of its very low sweetening effect, it must be supplemented with an intense sweetener. The choice of sweetener, or blend of sweeteners, is then of the utmost importance, if time-intensity sweetness and bitterness curves similar to those of sucrose are to be attained (Lim et al., 1989). When selecting a sugar substitute, it is equally important to test its effect on the shelf life of the product, as different bulking agents impact differently on the stability of fats present in cookies (Ochi et al., 1991).

Evidence provided by different studies on the manufacture of sugar-free cookies is valid under a particular set of conditions and must, therefore, be carefully tested when applied to new formulations, types of cookies, or processing conditions (commercial scale). In any case, all this experimentation has in common the substitution of sugar by bulking agents and, sometimes, intense sweeteners, a careful monitoring of the physical characteristics of doughs and cookies, and the sensory acceptability of the newly developed product.

12.5.2 FAT SUBSTITUTION

12.5.2.1 Cakes

Obtaining fat-free baking products with sensory characteristics similar to the regular products is a formidable challenge, because no fat replacer currently in use exhibits perfect structural similarity (Shukla, 1995). To overcome this obstacle, it has been necessary to develop blends of fat replacers, change the doses (1:1 ratios being not always advisable), and even reformulate the product.

Broadly speaking, carbohydrate-based fat replacers are the most widely used in the processing of fat-free cakes, owing to their capacity to provide products with good organoleptic properties (Bath et al., 1992; Swanson et al., 2002), which can even be improved by the addition of emulsifiers (Khalil, 1998). The substitution of fats by maltodextrins in cakes causes a reduction in batter viscosity, which, in turn, leads to a smaller final volume (Lakshminarayan et al., 2006). The same study reported that the best results were obtained when the amount of fat removed was replaced by a smaller amount of maltodextrin, approximately 50%, although cakes thus manufactured exhibited moderately sticky texture and mouthfeel. Moreover, it was shown that the addition of small amounts of glycerol monostearate (GMS) improved volume, entrapment of air in the batter, and the sensory quality of the cake, with less dense crumb grain and less sticky texture. The addition of sodium steroyl lactylate, however, failed to yield the same beneficial effects, proving that the choice of an appropriate emulsifier is essential. In cakes processed with maltodextrin as fat replacer, their physical and sensory characteristics can also be improved by adding corn amylodextrin (Kim et al., 2001).

A number of significant fat replacers useful in the processing of cakes with lower fat contents are included in the same group as fibers. Good results can be obtained with inulin and oligofructose (Devereux et al., 2003), polydextrose (Frye and Setser, 1992), fiber products derived from corn and oats (Warner and Inglett, 1997), and hydrocolloids, alone or in combination with emulsifiers (Kaur et al., 2000). These products provide a significant reduction in caloric content, negligible caloric effect, and, in some instances, in addition to the caloric reduction, a lower flour content becomes possible. However, products including these substitutes tend to score lower values in the hedonic scale if compared with the original products, on account of the changes that occur in the texture of the finished product. Furthermore, excessive consumption of these products may have adverse physiological effects.

12.5.2.2 Cookies

Although several studies about the processing of fat-free cookies have been published, no comprehensive comparison of all the different substitutes is yet available. Fat-free formulations for cookies yield doughs drastically changed in texture and finished products with different physical and sensory characteristics. These changes can be minimized by including in the formulation different fat replacers, although in all cases, acceptability is not as wide as with the regular products. Cookies with low fat content tend to exhibit less surface cracking, fewer surface protusions, more uniform cells, and more mouthcoating, apart from a different flavor (Armbrister and Setser, 1994). Carbohydrate-, protein-based, and fiber-derived replacers have been tested with varying results. Sudha et al. (2007) established that the addition of maltodextrin or polydextrose reduced the effects of fat removal on the processing of cookies, and with the addition of emulsifiers or guar gum, the quality of the finished product approached that of the original but was never quite the same. Sanchez et al. (1995) optimized a formulation of reduced-fat shortbread cookies with different fat replacers. This study showed that cookies with low fat content exhibited increased moisture and toughness, and less specific volume; the addition of emulsifiers led to improved results. A modification of the processing of the dough was introduced to obtain low-fat shortbread cookies. Zoulias et al. (2002b) compared different carbohydrate- and protein-based fat replacers and observed significant differences among the various replacers; the best results were obtained with inulin (Raftiline), a blend of microparticulated whey proteins and emulsifiers (Simplesse), and an oat-derived product rich in ß-glucans (C*Light). Among all these, Simplesse yielded cookies very similar in diameter to the original cookies but with higher values of moisture content (Zoulias et al., 2002a).

Additional fat replacers in the manufacture of different types of cookies have been tested with similar results, including oat-derived fiber products (Charlton and Sawyer-Morse, 1996; Conforti et al., 1997; Inglett et al., 1994; Lee and Inglett, 2006), prune paste (Charlton and Sawyer-Morse, 1996), hydrocolloid-based products (Conforti et al., 1997), or okra gum (Romanchick-Cerpovicz et al., 2002). All of them are capable of minimizing the effects of fat removal, but they fail to match the physicochemical properties of the regular products. On the other hand, the strategy for the manufacture of low-fat cookies cannot focus on an approach based on a single fat replacer when a blend of replacers is more likely to be successful in improving the final result (Zoulias, 2000b). Finally, the importance of reducing fat content in fillings and icings should not be overlooked—this could be obtained with the right blend of whey protein and hydrocolloids (Laneuville et al., 2005).

12.6   CONCLUSIONS

The recent growing awareness of the role of diet in health throughout most Western countries has prompted an increase in the consumption of low-calorie, low-fat, and sugar-free products. This holds true not only for comparatively small segments of the population affected by certain conditions, such as diabetes, requiring a mandatory restraint in the consumption of fats and sugar, but also for a growing number of consumers opting for "a healthy lifestyle" as a way of preventing future ailments. As a result, significant attention has been focused on the development of bakery products such as cakes and cookies low in fat and sugar. For many manufacturers, the development and marketing of this kind of products must have helped diversify production and widen the market by reaching to new consumers and gaining a competitive advantage over rival firms.

The development of these dietary products is a complex task, because it may involve something else beyond the mere substitution of fats and sugar. It is often essential to reformulate these products and modify their processing variables. Most approaches, however, favor the utilization of sugar and fat substitutes. Information about the interactions between these substitutes and dough components is hard to come by, in many cases it remains unpublished. Drafting and publication of studies dealing with such interactions and the way they impact on technological processes would be highly useful for the formulation of baking goods low in fats and sugar. On the other hand, there is more information available on the development of this kind of products when sugar and fats are replaced by other types of ingredients. Yet each product should be dealt with independently; it might even be necessary to vary the formulation in response to the degree of mechanization in the processing of such goods. A successful development of baking products low in fats and sugar demands, first, a thorough understanding of the functions of fats and sugar in the processing and quality of the final product and, second, an equally thorough understanding of the alternative ingredients and the possible modifications in the production process aimed at improving the quality of the final product. Finally, it should always be kept in mind that the success of these dietary products is not just due to their nutritional composition but, most importantly, is due to their organoleptic acceptability among potential consumers. Hence, studies about the sensory acceptability of newly developed products become essential.

REFERENCES

Alexander, R.J. 1998. *Sweeteners: Nutritive.* Eagan Press, St. Paul, MN.

American Dietetic Association (ADA). 2005. Position of the American Dietetic Association: Fat replacers. *Journal of the American Dietetic Association* 2:266–275.

Archer, S.L., K. Liu, A.R. Dyer, K.J. Ruth, D.R. Jacobs, L. Van Horn, J.E. Hilner, and P.J. Savage. 1998. Relationship between changes in dietary sucrose and high density lipoprotein cholesterol: The CARDIA Study. Coronary Artery Risk Development in Young Adults. *Annals of Epidemiology* 8:433–438.

Armbrister, W.L. and C.S. Setser. 1994. Sensory and physical-properties of chocolate chip cookies made with vegetable shortening or fat replacers at 50 and 75-percent-levels. *Cereal Chemistry* 71:344–351.

Ascherio, A. 2006. Trans fatty acids and blood lipids. *Atherosclerosis Supplements* 7 (2): 25–27.

Ascherio, A. and W.C. Willett. 1997. Health effects of trans fatty acids. *American Journal of Clinical Nutrition* 66: S1006–S1010 Suppl.

Attia, E.A., H.A. Shehata, and A. Askar. 1993. An alternative formula for the sweetening of reduced-calorie cakes. *Food Chemistry* 48:169–172.

Baltsavias, A., A. Jurgens, and T. van Vliet. 1999. Fracture properties of short-dough biscuits: Effect of food composition. *Journal of Cereal Science* 29:235–244.

Bath, D.E., K. Shelke, and R.C. Hoseney. 1992. Fat replacers in high-ratio layer cakes. *Cereal Food World* 37:495–500.

Bean, M.M. and W.T. Yamazaki. 1978. Wheat starch gelatinization in sugar solutions. I. Sucrose:microscopy and viscosity effects. *Cereal Chemistry* 55:936–944.

Beereboom, J.J. 1979. Low calorie bulking agents. *Critical Reviews in Food Science and Nutrition* 11:401–413.

Bolton-Smith, C. and M. Woodward. 1994. Coronary heart disease: Prevalence and dietary sugars in Scotland. *Journal of Epidemiology and Community Health* 48:119–122.

Bray, G.A. and B.M. Popkin. 1998. Dietary fat intake does affect obesity. *American Journal of Clinical Nutrition* 68:1157–1173.

Bray, G.A., S. Paeratakul, and B.M. Popkin. 2004. Dietary fat and obesity: A review of animal, clinical and epidemiological studies. *Physiology and Behavior* 83:549–555.

Calorie Control Council. 2006. Trends and Statistics. www.caloriecontrol.org/lcchart.html.

Charlton, O. and M.K. Sawyer-Morse. 1996. Effect of fat replacement on sensory attributes of chocolate chip cookies. *Journal of the American Dietetic Association* 96:1288-1290.

Colditz, G.A, W.C. Willett, M.J. Stampfer, S.J. London, M.R. Segal, and F.E. Speizer. 1990. Patterns of weight change and their relation to diet in a cohort of healthy women. *American Journal of Clinical Nutrition* 51:1100–1105.

Conforti, F.D., S.A. Charles, and S.E. Duncan. 1997. Evaluation of a carbohydrate-based fat replacer in a fat-reduced baking powder biscuit. *Journal of Food Quality* 20:247–256.

Devereux, H.M., G.P. Jones, L. McCormack, and W.C. Hunter. 2003. Consumer acceptability of low fat foods containing inulin and oligofructose. *Journal of Food Science* 68:1850–1854.

Drewnowski, A., C.L. Kurth, and J.E. Rahaim. 1991. Taste preferences in human obesity: Environmental and familial factors. *American Journal of Clinical Nutrition* 54:635–641.

Esteller, M.S., A.C.O. Lima, and S.C.S. Lannes. 2006. Color measurement in hamburger buns with fat and sugar replacers. *LWT-Food Science and Technology* 39:184–187.

Feskens, E.J. and D. Kromhout. 1990. Habitual dietary intake and glucose tolerance in euglycemic men: The Zutphen Study. *International Journal of Epidemiology* 19:953–959.

Frayn, K.N. and S.M. Kingman. 1995. Dietary sugars and lipid metabolism in humans. *American Journal of Clinical Nutrition* 62 (suppl):250S–261S.

Freeman, T.M. 1989. Sweeting cakes and cake mixes with alitame. *Cereal Foods World* 34:1013–1015.

Frye, A.M. and C.S. Setser. 1992. Optimizing texture of reduced-calorie yellow layer cakes. *Cereal Chemistry* 69:338–343.

Frye, A.M. and C.S. Setser. 1993. Bulking agents and fat substitutes. In *Low-Calorie Foods Handbook,* Ed. A.M. Altschul, Marcel Dekker, New York.

Garcia-Palmieri, M.R., P. Sorlie, J. Tillotson, R. Costas, Jr., E. Cordero, and M. Rodríguez. 1980. Relationship of dietary intake to subsequent coronary heart disease incidence: The Puerto Rico Heart Health Program. *American Journal of Clinical Nutrition* 33:1818–1827.

Giese, J.H. 1993. Alternative sweeteners and bulking agents: An overview of their properties, function, and regulatory status. *Food Technology* 47:114–126.

Glueck, C.J., R.D. Streicher, E.K. Illig, and K.D. Weber. 1994. Dietary fat substitutes. *Nutrition Research* 14:1605–1619.

Guinard, J., H. Smiciklas-Wright, C. Marty, R.A. Sabha, I. Soucy, S. Taylor-Davis, and C. Wright. 1996. Acceptability of fat-modified foods in a population of older adults: Contrast between sensory preference and purchase intent. *Food Quality and Preference* 7:21–78.

Hanger, L.Y., A. Lotz, and S. Lepeniotis. 1996. Descriptives profiles of selected high intensity sweeteners (HIS), his blends and sucrose. *Journal of Food Science* 61:456–459.

Hicsasmaz, Z., Y. Yazgan, F. Bozoglu, and Z. Katnas. 2003. Effect of polidextrosa-substitution on the cell structure of the high-ratio cake system. *LWT-Food Science and Technology* 36:441–450.

Hill, J.O. and A.M. Prentice. 1995. Sugar and body weight regulation. *American Journal of Clinical Nutrition* 62 (Suppl. 1):264S–273S.

Howard, B.V., and J. Wylie-Rosett. 2002. Sugar and cardiovascular disease. *Circulation* 106:523.

Ikawa, Y. 1998. Effects of sucrose replacement on baking process of sponge cakes. *Journal of the Japanese Society for Food Science and Technology* 45:357–363.

Inglett, G.E., K. Warner, and R.K. Newman. 1994. Sensory and nutritional evaluations of Oatrim. *Cereal Foods World* 39:755–527.

Kaack, K. and L. Pedersen. 2005. Low energy chocolate cake with potato pulp and yellow pea hulls. *European Food Research and Technology* 221:367–375.

Kamel, B.S. and V.F. Rasper. 1988. Effects of emulsifiers, sorbitol, polidextrose, and crystalline cellulose on the texture of reduced-calorie cakes. *Journal of Texture Studies* 19:307–320.

Katan, M.B., P.L. Zock, and R.P. Mensink. 1995. Trans-fatty-acids and their effects on lipoproteins in humans. *Annual Review of Nutrition* 15:473–493.

Kaur, A., G. Singh, and H. Kaur. 2000. Studies on use of emulsifiers and hydrocolloids as fat replacers in baked products. *Journal of Food Science and Technology–Mysore* 37:250–255.

Khalil, A.H. 1998. The influence of carbohydrate-based replacers with and without emulsifiers on the quality characteristics of lowfat cake. *Plant Foods for Human Nutrition* 52:299–313.

Khosla, P. and K. Sundram. 1996. Effects of dietary fatty acid composition on plasma cholesterol. *Progress in Lipid Research* 35:93–132.

Kim, H.Y.L., H.W. Yeom, H.S. Lim, and S.T. Lim. 2001. Replacement of shortening in yellow layer cakes by corn dextrins. *Cereal Chemistry* 78:267–271.

Klug, C., G.W.V. Lipinski, and D. Bottger. 1992. Baking stability of acesulfame-K. *Zeitschrift fur Lebensmittel-Untersuchung und-Forschung* 194:476–478.

Kocer, D., Z. Hicsasmaz, A. Bayindirli, and S. Katnas. 2007. Bubble and pore formation of the high-ratio cake formulation with polydextrose as a sugar- and fat-replacer. *Journal of Food Engineering* 78:953–964.

Kritchevsky, D. 2000. Overview: Dietary fat and atherosclerosis. *Asia Pacific Journal of Clinical Nutrition* 9:141–145.

Lakshminarayan, S.M., V. Rathinam, and L. KrishnaRau. 2006. Effect of maltodextrin and emulsifiers on the viscosity of cake batter and on the quality of cakes. *Journal of the Science of Food and Agriculture* 86:706–712.

Laneuville, S.I., P. Paquin, and S.L. Turgeon. 2005. Formula optimization of a low-fat food system containing whey protein isolate-xanthan gum complexes as fat replacer. *Journal of Food Science* 70:S513–S519.

Lee, S. and G.E. Inglett. 2006. Rheological and physical evaluation of jet-cooked oat bran in low calorie cookies. *International Journal of Food Science and Technology* 41:553–559.

Lichtenstein, A.H. 2000. Trans fatty acids and cardiovascular disease risk. *Current Opinion in Lipidology* 11:37–42.

Lim, H., C.S. Setser, and S.S. Kim. 1989. Sensory studies of high potency multiple sweetener systems for shortbread cookies with and without polidextrose. *Journal of Food Science* 54:625–628.

Liu, S., W.C. Willett, M.J. Stampfer, F.B. Hu, M. Franz, L. Sampson, C.H. Hennekens, and J.E. Manson. 2000. A prospective study of dietary glycemic load, carbohydrate intake, and risk of coronary heart disease in US women. *American Journal of Clinical Nutrition* 71:1455–1461.

Lucca, P.A. and B.J. Tepper. 1994. Fat replacers and the functionality of fat in foods. *Trends in Food Science and Technology* 5:9-12.

Makinen, K.K., and P. Isokangas. 1988. Relationship between carbohydrate sweeteners and oral diseases. *Progress in Food and Nutrition Science* 12:73–109.

Marshall, J.A., S. Hoag, S. Shetterly, and R.F. Hamman. 1994. Dietary fat predicts conversion from impaired glucose tolerance to NIDDM: The San Luis Valley Diabetes Study. *Diabetes Care* 17:50–56.

Montijano, H., F.A. Tomas-Barberan, and F. Borrego. 1998. Technological properties and regulatory status of high intensity sweeteners in the European Union. *Food Science and Technology International* 4:5–16.

Newsome, R. 1993. Sugar substitutes. In *Low-Calorie Foods Handbook*, Ed. A.M. Altschul. Marcel Dekker, New York.

Ngo, W.H. and M.V. Taranto. 1986. Effects of sucrose level on the rheological properties of cake batters. *Cereal Foods World* 31:317–322.

Ochi, T., K. Tsuchiya, Y. Ohtsuka, M. Aoyama, T. Maruyama, and I. Niiya. 1991. Effects of sweetener on fat stability of cookies. *Journal of the Japanese Society for Food Science and Technology* 38:910–914.

Odegaard, A.O. and M.A. Pereira. 2006. Trans fatty acids, insulin resistance, and type 2 diabetes. *Nutrition Reviews* 64:364–372.

Olinger, P.M. and V.S. Velasco. 1996. Opportunities and advantages of sugar replacement. *Cereal Foods World* 41:110–117.

Pateras, I.M.C., K.F. Howells, and A.J. Rosenthal. 1994. Hot-stage microscopy of cake batter bubbles during simulated baking: Sucrose replacement by polidextrosa. *Journal of Food Science* 59:168–170, 178.

Penna, E.W., P. Avendaño, D. Soto, and A. Bunger. 2003. Chemical and sensory characterization of cakes enriched with dietary fiber and micronutrients for the elderly. *Archivos Latinoamericanos de Nutrición* 53:74–83.

Perry, J.A., R.B. Swanson, B.G. Lyon, and E.M. Savage. 2003. Instrumental and sensory assessment of oatmeal and chocolate chip cookies modified with sugar and fat replacers. *Cereal Chemistry* 80:45–51.

Peterson, S., M. Sigman-Grant, B. Eissenstat, and P. Kris-Etherton. 1999. Impact of adopting lower-fat food choices on energy and nutrient intakes of American adults. *Journal of the American Dietetic Association* 99:177–183.

Pyler, E.J. 1988. *Baking Science and Technology*. Sosland, Merriam, KS.

Ponte, Jr., J.G. 1990. Sugar in bakery foods. In *Sugar: A User's Guide to Sucrose*, Ed. N.L. Pennington and C.W. Baker. Van Nostrand Reinhold, New York.

Rapaille, A., M. Gonze, and F. Van der Schueren. 1995. Formulating sugar free chocolate products with maltitol. *Food Technology* 49:51–54.

Romanchik-Cerpovicz, J.E., R.W. Tilmon, and K.A. Baldree. 2002. Moisture retention and consumer acceptability of chocolate bar cookies prepared with okra gum as a fat ingredient substitute. *Journal of the American Dietetic Association* 102:1301–1303.

Ronda, F., M. Gomez, C.A. Blanco, and P.A. Caballero. 2005. Effects of polyols and nondigestible oligosaccharides on the quality of sugar-free sponge cakes. *Food Chemistry* 90:549–555.

Rosenthal, A.J. 1995. Application of aged egg in enabling increased substitution of sucrose by litesse (polidextrose) in high-ratio cakes. *Journal of the Science of Food and Agriculture* 68:127–131.

Rothstein, W.G. 2006. Dietary fat, coronary heart disease, and cancer: A historical review. *Preventive Medicine* 43:356–360.

Salmeron, J., J.E. Manson, M.J. Stampfer, G.A. Colditz, A.L. Wing, and W.C. Willett. 1997a. Dietary fiber, glycemic load, and risk of non-insulin-dependent diabetes mellitus in women. *JAMA-Journal of the American Medical Association* 277:472–477.

Salmeron, J., A. Ascherio, E.B. Rimm, G.A. Colditz, D. Spiegelman, D.J. Jenkins, M.J. Stampfer, A.L. Wing, and W.C. Willett. 1997b. Dietary fiber, glycemic load, and risk of NIDDM in men. *Diabetes Care* 20:545–550.

Sanchez, C., C.F. Klopfenstein, and C.E. Walker. 1995. Use of carbohydrate-based fat substitutes and emulsifying agents in reduced-fat shortbread cookies. *Cereal Chemistry* 72:25–29.

Shinsato, E. 1996. Confectionery ingredient update. *Cereal Foods World* 41:372–375.

Shukla, T.P. 1995. Problems in fat-free and sugarless baking. *Cereal Foods World* 40:159–160.

Sigman-Grant, M. 1997. Can you have your low-fat cake and eat it too? The role of fat-modified products. *Journal of the American Dietetic Association* 97(Suppl.):S76–S81.

Sluimer, P. 2005. Principles of breadmaking: Functionality of raw materials and process steps. American Association of Cereal Chemistry, St. Paul, MN.

Sreebny, L.M. 1982. Sugar and human dental caries. *World Review of Nutrition and Dietetics* 40:19–65.

Stauffer, C.E. 1996. *Fats & Oils.* Eagan Press, St. Paul, MN.

Sudha, M.L., A.K. Srivastava, R. Vetrimani, and K. Leelavathi. 2007. Fat replacement in soft dough biscuits: Its implications on dough rheology and biscuit quality. *Journal of Food Engineering* 80:922–930.

Swanson, R.B., J.M. Perry, and L.A. Carden. 2002. Acceptability of reduced-fat brownies by school-aged children. *Journal of the American Dietetic Association* 102:856–859.

Tuorila, H., F.M. Kramer, and A.V. Cardello. 1997. Role of attitudes, dietary restraint, and fat avoidance strategies in reported consumption of selected fat-free foods. *Food Quality and Preference* 8:119–123.

Warner, K. and G.E. Inglett. 1997. Flavor and texture characteristics of foods containing Z-Trim corn and oat fibers as fat and flour replacers. *Cereal Foods World* 42:821–825.

Wolraich, M., R. Milich, P. Stumbo, and F. Schultz. 1985. Effects of sucrose ingestion on the behavior of hyperactive boys. *Journal of Pediatrics* 106:675–682.

Yudkin, J. 1978. Dietary factors in atherosclerosis: Sucrose. *Lipids* 13:370–372.

Yudkin, J. and E. Evans. 1972. Low-carbohydrate diet in treatment of chronic dyspepsia. *Proceedings of the Nutrition Society* 31:A12.

Zoulias, E.I., V. Oreopoulou, and E. Kounalaki. 2002a. Effect of fat and sugar replacement on cookie properties. *Journal of the Science of Food and Agriculture* 82:1637–1644.

Zoulias, E.I., V. Oreopoulou, and C. Tzia. 2002b. Textural properties of low-fat cookies containing carbohydrate- or protein-based fat replacers. *Journal of Food Engineering* 55:337–342.

Zoulias, E.I., S. Piknis, and V. Oreopoulou. 2000a. Effect of sugar replacement by polyols and acesulfame-K on properties of low-fat cookies. *Journal of the Science of Food and Agriculture* 80:2049–2056.

Zoulias, E.I., V. Oreopoulou, and C. Tzia. 2000b. Effect of fat mimetics on physical, textural and sensory properties of cookies. *International Journal of Food Properties* 3:385–397.

Index

275

T - #0387 - 071024 - C7 - 234/156/14 - PB - 9780367387617 - Gloss Lamination